AF766525

Ecosystem Services of Headwater Catchments

Josef Křeček • Martin Haigh • Thomas Hofer
Eero Kubin • Catrin Promper
Editors

Ecosystem Services of Headwater Catchments

Springer

CP

Editors
Josef Křeček
Department of Hydrology
Czech Technical University
Prague, Czech Republic

Martin Haigh
Department of Social Sciences
Oxford Brookes University
Oxford, UK

Thomas Hofer
Forestry Department
FAO
Rome, Italy

Eero Kubin
LUKE
Oulu University
Oulu, Finland

Catrin Promper
Austrian Federal Ministry of Agriculture
Forestry, Environment and Water
Management
Vienna, Austria

Co-published by Springer International Publishing, Cham, Switzerland, with Capital Publishing Company, New Delhi, India.

Sold and distributed in North, Central and South America by Springer, 233 Spring Street, New York 10013, USA.

In all other countries, except SAARC countries—Afghanistan, Bangladesh, Bhutan, India, Maldives, Nepal, Pakistan and Sri Lanka—sold and distributed by Springer, Tiergartenstr. 15, 69121 Heidelberg, Germany.

In SAARC countries—Afghanistan, Bangladesh, Bhutan, India, Maldives, Nepal, Pakistan and Sri Lanka—printed book sold and distributed by Capital Publishing Company, 7/28, Mahaveer Street, Ansari Road, Daryaganj, New Delhi 110 002, India.

ISBN 978-3-319-57945-0 ISBN 978-3-319-57946-7 (eBook)
DOI 10.1007/978-3-319-57946-7

Library of Congress Control Number: 2017943102

© Capital Publishing Company, New Delhi, India 2017
This work is subject to copyright. All rights are reserved by the Publisher, whether the whole or part of the material is concerned, specifically the rights of translation, reprinting, reuse of illustrations, recitation, broadcasting, reproduction on microfilms or in any other physical way, and transmission or information storage and retrieval, electronic adaptation, computer software, or by similar or dissimilar methodology now known or hereafter developed.
The use of general descriptive names, registered names, trademarks, service marks, etc. in this publication does not imply, even in the absence of a specific statement, that such names are exempt from the relevant protective laws and regulations and therefore free for general use.
The publisher, the authors and the editors are safe to assume that the advice and information in this book are believed to be true and accurate at the date of publication. Neither the publisher nor the authors or the editors give a warranty, express or implied, with respect to the material contained herein or for any errors or omissions that may have been made. The publisher remains neutral with regard to jurisdictional claims in published maps and institutional affiliations.

Printed on acid-free paper

This Springer imprint is published by Springer Nature
The registered company is Springer International Publishing AG
The registered company address is: Gewerbestrasse 11, 6330 Cham, Switzerland

Preface

Changes in headwater and mountain watersheds have major impacts on both the sustainable development of their headwater regions and on those who live downstream. New and effective responses to problems related to upland conservation and mountain watershed management are as necessary today as they were a hundred years ago when the oldest government services for torrent and avalanche control, and forest protection, were created. Water yielded from 'protective lands' cannot be considered as being of a good quality by definition. In addition, the impacts of the over-exploitation of resources, developments related to transport, tourism, forestry, agriculture, water supply, power supply, mining, etc., as well as global climate change and non-point pollution from various sources, can and have seriously degraded many fragile headwater environments.

Headwater control aims to promote grounded, better integrated and more self-sustainable development in headwater environments. It is constructed upon three principles: First, it recognises that headwaters are vulnerable habitats much threatened by environmental change, both climatic and more directly anthropogenic, such as the anthropogenic degradation of forests, biodiversity, ecosystem health, waters and soils and the damaging effects of air pollution, agriculture and economic development. Second, it argues that direct intervention can secure environmental quality in headwater environments through pollution control, forest engineering, water management, soil conservation, torrent control, landslide mitigation, land reclamation, bioengineering, applied environmental education and action-oriented community participation. Finally, it emphasises the practical application of holistic integrated environmental management, both in its biophysical and social components.

Short-term economic criteria are not able to guarantee ecological stability in headwater catchments. This book takes this agenda forward by analysing the environmental benefits of headwaters on a broader scale and by focussing on the problems of evaluating and conserving the ecosystem services they provide.

This volume contains 24 papers selected from contributions to recent meetings of the European Forestry Commission Working Party on the Management of Mountain Watersheds, which is managed by FAO (Food and Agriculture Organization of the

United Nations). The Working Party—its biennial sessions and intersession activities—focusses on the continuous exchange of knowledge and experience between professionals in Europe and other regions of the world confronted with similar issues.

This volume is cosponsored by the International Association for Headwater Control (NGO founded in 1989). It aims to promote inclusive stakeholder dialogue in headwater regions and to bring together the perspectives of applied science practitioners, researchers, policymakers and community groups.

Prague, Czech Republic Josef Křeček
Oxford, UK Martin Haigh
Rome, Italy Thomas Hofer
Oulu, Finland Eero Kubin
Vienna, Austria Catrin Promper

Contents

About the Editors

Josef Křeček, Ph.D. is the founder and managing codirector of the UNESCO-recognised *International Association for Headwater Control*, management committee member of EU COST Action 15226 *Climate-Smart Forestry in Mountain Regions* and former president of *EFC/FAO Working Party on the Management of Mountain Watersheds*. He is associate professor of the Czech Technical University in Prague. His research is oriented on hydrological processes related to impacts of the climate change and acid atmospheric deposition, as well as the participatory watershed management. He is experienced with several international projects in Europe and Asia and coordinated a number of publications on watershed management and headwater restoration.

Martin Haigh, Ph.D. is emeritus professor of geography at Oxford Brookes University, Oxford, UK. Formerly, he was (vice) president of the 'World Association for Soil and Water Conservation'. He is a co-founder of the UNESCO-recognised *International Association for Headwater Control* and co-organiser of most of its international conferences. He is a UK national teaching fellow and a senior fellow of the Higher Education Academy. In 2010, he was awarded the Royal Geographical Society's *Taylor and Francis Award for Excellence in Geography Teaching (Higher Education)*. Martin's current work includes environmental education as well as technical research in landslide prediction and landscape reconstruction in mining and mountainous areas.

Thomas Hofer is forestry officer and leader of the Watershed Management and Mountains Team at the Food and Agriculture Organization of the United Nations (FAO). Since 2006, he served as the secretary of the *EFC/FAO Working Party on the Management of Mountain Watersheds*. He has vast field project experience in Asia, Eastern Europe, Africa and Latin America. He has coordinated the development of a number of flagship publications on watershed management, sustainable mountain development and forest hydrology.

Eero Kubin, Ph.D. is former president of *EFC/FAO Working Party on the Management of Mountain Watersheds* and management committee member of EU COST Action 725. Over 15 years, he served as director of the Muhos Research Unit of the Finnish Forest Research Institute, and he is leader of long-term research projects on phenology and environmental aspects of forestry practices. As docent of Oulu University and Helsinki University, he is lecturer on forest ecology and supervisor of several doctoral thesis.

Catrin Promper, Ph.D. works for the Austrian Federal Ministry of Agriculture, Forestry, Environment and Water Management, in the Directorate for Torrent and Avalanche Control. She holds a PhD in geography from the University of Vienna and gained international work experience in the Watershed Management and Mountains Team at the Food and Agriculture Organization of the United Nations (FAO). Since 2015, she supports the secretariat of the EFC/FAO Working Party on the Management of Mountain Watersheds.

Part I
Headwater Environment and Natural Resources

Chapter 1
Headwater Catchments: Foundation Pillars for Ecosystem Services

Elaine Springgay

Recognizing the multi-functionality of landscapes and forested catchments beyond their ability to provide goods, such as timber, non-wood forest products and minerals, is increasingly more important due to continuous land-use and climate changes. Growing populations have led to increased pressure on land, and diminishing natural resources has forced us to face the fact that land and many of its resources are finite, or regenerate significantly slower than they are being utilized. It is not only about the over-consumption of resources, but impacts on the processes that generate these resources. Although the multi-functionality of landscapes is increasingly acknowledged, the inclusion of ecosystem services in policy and practice has been slow, thus moderating the potential for improved integrated landscape approaches.

'Ecosystem services' is a collective term used to describe the plethora of functions provided by ecosystems or landscapes. The Millennium Ecosystem Assessment (MEA 2005) defines ecosystem services as "the functions and products of ecosystems that benefit humans, or yield welfare to society". These include soil fertility, erosion and avalanche control, water regulation and purification, groundwater recharge, carbon storage, wildlife habitat, aesthetic beauty, as well as many others. The term, which gained momentum in the late 1990s, was coined in the mid-1980s (Ehrlich and Mooney 1983) and modified from the term 'environmental services' described by Wilson and Matthews (1970 in Lele et al. 2013). It is often used to encourage integrated landscape management, based on the premise that an ecosystem or landscape that is managed for the conservation of its functions is a healthy one that is resilient to shocks (natural and human-made) and will continue to provide a multitude of goods and services for current and future generations.

E. Springgay (✉)
Food and Agriculture Organisation of United Nations, Rome, Italy
e-mail: Elaine.Springgay@fao.org

© Capital Publishing Company, New Delhi, India 2017

J. Křeček et al. (eds.), *Ecosystem Services of Headwater Catchments*,
DOI 10.1007/978-3-319-57946-7_1

A holistic, integrated landscape approach that includes ecosystem functions is not a new concept. Throughout the first half of the twentieth century, Aldo Leopold argued that the human-centric approach to land management was unsustainable, and therefore, an evolution in ethics was required in order to properly care for land. Leopold (1949) called for a philosophical change in ideals, referred to as the 'land ethic'; it was suggested that by enlarging our sense of community to include soils, waters, plants and animals, an intrinsic value for the environment could be established, resulting in improved conservation and protection. In other words, we should care for our environment the way we care for our neighbours, and society, at large.

Instead of a philosophically based land management approach, a more economic (or market-based) approach evolved. The modern-day concept of including ecosystem function in land management recognizes that the ecosystem or landscape provides services in addition to goods. These services can be valued or monetized, in order to quantify the benefits provided by ecosystems and/or incentivize the mainstreaming of improved landscape management and the inclusion of ecosystem services in practices and policies. However, calculating a market value for these services and determining who pays for such services and to whom, can be complex, which has contributed to the slow adoption of ecosystem services in policy and practice. Similarly, Leopold's 'land ethic', a market-based approach also requires a shift in perspective as traditionally many of these ecosystem services are seen as 'free'. Arguably, many of these services are invaluable.

While ecosystem services are universally important, across all landscapes and at different scales, they are particularly important in headwater catchments: upland areas reputed as the sources of our water supplies, and refuges for biodiversity. As such, it can be said that upland catchments are a main life support, the foundational pillars for our planet. For example, upland areas support approximately one-quarter of terrestrial biological diversity, with half of the world's biodiversity hotspots concentrated in mountain regions (Spehn et al. 2010).

Mountains provide freshwater to over 50% of the world's population (Ariza et al. 2013), which is why they are often referred to as the Earth's "water towers". It is estimated that 75% of water is derived from forested watersheds (MEA 2005), and as much as 90% of a river's flow may originate from the river's headwater catchment (Kirby 1978; Saunders et al. 2002); therefore, headwater catchments are vital for our water supply used for agricultural, domestic, industrial and environmental purposes. As such, these catchments also influence the flow of sediments, nutrients and organic matter downstream, influencing production systems, water quality and energy production.

Due to their altitude and slope, headwater catchments are comparatively fragile ecosystems susceptible to long-term impacts due to natural and/or human-induced shocks, as well as climate change. The remoteness of these areas can also mean they are overlooked, or passively addressed in management and policy, despite their vital role of providing ecosystem services locally and further downstream (Ariza et al. 2013).

The important links between forests and water resources were internationally acknowledged in 2002 by the Shiga Declaration on Forests and Water. However, since then the adoption of integrated policies and practices to address water regulation (quality and quantity), as well as disaster-risk mitigation/management has yet been inadequate to non-existent. Since the signing of the Paris Agreement (2016) by over 190 countries and the adoptions of Agenda 2030 whose Sustainable Development Goals (SDGs) set universal targets, applicable to both developing and developed countries, the push for integrated management and increased recognition of ecosystem services in policy and practice is timely. Agenda 2030 includes forests and mountains in both SDGs 6 (water) and 15 (land), with both goals recognizing the link between forests and mountains to water resources, thus creating renewed momentum and incentives for improved policies and practices.

Collaborative efforts, such as the International Forests and Water Agenda, a process which includes organizations such as the Food and Agriculture Organization of the United Nations (FAO), the International Centre for Integrated Mountain Development (ICIMOD), the World Agroforestry Center (ICRAF), the International Network for Bamboo and Rattan (INBAR) and the International Union for Forest Research Organizations (IUFRO), have strongly advocated for integrative forest-water policies and practices based on sound scientific understanding. The Forests and Water Agenda has been successful at establishing a network of stakeholders representing both forest and water sectors, as well as international organizations, academic institutions, civil society, government and non-government actors, who are actively engaged in sharing knowledge and experiences.

At the XIV World Forestry Congress in Durban, South Africa (September 2015), the Agenda launched *Forests and Water: A five-year action plan* (2016) to guide international concerted efforts in improving forest management for water-related ecosystem services. The launch was a major milestone in the Forest and Water Agenda process as it marked a clear transition from discourse to action.

The Action Plan precipitated the establishment of the Forest and Water programme at FAO, which aims to assist countries in the development and implementation of scientifically-based integrated forest-water policies and practices that will contribute to meeting their Agenda 2030 targets. This will be achieved through advocacy, as well as the sharing of scientific knowledge and the lessons learned regarding policies and practices between regions and countries.

Since 1950, the European Forestry Commission (EFC) has had a group dedicated to addressing upland watershed management, including soil conservation, water management, disaster risk reduction/management and restoration of degraded lands in upland watersheds. The EFC Working Party on the Management of Mountains and Watersheds is tasked to engage EFC member countries in an information exchange on forest and water policies, and watershed and risk management practices, as well as to highlight gaps in research, policy and practice (Hofer and Ceci 2012). The Working Party has remained relevant for over 60 years by continuously engaging in emerging issues of global importance, including the recent issues of climate change and ecosystem services. For example, a Working Group on Forests and Water was established in 2014 to address the water-related ecosystem services provided by mountains and their forests; the Working Group collaborates with the FAO Forest and Water Programme to share scientific knowledge and

European policies and practices with the international community. As such, the EFC Working Party on the Management of Mountains and Watersheds, along with collaborating institutions, organizations and researchers, have contributed to this publication to improve the understanding of ecosystem services in a changing world and the implications this has in practice, policy and society.

References

Ariza, C., Maselli, D., & Kohler, T. (2013). *Mountains: Our life, our future. Progress and perspectives on Sustainable Mountain development from Rio 1992 to Rio 2012 and beyond.* Bern: Swiss Agency for Development and Cooperation (SDC), Centre for Development and Environment.

Ehrlich, P., & Mooney, H. (1983). Extinction, substitution, and ecosystem services. *Bioscience, 33*, 248–254.

FAO. (2016). *Forests and water: A five-year action plan.* Available online: www.fao.org/2/ForestsAndWater. Accessed on 16 Mar 2016.

Hofer, T., & Ceci, P. (2012). Mission and history of the European Forestry commission Working Party on the Management of Mountain Watersheds. In J. Křeček, H. Haigh, T. Hofer, & E. Kubin (Eds.), *Management of mountain watersheds*. Dordrecht: Springer.

Kirby, M. J. (1978). *Hillslope hydrology*. Chichester: Wiley.

Lele, S., Springate-Baginski, O., Lakerveld, R., Deb, D., & Dash, P. (2013). Ecosystem services: Origins, contributions, pitfalls, and alternatives. *Conservation and Society, 11*, 343–358.

Leopold, A. (1949). *A sand county almanac and sketches here and there*. Oxford: Oxford University Press.

MEA. (2005). *Ecosystems and human well-being, Millennium Ecosystem Assessment.* Washington, DC: Island Press.

Saunders, D., Meeuwig, J., & Vincent, A. (2002). Freshwater protected areas: Strategies for conservation. *Conservation Biology, 16*, 1523–1739.

Spehn, E., Rudmann-Maurer, K., Körner, C., & Maselli, D. (Eds.). (2010). *Mountain biodiversity and global change*. Basel: GMBA-DIVERSITAS.

Wilson, C., & Matthews, W. (Eds.). (1970). *Man's impact on the global environment: Report of the study of critical environmental problems (SCEP).* Cambridge, MA: MIT Press.

Chapter 2
Technical and Ecological Methods to Control the Water Cycle in Mountain Watersheds

Edward Pierzgalski

1 Introduction

In the Polish part of the Sudeten and Carpathian Mountains, the average annual precipitation ranges from 800 to 1500 mm. Runoff coefficient in mountain catchments there varies from 0.4 to 0.9. Seasonal variation in runoff, and, particularly, extreme events (floods and droughts), cause series problems in the down-hill urban areas, as well as in agriculture and forestry. Namely, heavy rainstorms are harmful, producing devastating surface runoff, soil erosion and stream channel instability. But, in the last years, the periods of water deficit are more frequent.

Nowocień (2008) reported the annual loss of soil in catchments of the Carpathian 280 t/km^2. Mountain streams with slopes of above 40% are very dangerous in forming flash floods by storms exceeding 40 mm of rainfall. The rapid velocities in streams initiate transport of large amounts of debris-flow including boulders and rock blocks. The debris come primarily from the erosion of channel bottoms and banks, but some is derived also by landslides. These processes lead to sedimentation in rivers and water reservoirs (from a few to several tens of centimetres annually). Sheet erosion causes the loss of surface soil layers, particularly, the organic particles, fertilizers and various chemicals. Thus, the soil erosion processes contribute also to the transfer of pollutants and increased risk of flooding. Gully erosion is still a major factor contributing to degradation of the environment in mountain watersheds in Poland. Despite the fact that gully networks in the mountains are already strongly developed, they are still growing due to the forest logging practices.

The aim of this paper is to present results of long-term hydrological studies on forest hydrology conducted by the Polish Forest Research Institute, and to assess the eco-services (technical and ecological measures) adopted in mountain basins by the Coordination Center for Environmental Projects (2010) in 2007–2015. The

E. Pierzgalski (✉)
Forest Research Institute, Sękocin Stary, ul. Braci Leśnej 3, 05-090 Raszyn, Poland
e-mail: E.Pierzgalski@ibles.waw.pl

© Capital Publishing Company, New Delhi, India 2017

J. Křeček et al. (eds.), *Ecosystem Services of Headwater Catchments*,
DOI 10.1007/978-3-319-57946-7_2

specific objectives of this project were to mitigate environmental damages and to support: water resources recharge, conservation of soil, and control of torrent erosion, stream channels stability of biodiversity.

2 Material and Methods

In general, the project has been implemented in 55 Forest Districts in the mountains. Approx. 3500 different measures were applied, including 130 reconstructed and new ponds, rehabilitation of 53 km of skid-roads, and revitalization of 173 km of streams. The partial tasks were divided into three main groups:

1. To increase the retention capacity
2. To reduce erosion processes (sheet, rill, gully and stream channel erosion)
3. To restore biological corridors in streams

Methods for enhancing the water retention capacity are shown in Fig. 2.1, to reduce erosion processes in Fig. 2.2, and to restore biota in mountain streams and rivers in Fig. 2.3.

Environmentally sound forestry practices were adopted in mountain watersheds to control extreme hydrologic events. The considered interactions between water and forest resources are shown in Table 2.1

The effects of forest harvest on runoff have been studied in two forest catchments located in the Sudeten Mountain and harvested in the 1980s: *Czerniawka*, with only 4% area harvested, and *Ciekoń*, where the clear-cut extended on 40% of the basin area (Pierzgalski et al. 2007, 2009).

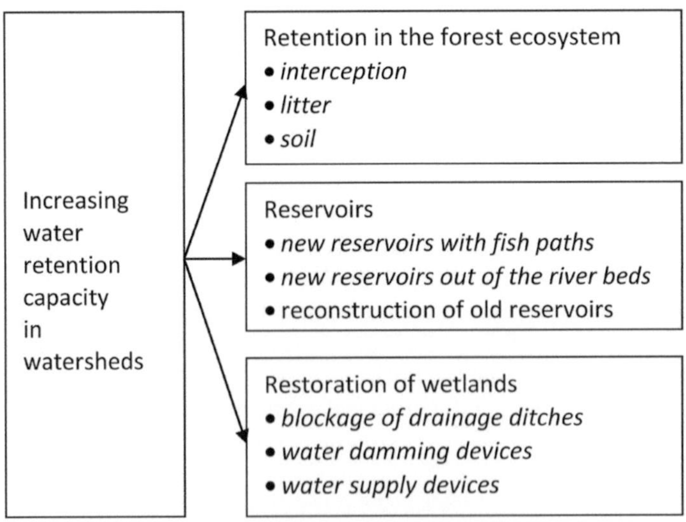

Fig. 2.1 List of measures to increase water retention capacity

Fig. 2.2 List of measures reducing the erosion

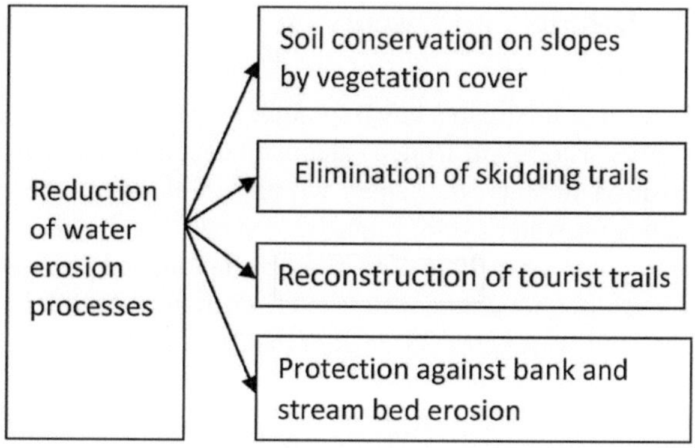

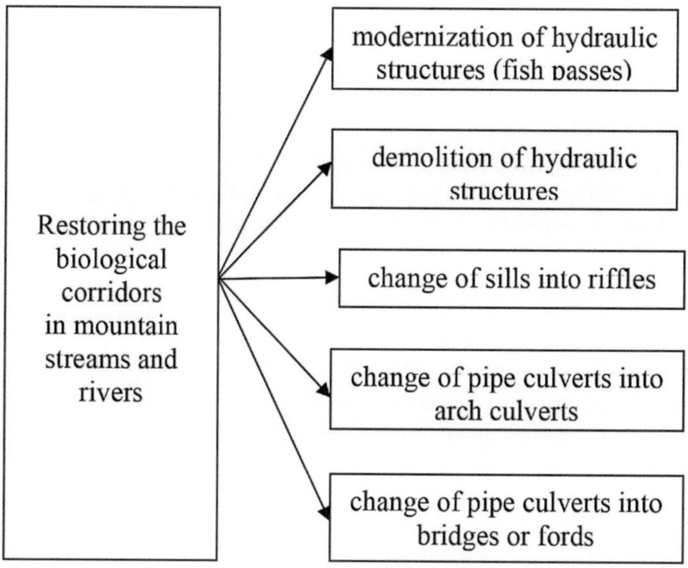

Fig. 2.3 Practices to support revitalization of streams and rivers

Table 2.1 Interaction between forest and water

Influence of forests on water resources	Influence of water on forest resources
Interception	Forest habitat type
Evapotranspiration	Flora and fauna
Retention	Timber growth
Discharge variability	Resistance to disturbances
Erosion and sedimentation	Adaptation to the climate change
Water quality	
Water habitat	Carbon sequestration

3 Results and Discussion

Discharge in streams has been affected by a complex of natural and anthropogenic factors. The runoff from a catchment depends on the area harvested, and also on the position of harvested spots within a watershed. Analyzing the impact of forest stands on water cycle; it has been stated already that, particularly the interception loss ranges from 20 to 40% of precipitation, and plays a significant role in the runoff genesis of a mountain watershed (Lenart et al. 2003). In deciduous forests, the interception loss is generally smaller than in coniferous stands (Osuch 1998). The impact of harvested spruce forests (*Picea abies*) in the Sudeten Mountains on the flood hydrographs is shown in Fig. 2.4: the significantly higher peak and sharper flood hydrograph were found in the Ciekoň basin with more extended intensive forest harvest.

The harvest of forests affected also the seasonal distribution of water yield (Fig. 2.5). Summarizing hydrological observations in Poland (Pierzgalski et al. 2007) the effect of the age of forests on the specific discharge in mountain streams is demonstrated in Fig. 2.6.

To control impacts of extreme hydrological events, at present, the prevailing view prefers ecological oriented measures against the technical approach (Shields et al. 2003). In comparison with commercial forestry, the retention capacity of studied forests was increased by environmental logging technologies, and also by the alternative soil preparation before reforestation. The current environmental forestry includes:

- Restriction of a clear-cut on slopes above 40%,
- Ecological silviculture to produce more stable forest stands with a rich species composition to be resistant against insect or wind damages,

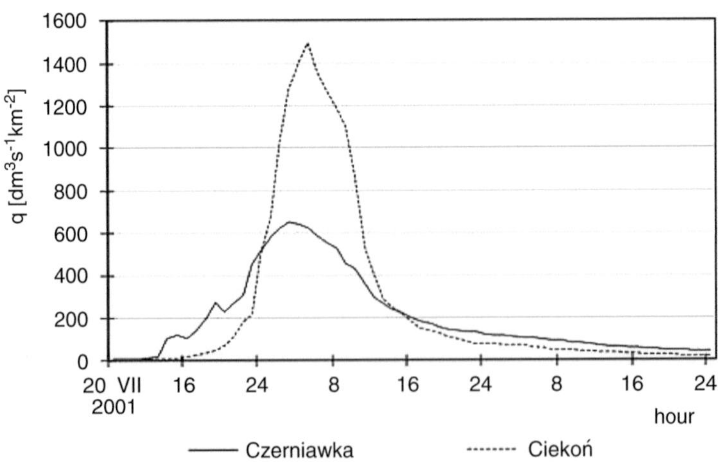

Fig. 2.4 Influence of forest harvest on flood hydrograph (Ciekoň: 40% harvested)

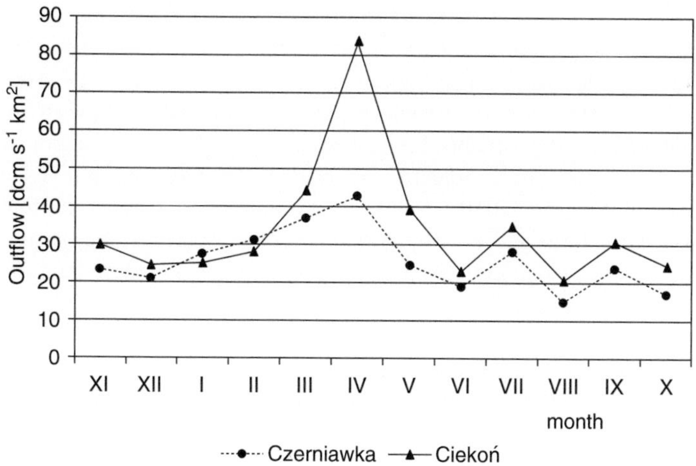

Fig. 2.5 Effect of forest cover on seasonal runoff (Ciekoň: 40% harvested)

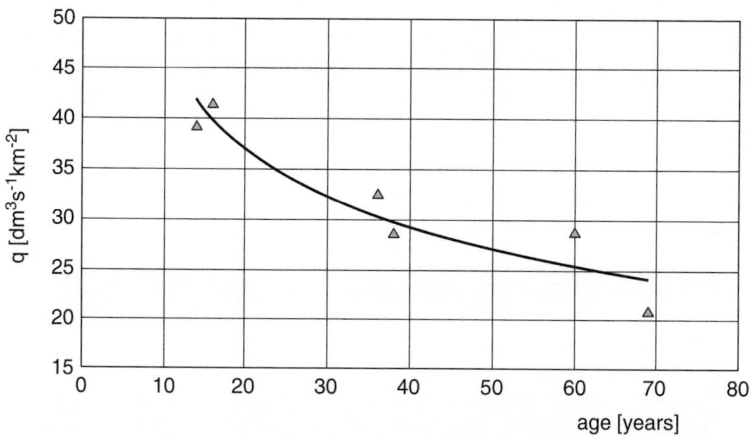

Fig. 2.6 Influence of forest age on specific discharge in mountain streams

- Rehabilitate of skidding roads or trails after the logging, and
- Priority reforestation of slopes above 15%.

In mountain watersheds, all the damages associated with extreme hydrologic events might not be eliminated by adequate forestry practices. The application of traditional technical measures is still important. In the framework of this project, the technical measures included: water reservoirs, hydraulic structures in streams, water damming devices and measures of road protection.

Small water reservoirs represent the primary way to control surface water dynamics. These are usually objects with multiple tasks, clearly defined for their proper design and operation. The reservoirs in mountain forests should be designed comprehensively in terms of catchments despite the fact that they are usually small. Their impact is not determined by size, but by the number of devices in the watershed, which corresponds to the principle of distributed risk. These facilities primarily serve ecological functions, but each project must be assessed for the environmental impact and effective investments. Special environmental impact assessments should be made at the locations protected by Natura 2000 (Coordination Center for Environmental Projects 2010). Small-scale retention structures in forest areas should use the natural and local materials (wood, fascines, stone, sand, cohesive soils). The main goal is to limit the transport of materials, to reduce investment costs and to minimize damages in the forest sector during construction. Devices on the water reservoirs should be durable with a minimum operational service. The philosophy of hydraulic structures in a mountain watershed has been changing. In the past, they were constructed mainly with concrete and rock, blocking the connectivity of the biological corridors. Currently, the biodiversity in stream waters is preferred; therefore, the accepted design has to stabilize stream channels while allowing movement of aquatic organisms (Bojarski et al. 2005). Among the solutions used in this project, stabilization of forest roads by local materials (Fig. 2.7) was used to reduce the risk of sheet erosion and direct flow acceleration.

In the recent period of 1990–2015, the annual amount of direct flood damages in mountain watersheds of Poland ranged from 30 million EUR (2010) to 50 million EUR (1997) (Lenart et al. 2003; Pierzgalski et al. 2011), not including consequent damages (for example reduced timber production, impacts fungal diseases or insect degradation). The hope of the investment in eco-technical services introduced by this project is to reduce significantly those amounts.

Fig. 2.7 Reconstruction of forest roads with local materials (Photo: Edward Pierzgalski)

4 Conclusions

In mountain regions, the control of water resources is essential for the sustainable existence of human society in large downhill areas. Therefore, the proper maintenance of ameliorative infrastructures in forest catchments is urgently needed for enhancing the resilience of forest ecosystems and prevention of their disturbances.

In Poland, hydrological extremes threaten forest catchments, particularly, in the Sudeten and Carpathian Mountains. The most harmful are the effects of soil and stream channel erosion. In 2007–2015, more than 3500 ameliorative measures were realized in the framework of the project "Counteracting the effects of rainwater runoff in mountainous areas". Both technical and ecological techniques were applied to increase the retention capacity, and to reduce sheet, rill, gully, and stream channel erosion.

The environmental assessment concentrated namely on the landscape stabilization, water retention and retardation, and biodiversity protection. Special attention has been paid to the restoration of biological corridors in mountain streams. Combination of both ecological oriented and technical acquisitions showed a good evaluation in term of eco-services.

References

Bojarski, A., Jeleński, S., Jelonek, M., Litewka, T., Wyżga, B., & Zalewski, Z. (2005). *Good practices in maintaining of mountainous torrents and rivers* (in Polish). Warsaw: Ministry of Environment.

Coordination Center for Environmental Projects. (2010). Program of project: Counteracting the effects of rainwater runoff in mountainous areas. Increasing retention and maintaining streams and associated infrastructure in good condition. Warsaw.

Lenart, E., Pierzgalski, E., & Niemtur, S. (2003). *National report on forest watershed management in mountain regions in Poland* (pp. 155–165) (Swiss Agency for the Environment, Forest and Landscape SAEFL, Berne Environmental Documentation, No. 165).

Nowocień, E. (2008). Selected issues of soil erosion in Poland. (In Polish). In*:* Study and reports IUNG-PIB, Puławy, 10.

Osuch, B. (1998). An averaged depth of the water layer intercepted by plant surface as the three stand interception reservoir volume coordinate. In *Proceedings of the international conference "Forest and Water"*, Cracow.

Pierzgalski, E., Tyszka, J., & Stolarek, A. (2011). Floods and droughts in the forests (in Polish). In Lorenc H. (Ed.), *Natural disasters: A safety inside the country* (Monographs IMGW-PIB: Part III).

Pierzgalski, E., Janek, M., Kucharska, K., Tyszka, J., & Wróbel, M. (2007). *Hydrological research in Sudeten forest watersheds (in Polish).* Warsaw: Forest Research Institute.

Pierzgalski, E., Janek, M., Kucharska, K., Niemtur, S., Stolarek, A., Tyszka, J., & Wróbel, M. (2009). *Hydrological and erosion processes in the mountain forest catchments (in Polish).* Warsaw: Forest Research Institute.

Shields Jr., F. D., Copeland, R. R., Klingeman, P. C., Doyle, M. W., & Simon, A. (2003). Design for stream restoration. *Journal of Hydraulic Engineering, 129*(8), 575–584.

Chapter 3
Ecosystem Services Supporting Water Supply Systems

Özden Görücü

1 Introduction

Forests can provide human societies with series of benefits and services, generally called as ecosystem services. The ecosystem services are in the form of separated or integrated forms. In general, the ecosystem services could be classified as:

- Provisioning services: Materials that ecosystems provide such as food, water, wood, grass, medicinal plants and other raw materials.
- Regulating services: Services that ecosystems provide regulators such as soil and air quality, carbon storage, flood, erosion and disease control.
- Supporting services: Services that ecosystems provide sustained space such as biodiversity of flora and fauna, nutrient cycling.
- Cultural services: Services that ecosystems provide humanity quality such as recreation, aesthetic values, spiritual inspiration and so mental health.

In recent years, there are many efforts and scientific studies to assign economic values to the ecosystem services. This is necessary to reflect the economic values of the ecosystem services to the national account. For this, the theories and application ways of forest economics are being followed (Price 1989). On this topic the crucial question is why to assign economic values to ecosystems? Because there is a failure in environmental management and sustainability to account and to create insource or outsource financing for the full economic values of ecosystems and biodiversity has been a significant factor in their continuing loss and degradation. So, optimal harvest scheduling (Armstrong et al. 1992) and consistent economic value assignment mechanisms to ecosystems (Price 1989) were put forward

Ö. Görücü (✉)
Department of Forest Engineering, Faculty of Forestry, Kahramanmaraş Sütçü İmam University, Kahramanmaraş, Turkey
e-mail: ogorucu@ksu.edu.tr

© Capital Publishing Company, New Delhi, India 2017

J. Křeček et al. (eds.), *Ecosystem Services of Headwater Catchments*,
DOI 10.1007/978-3-319-57946-7_3

in practice. Therefore, ecosystem valuation approaches are being indicated by Görücü (1998) as:

- Direct market valuation approaches
- Revealed preference approaches
- Stated preference valuation approaches
- Multi criteria analysis

Multi criteria analysis allows formal integration of multiple values of the ecosystems by assigning relative weights to each criterion. On the other hand, this approach has the potential to be extended to cover ecosystem services. The outcomes obtained by multi criteria analysis are a ranking of preferences which supply a basis for taking decisions among various options. In this article, multi criteria analyses were used and operated for the ecosystem services supporting water supply systems.

Turkey has a total area of 78 million hectares; about one quarter is designated as forest. Mountain ranges run generally parallel to the northern and southern coasts, surrounding the central Anatolian Plain, which rises from 500 m elevation in the west to over 2000 m in the east. About 80% of soils in Turkey suffer from moderate to severe sheet and gully erosion and most rivers carry heavy loads of sediment. The mean annual precipitation varies from 250 (central and southeastern plateaus) to 2500 mm (northern coastal plains and mountains). In the western and southern coastal zones, a subtropical Mediterranean climate predominates, with short, mild and wet winters and long, hot, dry summers. Arid and semiarid continental climates prevail in central regions with frequent and heavy snowfall in the higher parts of the Anatolian Plain (Görücü 2009). The extend of forest land has gradually decreased as a result of long-term improper and extensive use, and currently, it has become necessary to take preventive measures for the protection of forests (Görücü 2002). Next to the timber economy, forests have collective benefits that cannot be evaluated monetarily, such as regulations of the climate, controlling the water regime, prevention of erosion, as well as their potential contribution to the defence of the country (Görücü 1998).

The aim of this study is to analyse and evaluate the sectoral relationships between forests and water under the umbrella of ecosystem services. Series of scenarios were designed to investigate the application of multi-criteria analysis in the management of forests. These scenarios are based on the analytical hierarchy process, using weighting and scoring of ecosystems. The output of this study leads to the development of guidelines for the management and practices of the Kahramanmaraş Suçatı forest districts.

2 Material and Methods

This study was carried out in the Ceyhan watershed, located in the upland and semiupland landscapes of the Suçatı Mountains. The altitude alters from 1036 to 2100 m, and forest stands of Turkish pine (*Pinus brutia*) cover 58.6% of the catchment. This study was designed to solve problems of Kahramanmaraş Suçatı Forest Districts, where the management of evenaged forest stands has been discussed, and several levels of clear cutting applied. The management unit of Turkish pine was planned by linear programming with respect of FORPLAN software, as only a timber production forest in one scenario (Görücü 1996). There are 224 compartments in the Suçatı district, but in this study, they were aggregated (according to the criteria of slope, age, exposure, site index etc.) in 95 activity areas. Altogether, the Suçatı forest district of Turkish pine (15,256 ha) consists of productive forests (8183 ha), unproductive forests (3714 ha), and the deforestation area (3359 ha). The data of the selected 95 activity areas were used in the matrix of the FORPLAN software, the linear programming was used for all models in given objectives of the study. These different possible schedules for timber production are based on scenarios:

(a) Four rotation ages: 30, 40, 50 and 60 years
(b) Three different interest rates: 4, 5 and 6%
(c) Long-term sustained yield capacity, present net value as the criterion of efficiency to measure cut and inventory level
(d) Three different objective functions as maximize ending inventory, maximize the sum of cutting and maximize present net value
(e) Three allowable cut levels as 85,000 m^3, 102,000 m^3 and 140,000 m^3

The solutions were produced which test the sensitivity of the harvest schedules to changes in the limit on volume that may be harvested in any of 10 years, and the planning horizon was 100 years (Davis and Johnson 1987). A hundred-year planning period was used with harvesting operations scheduled for each ten-year subperiod within the period 1990–2089. Some constraints which were timber market price, costs of afforestation and road constructions, and administrative expenses had been included in the models. Linear programming was run to get optimal timber production according to present net value of each activity area in the context of multiple constraints on scarce resources. According to the linear programming proposed by Taha (2007):

$$\begin{aligned}
&\text{Objective function:} && maxZ = C_1X_1 + C_2X_2 + \ldots + C_nX_n \\
&\text{Constraints:} && a_{11}X_1 + a_{12}X_2 + \ldots + a_{1n}X_n \le b_1 \\
& && a_{21}X_1 + a_{22}X_2 + \ldots + a_{2n}X_n \le b_2 \\
& && a_{m1}X_n + a_{m2}X_n + \ldots + a_{mn}X_n \le b_m \\
&\text{Positiveness:} && X_1 \ge 0, X_2 \ge 0, \ldots, X_n \ge 0
\end{aligned}$$

In the next step, the ecosystem appropriateness criteria were selected to set up and weightings at forest level as well. For this, on the experimental site, the criteria of the ecosystems were decided as water production, timber production, recreation (and ecotourism) and also carbon storage. According to potentials of the site productivity which was obtained from actual forest management plans, the scores (points) and weightings (percentage) were assigned and included in the model. The cost of water as the ecosystem output must be calculated to reflect the value of the water produced by forest ecosystems to the national income balances. Eker (2005) analysed water economy related to forests in the Darlık Watershed. The costs of general administration, expropriation, maintenance and afforestation were included in this study. To evaluate forest resources, Görücü and Eker (2009) included also the carbon economy in the watershed of Kahramanmaraş Ayvalı Dam. In this study, stands of Turkish pine with site class 3rd in three compartments were analyzed with regard to carbon storage and emission economics such as the monetary value of the carbon content in biomass. The most important parameters used in these calculations were carbon amount in the dry mass of pine forests (amounts of CO_2 fixed in 1 kg of timber) and the default price of carbon per tonne. These parameters were used to calculate the present net value of the carbon storage in the pine biomass.

Subsequently, new evaluation methods on the water economics were tried and made progress in Turkey. The latest one was sophistically carried out and explained here in the Suçatı watershed to measure the ecosystems services which include water as well, by using multi criteria analysis in water centred forest management regime under the analytical hierarchy processing, by weighting and scoring of the selected ecosystem services.

3 Results and Discussion

For only timber production some important scenarios and results were obtained by using clear cutting and thinning-clear cutting, while interest rate was 5%, based on allowable cut level respectively 85,000 m^3, 102,000 m^3 and 140,000 m^3 and also for four different rotation (Görücü 2004). As an example, solution of periodical allowable cut for various rotation ages by clear cutting is given in Table 3.1.

Assuming four ecosystem services (water production, timber production, recreation-ecotourism and carbon storage), the solution is given in Table 3.2. All points and weights were assigned due to the actual forest structure and the watershed landscape, and the multi criteria analysis under the analytical hierarchy process applied.

The derived data according to scenarios 1 and 2, based on the model of Görücü (2011) are in Table 3.3.

In the Suçatı watershed, the forest practices are based on clear cutting and reforestation in ten periods. Both natural and cultivated stands are considered in the schedule (Table 3.1). The growth there is decelerating with the age of a stand. For the 85,000 m^3 of allowable cut the best alternative is the maximum present net value of \$16,013 and rotation age 30 years, in the tenth period by long-term sustained yield capacity. For the 102,000 m^3, the best alternative is the maximum present net value of \$18,729 and rotation age 40 years, in the nineth period by long-term sustained yield capacity. For the 140,000 m^3 of allowable cut the best alternative is the maximum present net value \$19,440 and rotation age 30 years, in the nineth period by long-term sustained yield capacity.

As the next step of this study (Table 3.2), the evaluation of four ecosystem services (water production, timber yield, recreation-ecotourism and carbon storage) was applied. All the used points and weights in the analysis were assigned due to the

Table 3.1 Periodical mean increment for various rotation ages and allowable cut by clear cutting within the Suçatı watershed

Rotation age = 30		Rotation age = 40		Rotation age = 50		Rotation age = 60	
Age (years)	Periodical increment (m^3/ha)	Age (years) (m^3/ha)	Periodical increment	Age (years) (m^3/ha)	Periodical increment	Age (years) (m^3/ha)	Periodical increment
Periodical allowable cut = 85,000 m^3							
25	55	35	54	45	48	55	43
30	63	40	65	50	60	60	40
35	54	45	60	55	52	65	30
40	50	50	58	60	40	70	28
45	48	55	50	65	32		
50	45	60	40	70	30		
55	43	65	35				
60	40	70	33				
65	38						
70	35						
Periodical allowable cut = 102,000 m^3							
25	73	35	72	45	68	55	62
30	76	40	77	50	74	60	60
35	72	45	70	55	65	65	55
40	70	50	66	60	60	70	46
45	68	55	63	65	51		
50	65	60	60	70	47		
55	62	65	55				
60	60	70	50				
65	58						
70	56						
Periodical allowable cut = 140,000 m^3							
25	43	35	50	45	40	55	35
30	52	40	46	50	45	60	33
35	50	45	42	55	40	65	27
40	45	50	40	60	33	70	24
45	40	55	38	65	29		
50	38	60	33	70	27		
55	35	65	30				
60	33	70	28				
65	31						
70	40						

actual forest structure and watershed landscape. The criteria, criterion points (pts) and weightings of the ecosystem services were used in the multi criteria analysis of the Suçatı watershed. Two scenarios were considered in context of the ecosystem services weightings and pointings within the analytical hierarchy process. The scenario 1 means, that by the timber production of 6–10 m^3, recreation (including eco-tourism) includes 21 visitors, and the carbon storage is 91–120 tonnes; so, the water

Table 3.2 Multi criteria analysis in water centered forest regime: Analytical hierarchy process, weighting and scoring on ecosystem services in the Suçatı watershed

Water production (1–5 pts)	Wood production (1–5 pts)	Recreation + ecotourism + (1–5 pts)	Carbon storage (1–5 pts)
Ecosystem appropriateness criteria and weightings at forest level (hectare/year)			
Criterion weight: 50%	Criterion weight: 30%	Criterion weight: 10%	Criterion weight: 10%
Criterion lowest point: 0.50	Criterion lowest point: 0.30	Criterion lowest point: 0.10	Criterion lowest point: 0.10
Criterion highest point: 2.50	Criterion highest point: 1.50	Criterion highest point: 0.50	Criterion highest point: 0.50
Criterion average point: 1.50	Criterion average point: 0.90	Criterion average point: 0.30	Criterion average point: 0.30
Subcriteria and points assigned			
1–4000 m3 1	1–5 m3 1	1–5 visitors 1	1–30 tonnes 1
4001–8000 m^3 2	6–10 m^3 2	6–10 visitors 2	31–60 tonnes 2
8001–12,000 m^3 3	11–15 m^3 3	11–15 visitors 3	61–90 tonnes 3
12,001–16,000 m^3 4	16–20 m^3 4	16–20 visitors 4	91–120 tonnes 4
16,001+ m^3 5	21+ m^3 5	20+ visitors 5	121+ tonnes 5
Weighted criterion point (criterion weight × subcriterion point) and criteria total appropriateness point (CTAP)			
1–4000 m^3 $0.50 \times 1 = 0.50$	1–5 m^3 $0.30 \times 1 = 0.30$	1–5 visitors $0.10 \times 1 = 0.10$	1–30 tonnes $0.10 \times 1 = 0.10$
4001–8000 m^3 $0.50 \times 2 = 1.00$	6–10 m^3 $0.30 \times 2 = 0.60$	6–10 visitors $0.10 \times 2 = 0.20$	31–60 tonnes $0.10 \times 2 = 0.20$
8001–12,000 m^3 $0.50 \times 3 = 1.50$	11–15 m^3 $0.30 \times 3 = 0.90$	11–15 visitors $0.10 \times 3 = 0.30$	61–90 tonnes $0.10 \times 3 = 0.30$
12,001–16,000 m^3 $0.50 \times 4 = 2.00$	16–20 m^3 $0.30 \times 4 = 1.20$	16–20 visitors $0.10 \times 4 = 0.40$	91–120 tonnes $0.10 \times 4 = 0.40$
16,001+ m^3 $0.50 \times 5 = 2.50$	21+ m^3 $0.30 \times 5 = 1.50$	21+ visitors $0.10 \times 5 = 0.50$	120+ tonnes $0.10 \times 5 = 0.50$
CTAP 7.50	4.50	1.50	1.50
GTAP (grand total appropriateness point)		15.00 (7.50 + 4.50 + 1.50 + 1.50)	

production is expected between 4001 and 8000 m^3 per hectare. Similarly, the scenario 2 explains that by the water production of 8001–12,000 m^3, the recreation includes 11–15 visitors, and carbon storage is 31–60 tonnes. The solution is based on following conditions: (1) As a correct guide for allocation of the ecosystem services to functions, (2) Capacity building on various goods and services, (3) Costing of the functions according to their weighting and pointing, (4) Estimating the value of the marketable and unmarketable ecosystem services, (5) Handling of the integrated watershed management, (6) Analyzing the watersheds in the holistic manner, and (7) Easier but also analytical valuing of the ecosystem services in the watersheds.

Table 3.3 Scenarios 1 and 2 based on the model by multi criteria analysis for forest related ecosystem services

Scenario 1	Scenario 2
Scenario 1. Data belong to appropriateness criteria:	*Scenario 2. Data belong to appropriateness criteria:*
Wood production: 6–10 m^3	Water production: 8001–12,000 m^3
Recreation + ecotourism: 21+ visitors	Recreation + ecotourism: 11–15 visitors
Carbon storage: 91–120 tonnes	Carbon storage: 31–60 tonnes
To be found:	*To be found:*
Water production:?	Wood production:?
Points assigned to appropriateness criteria:	*Points assigned to appropriateness criteria:*
Wood production: 2 pts	Water production: 3 pts
Recreation + ecotourism: 5 pts	Recreation + ecotourism: 3 pts
Carbon storage: 4 pts	Carbon storage: 2 pts
Weighted criterion point of appropriateness criteria:	*Weighted criterion point of appropriateness criteria:*
Wood production: 2 × 0.30 = 0.60	Water production: 3 × 0.50 = 1.50
Recreation + ecotourism: 5 × 0.10 = 0.50	Recreation + ecotourism: 3 × 0.10 = 0.30
Carbon storage: 4 × 0.10 = 0.40	Carbon storage: 2 × 0.10 = 0.20
Criteria total appropriateness point (CTAP): 0.60 + 0.50 + 0.40 = 1.50	*Criteria total appropriateness point (CTAP):* 1.50 + 0.30 + 0.20 = 2.00
Grand total appropriateness point (GTAP): 15.00	*Grand total appropriateness point (GTAP):* 15.00
Water production:	*Wood production:*
7.50 × 1.50/15.00 = 0.75 and goes to 4001–8000 m^3	4.50 × 2.00/15.00 = 0.60 and goes to 6–10 m^3

4 Conclusions

Besides the timber production, forests serve in providing human societies with water, erosion control, recreation, game and wildlife shelter, and also reducing carbon emissions and decreasing the negative effects of green house gases and global climate change. To evaluate forest ecosystem services, the optimization methods and quantitative approaches are used worldwide (Görücü and Eker 2009). In forest watersheds with concern of water supply, four ecosystem services should be considered: water production, timber yield, recreation-ecotourism and carbon storage. The scenarios used in the Suçatı watershed confirmed that the applied approach of ecosystem service evaluation is available in the watershed planning worldwide.

In the light of the ecosystem services evaluation processes, the missions to be undertaken about the forest and water agenda in Turkey are being seen in:

- Focusing on wider water cost approach in a catchment (not only water prices).
- Developing integrated resource management concept between forest and water sectors and extending the cooperation portfolio among stakeholders.
- Introducing the Water Framework Directive and International Water Convention to the public compromise towards an integrated watershed management.

- Adding the forestry sector into bilateral agreement of water use (currently between the State Water Affairs and the Juristic Person).
- Supporting the applied research on quantitative interactions between forest and water sectors.
- Therefore, the most important recommendations addressing the forest-water issues at the national level in Turkey are:
- Political support should be created and maintained at all levels and across all relevant sectors to promote the system of payment for ecosystem services.
- Political support is needed to adapt the effective legislation, and the principles of green economy (to combat global economy crisis).

References

Armstrong, G., Philips, W., & Beck, J. (1992). Optimal harvest scheduling under harvest volume constraints. *Canadian Journal of Forest Research, 22*, 497–503.

Davis, L. S., & Johnson, K. N. (1987). *Forest management* (3rd ed.). New York: McGraw Hill.

Eker, Ö. (2005). *Economic analysis of the water production function of forests*. PhD thesis, University of Istanbul, Faculty of Forestry, İstanbul.

Görücü, Ö. (1996). *Researches and improving of timber production planning in forest districts*. Publication of IUFRO/Danish Forest Landscape Research Institute, Copenhagen.

Görücü, Ö. (1998). Criteria for sustainable Forest management and studies in Turkish forestry. In: Future forest policies in Europe balancing economic and ecological demands, EFI Proceedings No. 22, Joensuu.

Görücü, Ö. (2002, August 27–30). Socioeconomic aspects of outdoor recreation in Turkey. In: Urban forestry and urban greening forestry serving urbanised societies, IUFRO European regional conference, Copenhagen.

Görücü, Ö. (2004). Improving timber production planning for forest districts in Calabrican pine region. *Journal of Faculty of Forestry*, Univeristy of İstanbul, Series A 54: 25–49.

Görücü, Ö. (2009). *Forestry based ecotourism possibilities and its role on economical development in Turkey*. Poland: University of Business in Wroclaw.

Görücü, Ö., & Eker, Ö. (2009). The research on carbon emission and economics in the watershed of Kahramanmaraş Ayvalı Dam. In: *Proceedings of the third congress of socioeconomic problems in Forestry*. Süleyman Demirel University, Isparta.

Price, C. (1989). *The theory and application of forest economics*. Oxford: Basil Blackwell.

Taha, H. A. (2007). *Operations research: An introduction* (8th ed.). Upper Saddle River: Pearson Prentice Hall.

Chapter 4
Ecosystem Services in Headwaters of the Boreal Environment

Eero Kubin and Josef Křeček

1 Introduction

The boreal forest (taiga or snow forests) is the world's largest terrestrial biome characterized by a subarctic climate and coniferous tree species (spruce, pine and larch) (Hämet-Ahti 1981). A long and cold winter is the dominant feature there, and snow processes are principal hydrological phenomena. In Finland, boreal forests cover 74% of the land area; therefore, forestry practices can affect nature capital of headwater catchments. These benefits can address all four groups of ecosystem services (MEA 2005): provisioning (timber yield, water resources recharge, fisheries), regulating (control of floods, carbon and bio-geo-chemical cycles), supporting (biodiversity) and cultural (recreation or aesthetic values).

Nowadays, the main forestry practice in the boreal forest is clear-cutting with replanting the forest plantations (monocultures) (Kubin and Kemppainen 1991). Forestry practices affect forest ecosystems and cycling of energy, water and nutrients (Tamm et al. 1974). The impacts on ground waters, watercourses or lakes depend, particularly, on the method of forest regeneration. In Finland, the impacts of forestry practices on runoff genesis and water quality has been monitored since 1974 (Kubin 1995, 1998). The aim of this paper is to summarize results of the long-term studies on environmental consequences of clear-cutting (including whole-tree and stump harvest) and side preparation.

E. Kubin (✉)
Natural Resources Institute Finland, Paavo Havaksen tie 3, FI -90014 Oulu, Finland
e-mail: eero.kubin@luke.fi

J. Křeček
Department of Hydrology, Czech Technical University, Prague, Czech Republic

© Capital Publishing Company, New Delhi, India 2017

J. Křeček et al. (eds.), *Ecosystem Services of Headwater Catchments*,
DOI 10.1007/978-3-319-57946-7_4

2 Material and Methods

Environmental impacts of forestry practices were studied in the middle-boreal zone of central Finland, represented by sand to sandy till soils and coniferous forests. The groundwater table was monitored at Pahalouhi (64°28′ N, 27°33′ E, Fig. 4.1), and stream-waters at Oijusluoma (65°50′ N, 29°0′ E, Fig. 4.2) research sites. Additionally, detailed snow observations were carried out at the Paljakka research area (65° 26′ N, 26° 26′ E) in mature spruce (*Picea abies*) stand and in an open grassland.

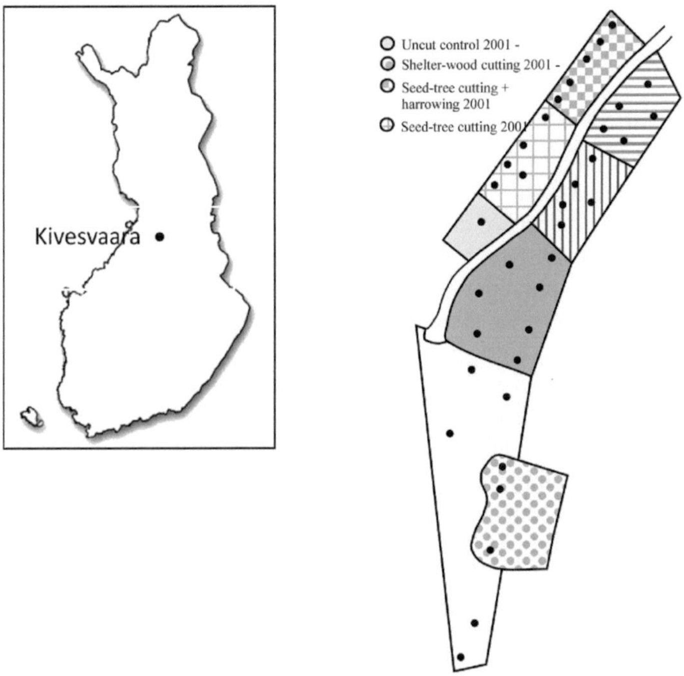

Fig. 4.1 Kivesvaara: Groundwater monitoring at the Pahalouhi research site

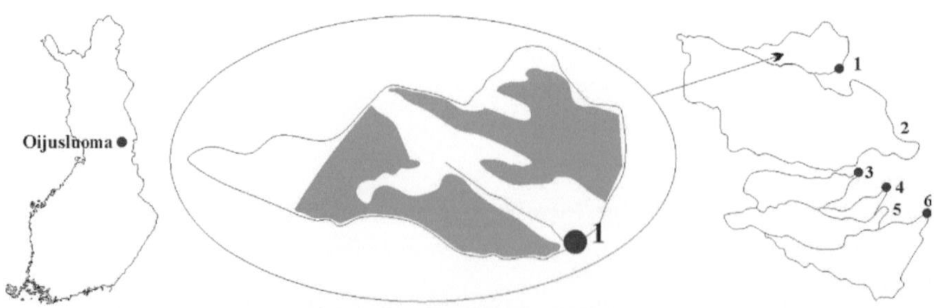

Fig. 4.2 Oijusluoma: A small catchment experiment

In the Pahalouhi experimental area, the continuous groundwater observation started in 1985: 24 observation wells were installed there, and water was sampled by the low-pressure pump at monthly intervals. The water samples were then analysed in the laboratory of the Muhos Research Station; concentrations of nitrate- and ammonia-nitrogen were estimated by standard methods. The experimental area included one control (no treatment) and three experimental plots (ploughed and planted manually, clear-cut without collected waste, and clear-cut with collected waste). In 1986, the clear-cut by a harvester reached 141 m^3/ha (Scots pine – 51%, Norway spruce – 46%, and birch – *Betula pendula*, *Betula pubescens* – 3%). One year after, in 1987, the harvested spots were reforested with plantations of Scots pine *(Pinus sylvestris)*.

In 2001, the Pahalouhi experiment was extended with the natural regeneration by the shelter-wood cut and seed-tree cuttings. The shelter-wood area of spruce was not treated with any site preparation while the Scots pine seed-tree area was harrowed. In the shelter-wood area, there were identified 300 stems per ha, and in the seed-tree plot only 50 stems per ha.

In the Oijusluoma research site (Fig. 4.2), six small catchments covered by mature spruce stands were instrumented in 2007. At the outlets (Fig. 4.2), streamflow has been registered by the water level recorder (120° V-notch weirs). The stream-water sampling has been provided twice a month, chemical analyses were carried out in the Finnish Forest Research Institute and concentrated on the mobility of nitrogen. The daily meteorological data of nearby climate stations (standard observation network) of the Kuusamo region were used.

After 2 years of the calibration (2007–2009), the experimental basin No. 1 was treated: the clear-cut (whole-tree and stump harvest) extended on the area of 12.6 ha, only untreated buffer zones along the stream-channel were respected. The stumps with diameter over 15 cm were harvested, approximately, one third of the logging residues were left uncollected. The paired catchment analysis (Bosch and Hewlett 1982) was used to compare hydrological processes in the pre-treatment (2007–2009) and treatment (2010–2011) periods, compared with the control basin. Simultaneously, the conceptual hydrological model HBV (Bergström 1992) was employed to simulate the hypothetical runoff from the experimental basin not being affected by the treatment.

3 Results

3.1 Groundwater Recharge

3.1.1 Impacts of Clear-Cut and Waste Wood Harvest on Groundwater

At the Pahalouhi site, nitrate-nitrogen concentrations were initially 30–50 μg/l. After the clear-cut, these concentrations were rising for 4–5 years to the peak 500 μg/l in 1990 (Fig. 4.3).The long-lasting effect is evident: still after 17 years the

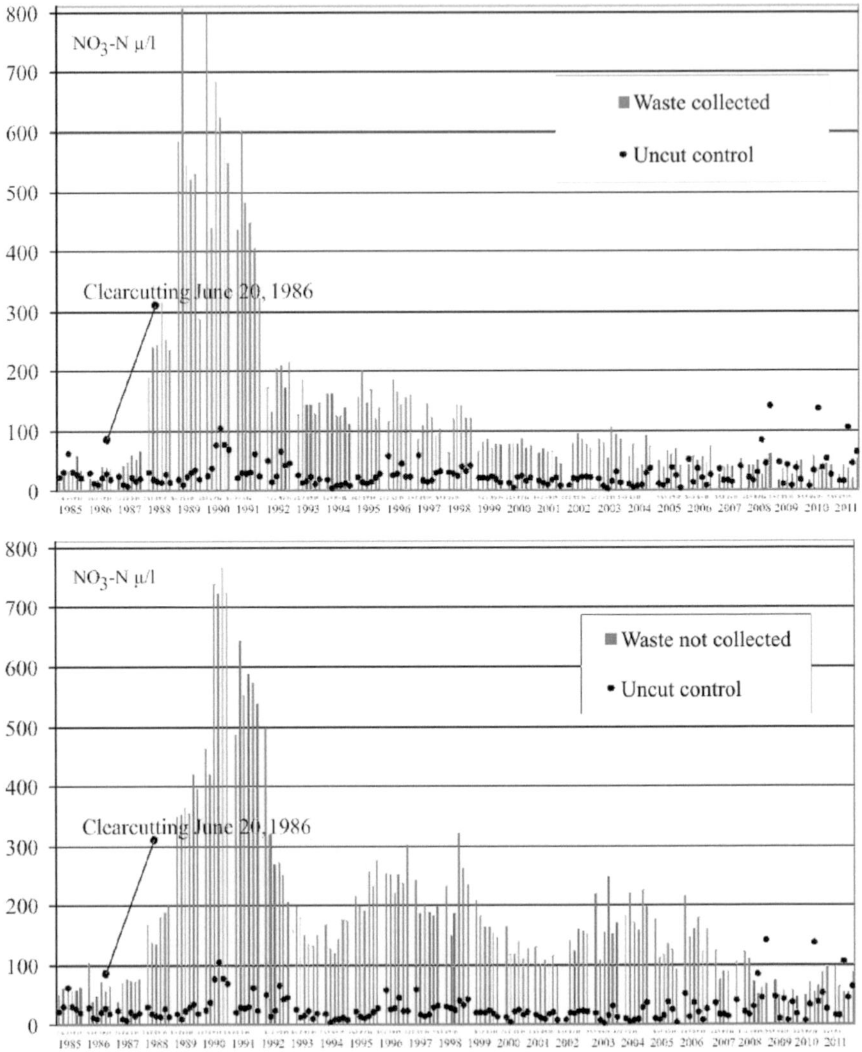

Fig. 4.3 Leaching of nitrate-nitrogen after clear-cutting; waste wood collected (*left*) or not (*right*)

content of nitrate-nitrogen exceeded the pre-treatment values. Significantly lower values were found by collected waste. Contrary to nitrates, contents of ammonia did not show any evidence of increase.

This experiment indicates a long-term disturbance in forest ecosystem services controlling the groundwater quality.

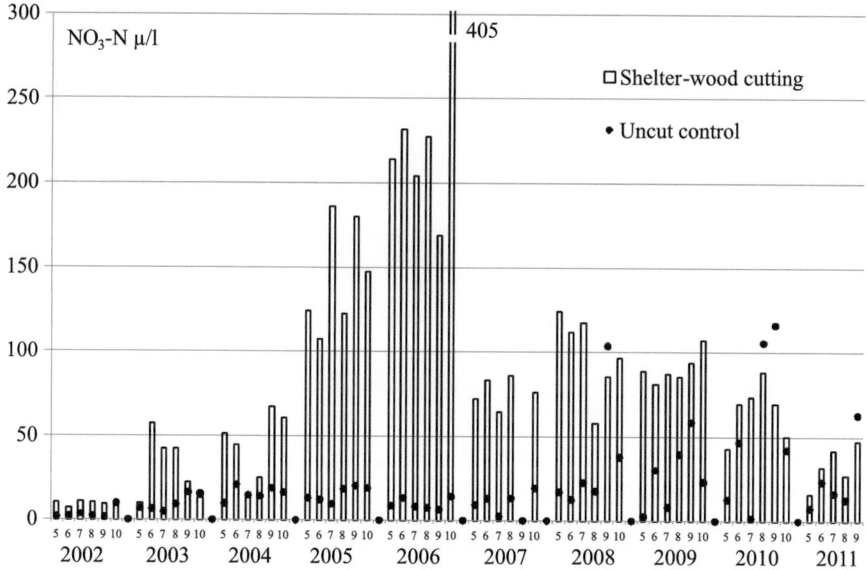

Fig. 4.4 Impacts of shelter-wood cut of spruce stands (followed by natural regeneration) on nitrate concentrations in groundwater (2002–2011)

3.1.2 Impact of Natural Forest Regeneration

In 2001, the shelter-wood cut with natural forest regeneration was performed in mature spruce stands. This treatment initiated increased leaching of nitrate-nitrogen during the period of next 5 years (Fig. 4.4), smaller effect in comparison with the clear-cut.

Similar effects were observed during the natural regeneration of Scots pine: increased nitrate leaching was also evident, but, lower in comparison with spruce stands. In the clear-cut area, the maximum content of nitrate-nitrogen was 800 µg/l, by the natural regeneration of spruce stands 400 µg/l, while in this case only 100 µg/l.

3.2 *Streamflow Processes*

3.2.1 Water Yield

Both, the paired-catchment analysis (Fig. 4.5) and the water balance simulation by HBV model, confirmed the drop in evapotranspiration after the clear-cut. The annual water yield increased by 85 mm (35%) in accordance with the shift in the Penman's potential evaporation (from 269 to 180 mm). The daily stream-flow correlation between the experimental (No. 1) and control (No. 2, 3 and 4) catchments

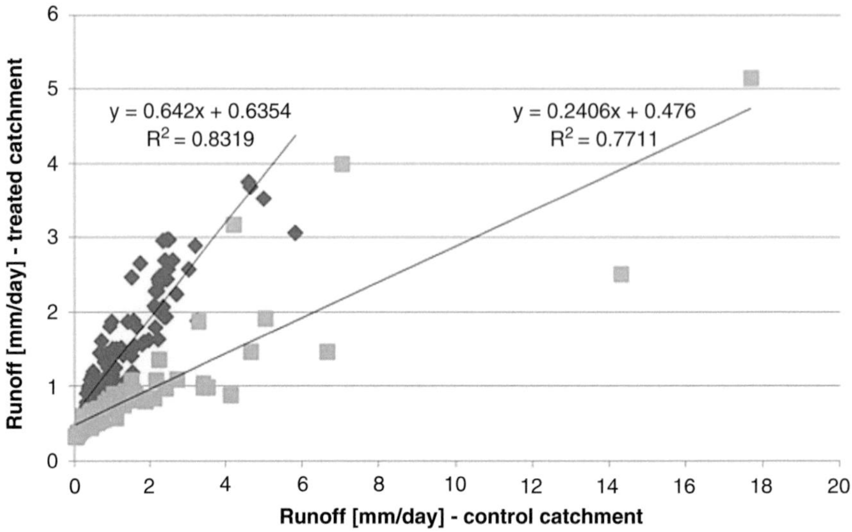

Fig. 4.5 Paired catchments analysis

was found significant, the highest correlation was found between basins 1 and 4 ($R^2 = 0.83$, $R_{crit} = 0.19$). Therefore, the pair of catchments 1 and 4 was accepted in the further hydrological analysis. The simulation done by the HBV model provided us with more detailed information concerning the runoff genesis within the experimental catchment: increasing groundwater table and base-flow discharge after the clear-cut.

3.2.2 Runoff Timing

After the clear-cut, peak-flows at the experimental catchment increased by 62–103% (Fig. 4.6) but minimum flows did not change significantly. From snow-melt and rain-storm hydrographs observed before and after the clear-cut, significant changes are evident there, including changes in the lag-time and the time of concentration. After the clear-cut, peak-flows are generated approximately 1 day earlier in comparison with the original mature spruce stands; and, volumes of the direct runoff changed from CN = 47 to CN = 68.

3.2.3 Water Quality

After the clear-cut, more significant relationship between the daily discharge and nitrate-nitrogen concentrations in the outlet of the experimental catchment (Fig. 4.7) corresponds to increasing mobility of nitrogen in soil water as well as with more intensive stream-flow responses. Thus, in stream-flow, concentrations of

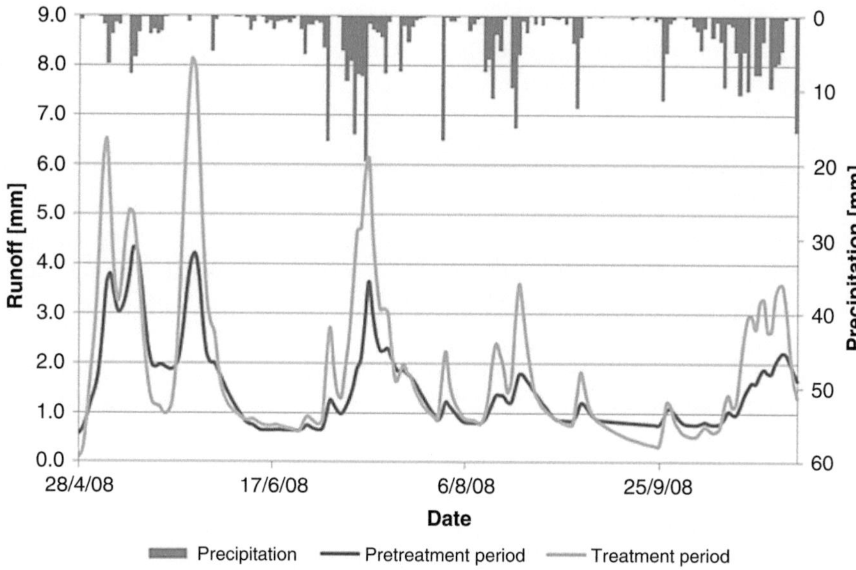

Fig. 4.6 Changes in runoff timing

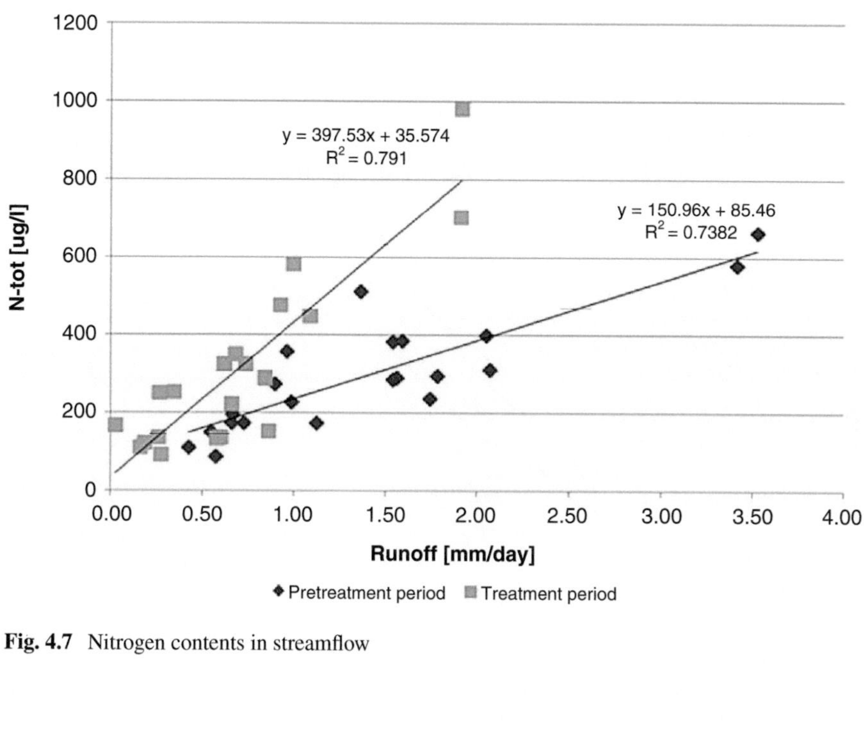

Fig. 4.7 Nitrogen contents in streamflow

nitrogen increased up to 100%. Those changes result mainly from significant rising content in nitrate-nitrogen, while the content of ammonium did not change significantly.

3.3 Runoff Genesis

During a snow season, the ratio between the observed snow-water equivalent and the snow-depth increased from 0.12 to 0.25. The final snowpack depletion in the mature spruce stand occurred over 44 days, compared to 25 days in grassland areas with a mean melt intensity (7 versus 10 mm day^{-1}) (Fig. 4.8). In the high melt period (April–May), the forest stand showed lower degree-day factors (2.3–3.5 mm °C^{-1} day^{-1}) compared to the more compact snowpack of the grasslands (3.8–6.0 mm °C^{-1} day^{-1}). Observed maximum daily snow-melt intensities varied between 30 and 50 mm.

It is evident that, in cold boreal forests, intercepted snow may be retained in the canopy much longer, compared with temperate forests. The mature boreal spruce stand maintains a higher snowpack (in average by 21 cm, 26%) compared to the grassland. The maximum snow-water content observed in the forest exceeds that in the open by 32%. The snow-water intercepted in the canopy reached a maximum of 27% of the snow stored in the forest ground, and the loss of intercepted snow by sublimation was approximately 26% of the annual snowfall.

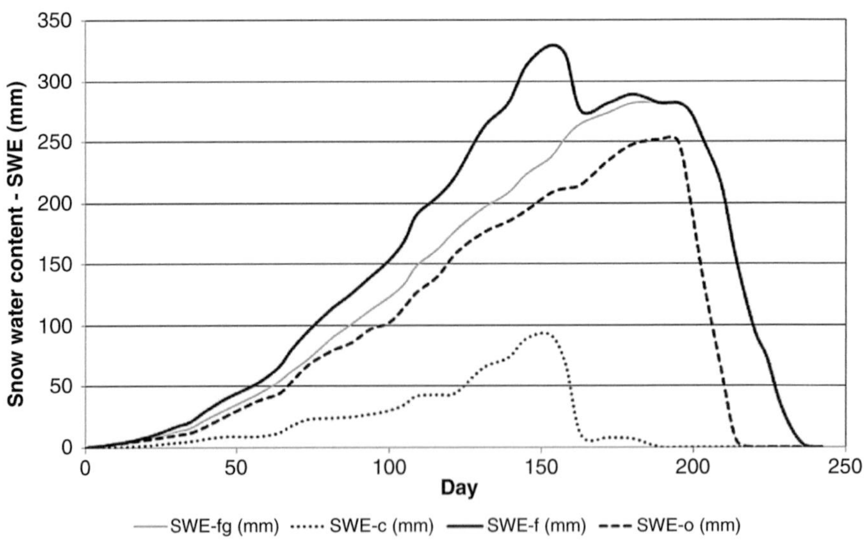

Fig. 4.8 Mean annual snow-water content in the boreal environment (Paljakka, 1992–2008): Spruce forest total (SWE-f), forest canopy (SWE-c), forest ground (SWE-fg) and open grassland (SWE-o) (Kremsa et al. 2015)

4 Conclusions and Discussion

The effects of forestry practices on mobility of nitrogen and contamination of water environments have been reported already by several studies worldwide (Tamm et al. 1974; Bormann and Likens 1979; Kubin 1987; Ahtiainen and Huttunen 1999; Piirainen et al. 2008; Kubin and Krecek 2008), but the long-term impact of increasing nitrogen concentrations in groundwater (persisted some 25 years after the clear-cut, and 10 years after the natural regeneration) have not been reported. By almost constant atmospheric deposition, the rising concentrations of nitrate are evidently due to the decomposition of the logging waste and humus (Kreutzweiser et al. 2008). Contrary to nitrate, there was not any evidence of increasing contents of ammonia (confirmed in boreal forests also by Rusanen et al. (2004). Therefore, ecosystem services of boreal forests are significantly affected by forestry practices. The impacts of forest regeneration on aquatic ecosystems can be prevented quite effectively by surface waters (Kubin et al. 2000), but more complicated in the groundwater protection.

The role of the boreal spruce forest in the local water cycle is particularly important for supporting the seasonal water resources recharge, and reducing the risk of spring floods by controlling the processes of snow accumulation and melt. However, to control the flood risk by watershed planning, the design with a certain percentage of open areas (or forest openings) might still help to prolong the snowmelt period. The clear-cut might increase water yield (a positive feedback) but also cause rising peak-flows, reducing lag-time, and increasing mobility of nitrate. The main reason of rising water yield is the drop in evapotranspiration (approx. 22% of the annual value). The estimated increase in water yield (85 mm, e.g. 35%) has been recognised reasonable with the canopy removed from the watershed. These findings correspond also to results reported by several studies on forest hydrology summarized by Bosch and Hewlett (1982). Of course, such an increase in stream-flow usually diminished with the forest re-growth and recovery of ground vegetation. In the Oijusluoma Research Area, the significant rise in water yield lasted for 5 years. On the other hand, the negative feedbacks in rising flood events and nitrogen concentrations might be critical in water quality and biota control. Hypothetically, the clear-cut not including logging residues increases the amount of decomposable organic matter only, but the soil disturbances (stump removal and site preparation) could increase the intensity of decomposition, mobilise nitrogen in the soil water and rise nitrogen concentrations in stream-water (Grip 1982; Ahtiainen and Huttunen 1999; Watmough and Dillon 2004).

Acknowledgements The study was supported by the Finnish Ministry of Agriculture and Forestry (Programme on Groundwater and catchment impacts of stump harvesting) and the Czech Technical University at Prague (SGS 14/128/OHK1/2 T/11). The experimental treatment was managed by Metsähallitus, field observations and chemical analyses carried out by LUKE (Natural Resources Institute Finland). The Oulanka Research Station of the Oulu University supported water sampling and run-off registrations.

References

Ahtiainen, M., & Huttunen, P. (1999). Long-term effects of forestry managements on water quality and loading in brooks. *Boreal Environment Research, 4*, 101–114.

Bergström, S. (1992). The HBV model—Its structure and applications. In *Institut reports hydrology 4*. Norrköping: Sweden.

Bormann, F. H., & Likens, G. E. (1979). *Pattern and process in a forested ecosystem*. New York\Heidelberg\Berlin: Springer.

Bosch, J. M., & Hewlett, J. D. (1982). A review of catchment experiments to determine the effect of vegetation changes on water yield and evapotranspiration. *Journal of Hydrology, 55*, 3–23.

Grip, H. (1982). *Water chemistry and runoff in forest streams at Kloten*. Uppsala University, PhD thesis, UNGI rapport 58.

Hämet-Ahti, L. (1981). The boreal zone and its biotic subdivisions. *Fennia, 159*, 69–75.

Kremsa, J., Křeček, J., & Kubin, E. (2015). Comparing the impacts of mature spruce forests and grasslands on snow melt, water resource recharge, and run-off in the northern boreal environment. *International Soil and Water Conservation Research, 3*, 50–56.

Kreutzweiser, D. P., Hazlett, P. W., & Gunn, J. M. (2008). Logging impacts on the biogeochemistry of boreal forest soils and nutrient export to aquatic systems: A review. *Environmental Reviews, 16*, 157–179.

Kubin, E. (1987). Site preparation and leaching of nutrients. *Metla Research Papers, 567*, 55–62.

Kubin, E. (1995). The effect of clear cutting, waste wood collecting and site preparation on the nutrient leaching to groundwater. In L. O. Nilsson, R. F. Hüttl, & U. T. Johansson (Eds.), *Nutrient uptake and cycling in Forest Ecosytems*. Dordrecht: Kluwer.

Kubin, E. (1998). Leaching of nitrate nitrogen into the groundwater after clear felling and site preparation. *Boreal Environmental Research, 3*, 3–8.

Kubin, E., & Kemppainen, L. (1991). Effect of clear cutting of boreal spruce forest on air and soil temperature conditions. *Acta Forestalia Fennica, 225*, 1–42.

Kubin, E., & Krecek, J. (2008). Impact of forestry practices on groundwater quality in the boreal environment. In M. Taniguchi, W. C. Burnett, Y. Fukushima, M. Haigh, & Y. Umezawa (Eds.), *From Headwaters to the Ocean*.

MEA. (2005). *Ecosystems and human well-being: Synthesis, Millenium Ecosystem Assessment*. Washington, DC: Island Press.

Piirainen, S., Finér, L., Mannerkoski, H., & Starr, M. (2008). Effect of clear-cutting and site preparation on the level and quality of groundwater in some headwater catchments in eastern Finland. *Forest Ecology and Management, 243*, 10–18.

Rusanen, K., Finer, L., Antikainen, M., Korkka-Niemi, K., Backman, B., & Britschgi, R. (2004). The effect of forest cutting on the quality of groundwater in large aquifers in Finland. *Boreal Environmental Research, 9*, 253–261.

Tamm, C. O., Holmen, H., Popovic, B., & Wiklander, G. (1974). Leaching of plant nutrients from soils as a consequence of forestry operations. *Ambio, 3*, 211–221.

Watmough, S. A., & Dillon, P. J. (2004). Major element fluxes from coniferous catchment in Central Ontario. *Biogeochemistry, 67*, 369–398.

Chapter 5
Climate Change Impacts on Water Resources in a Snow-Dominated Watershed of Northern Japan

Yoshinobu Sato

1 Introduction

The average air temperature over Japan during the period of 1981–2010 was approx. 10.3 °C, but the regional difference (from the northern to southern regions of Japan) was more than 10 °C (Fig. 5.1). Furthermore, since 1910, the mean annual air temperature over Japan has increased by 1.2 °C. The total annual precipitation in Japan during the same period was about 1745 mm, most of which fell over the coastal regions of the Sea of Japan and the Pacific Ocean (Fig. 5.2). In southwestern Japan, most of the precipitation on the Pacific Coast falls as rainfall delivered by tropical cyclones (typhoons) or frontal rainfalls during the summer. In contrast, most of the precipitation on the more northern coast of the Sea of Japan falls as snow, which is driven by cold westerly winds and warm ocean currents during the winter season. As shown in Fig. 5.3, the region of Japan that receives heavy snowfall occupies about 50% of its total land area and affects 16% of the Japanese population.

According to future climate projections from high-resolution global climate models (GCMs), it is expected that the average annual air temperature will rise more than 3 °C by the end of this century. With this increase in air temperature, snowfall is expected to decrease during the winter, and subsequently, snowmelt supplied to rivers will also decrease in early spring. These changes have the potential to seriously impact cold mountainous regions, as the water supply of these regions is typically dominated by snowmelt (Sato et al. 2013). Therefore, the impact of climate change on water resources will be clearer in areas in which the current average winter air temperature is less than 0 °C and predicted to be more than 0 °C in the future. In this study, we examined past trends in air temperature and precipitation in a snow-dominated watershed of northern Japan, and estimated future climate-driven impacts to the region through the use of high-resolution hydrological modelling.

Y. Sato (✉)
Faculty of Agriculture, Ehime University, Matsuyama, Japan
e-mail: sato@agr.ehime-u.ac.jp

© Capital Publishing Company, New Delhi, India 2017

J. Křeček et al. (eds.), *Ecosystem Services of Headwater Catchments*,
DOI 10.1007/978-3-319-57946-7_5

Fig. 5.1 Average annual air temperature (°C) in Japan (1981–2010)

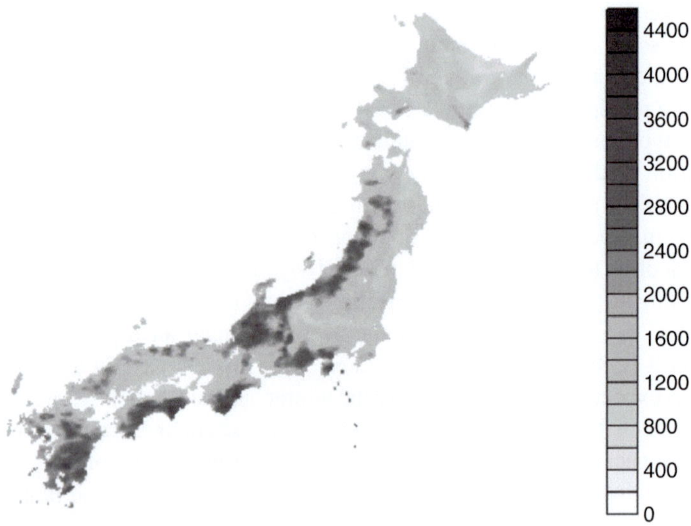

Fig. 5.2 Total annual precipitation (mm) in Japan (1981–2010)

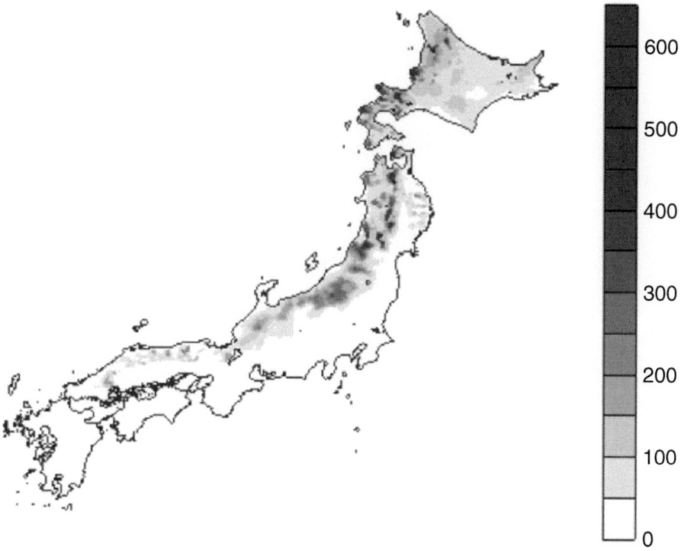

Fig. 5.3 Maximum annual snow depth (cm) in Japan (1981–2010)

2 Field Study

2.1 Tedori River Basin

The first part of this paper concerns the Tedori River basin located in the Ishikawa prefecture. The total basin area is 809 km^2 with a stem channel length of 72 km. The maximum elevation is approximately 2702 m at Mt. Hakusan, and 92% of the basin is covered by coniferous forest. The total annual precipitation is concentrated in the mountainous area (approx. 3000 mm) and gradually decreases along the lower reach of the river (approx. 2500 mm). This river supplies drinking water for 75% of the population of Ishikawa prefecture, as well as water for irrigation and hydroelectric power generation. There are two dam reservoirs (Tedori, 190.0 million m^3; Dainichi, 23.9 million m^3) used for flood control, as a water supply, and for river environment conservation. The channel network, location of dams, and the reference point for river discharge measurements are shown in Fig. 5.4.

2.2 Meteorological and Hydrological Observations

The meteorological data (precipitation, air temperature, wind speed, and sunshine duration) were obtained from the Automated Meteorological Data Acquisition System (AMeDAS) sites. The atmospheric and vapour pressure data were obtained from local observatories. Furthermore, the long-term mean daily data of discharge in

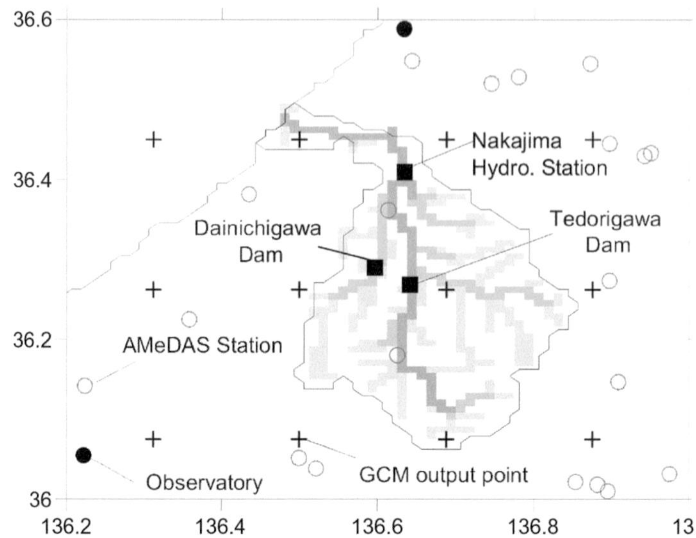

Fig. 5.4 Tedori River basin

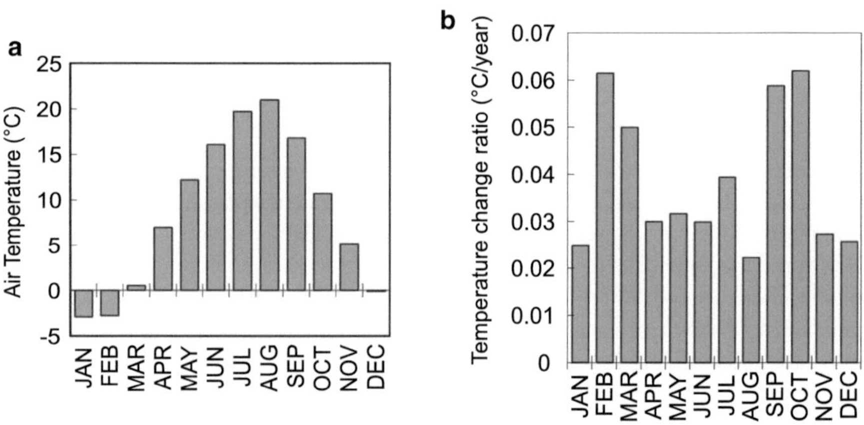

Fig. 5.5 **(a)** Average monthly air temperature in the Tedori River basin **(b)** Monthly temperature change ratio in the Tedori River basin

the reference profile of Nakajima at the Tedori River were received from the Japan River Association. Mean monthly air temperature in the Tedori watershed is shown in Fig. 5.5a. The mean monthly air temperature from December to February remained less than 0 °C during the period of 1980–2009. However, during these 30 years, the average annual air temperature increased continuously at a rate of approximately +0.03 °C/year, and the monthly temperature change was particularly significant (more than +0.05 °C/year) at the end of winter (February to March) and in early autumn (September to October; Fig. 5.5b). As shown in Fig. 5.6a, the total annual

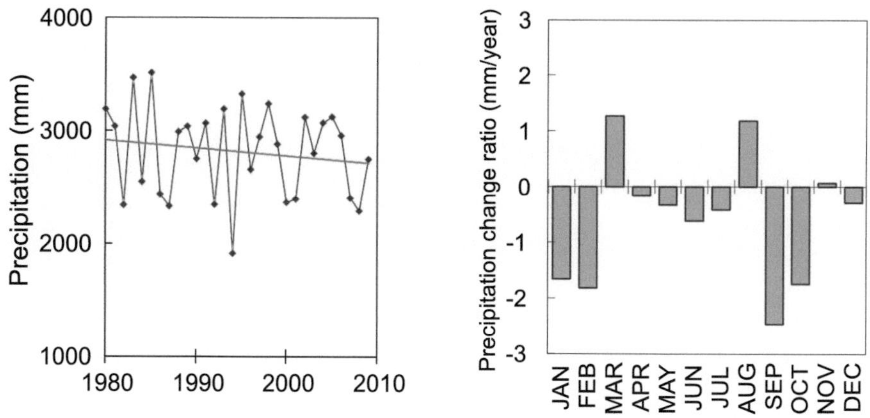

Fig. 5.6 **(a)** Total annual precipitation in the Tedori River basin **(b)** Monthly precipitation change ratio in the Tedori River basin

precipitation in the Tedori River basin showed a decreasing trend (−6.99 mm/year), which suggests that the amount of water available from the Tedori River basin must be gradually decreasing. The changes in monthly precipitation are shown in Fig. 5.6b. Monthly precipitation during mid-winter (January to February) and in early autumn (September to October) showed a significant decreasing trend (> −1.0 mm/year; Fig. 5.6b). The depth of 1 mm of precipitation represents here (the Tedori basin of 809 km^2) a volume of 0.809 million m^3. Therefore, a decrease of −6.99 mm/year corresponds to a loss of 5.65 million m^3 of available water per year, which is 2.64% of the total storage capacities of the two reservoirs in the Tedori River.

The monthly discharge observed at Nakajima (Tedori river) during 1980–2010 is shown in Fig. 5.7. Although monthly precipitation in the mid-winter is decreasing (Fig. 5.6b), the river discharge showed an increasing trend. This may be due to an increase in rainfall rather than snowfall due to the rising air temperature (Fig. 5.5b), which would result in higher discharge during the winter. In contrast, monthly river discharge showed a significant decreasing trend during the spring and early summer (April–June). This is likely due to decreasing snowfall during the winter, and subsequently also decreasing snowmelt freshet during the spring and early summer. Figure 5.8 shows the monthly change in the river discharge (m^3/s/year) at the Nakajima station in the Tedori River. From these data, it would appear that the river flow regime is already shifting due to climate change. Figure 5.9 shows the minimum value of the mean monthly discharge of the Tedori River during the period of 1980–2010. The amount of monthly water allocations for irrigation is also shown. Over the past 30 years, the mean monthly discharge has never decreased to a level of irrigation allocations per month. However, when this trend in the river discharge will continue, severe water shortages will probably occur during the growing season (in the sensitive seedling stage) in the near future. Therefore, to adapt to these negative impacts of climate change, Japanese growers may have to adjust their cropping schedule, shift to more drought-hardy cultivars, and/or develop new water resources as soon as possible.

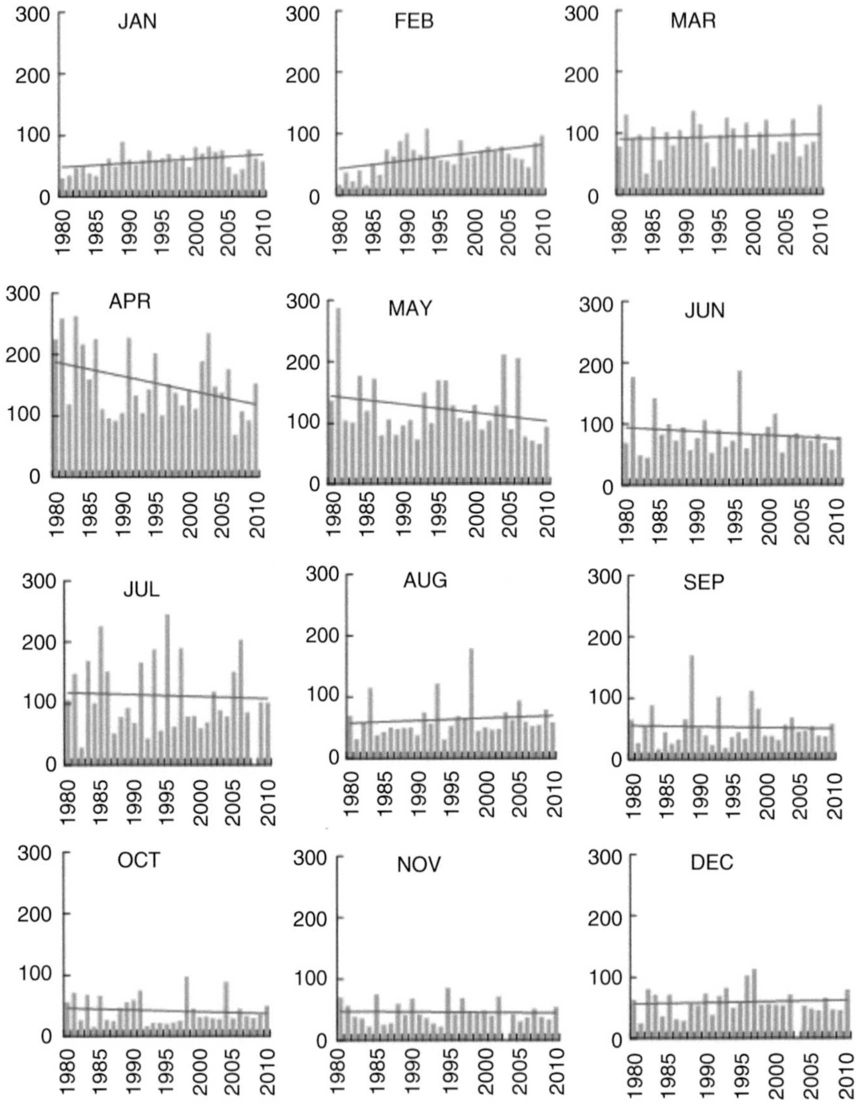

Fig. 5.7 Monthly river discharge (m^3/s) at Nakajima hydrological station in the Tedori River basin

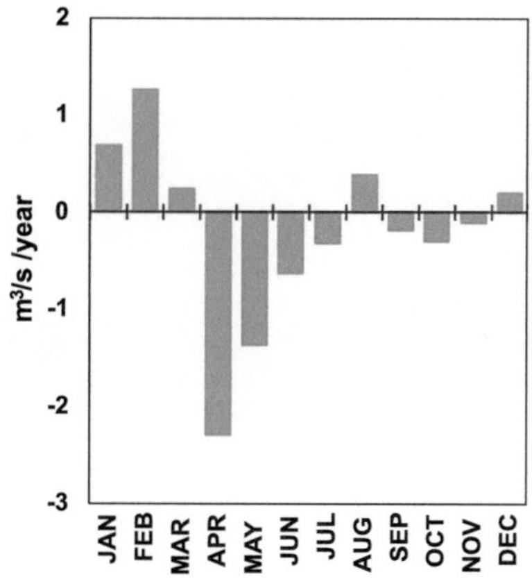

Fig. 5.8 Monthly change in river discharge in the Tedori River basin

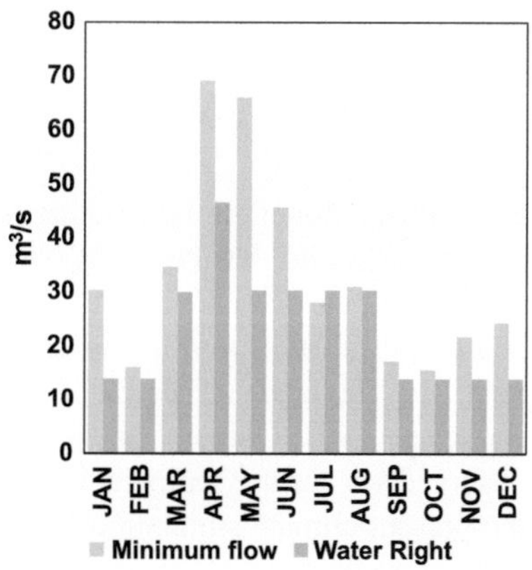

Fig. 5.9 Minimum monthly river discharge and irrigation water allocation (Water Right) in the Tedori River basin

3 Outline of the Model Simulation

3.1 Hydro-BEAM

The Hydrological River Basin Environment Assessment Model (Hydro-BEAM), developed by Kojiri (2006), was used for the hydrological simulation. As shown in Fig. 5.10, surface and subsurface flow from the upper soil layer and channel flow were calculated using a kinematic wave model. Baseflow from the lower soil layer was calculated using a multi-layer linear storage function model. However, reservoir operations, such as flood control, water supply, and water withdrawal from the river channel, were not considered in this analysis. All of the input variables for Hydro-BEAM were interpolated to a 1 km grid mesh using the inverse distance-weighting (IDW) method.

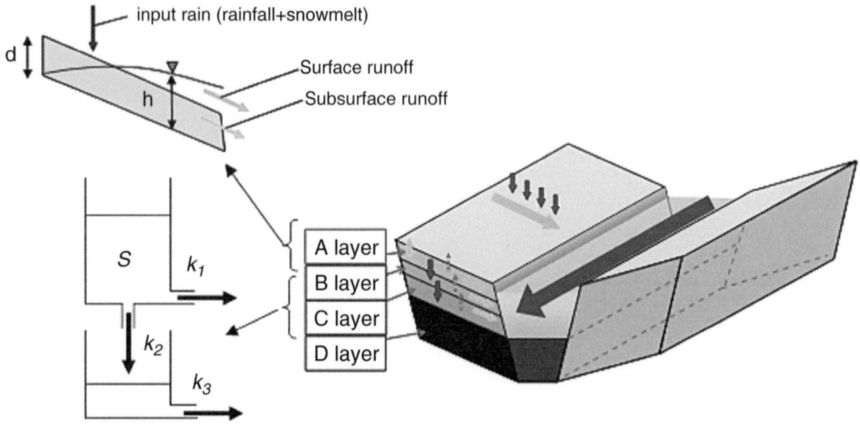

Fig. 5.10 Schematic structure of the hydrological model

3.2 *MRI-AGCM3.2S*

The climate model used for future projections was the MRI-AGCM3.2S. The MRI-AGCM3.2S was developed by the Japan Meteorological Agency (JMA) of the Meteorological Research Institute (Mizuta et al. 2000). The transform grid was composed of 1960 × 960 grid cells, corresponding to approximately 20 km grid intervals. The land surface model used was the new version of the JMA Simplified Biosphere (SiB; Ohizumi and Hosaka 2000). For the lower boundary conditions, observed sea surface temperatures (SSTs) and sea ice concentrations were used for the present climate simulation (1979–2003). For the climate simulations up to the end of the twenty-first century (2075–2099), the SSTs projected by the Coupled Model Intercomparison Project phase 5 (CMIP5) multi-model ensemble dataset were used. These data were used assuming the RCP8.5 scenario in the Fifth Assessment Report (AR5) of the Intergovernmental Panel on Climate Change (Mizuta et al. 2014). The input variables used for hydrological simulation are summarized in Fig. 5.11.

3.3 *Future Climate Projected via the MRI-AGCM3.2S*

According to the projection derived using the MRI-AGCM3.2S (RCP8.5 scenario), the air temperature in the future (2075–2099) will increase approximately 4.17 °C compared to the present climate (1979–2003) in the Tedori River basin. In contrast, the annual and monthly precipitation did not change significantly (Fig. 5.12). Figure 5.13 summarizes the change in water balance in the Tedori River basin between the present and future climate conditions. This comparison showed that although the amount of precipitation does not change markedly, the overall water

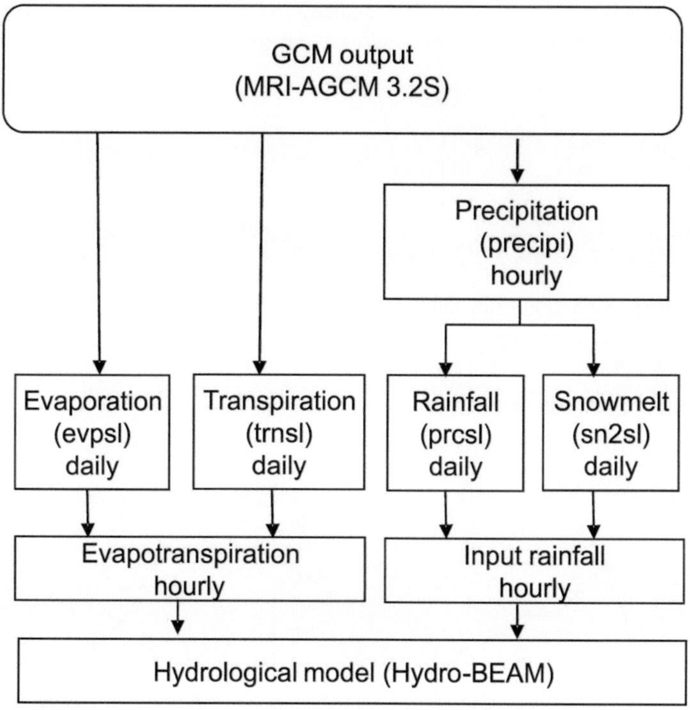

Fig. 5.11 Flow chart to input variables for the hydrological model simulation

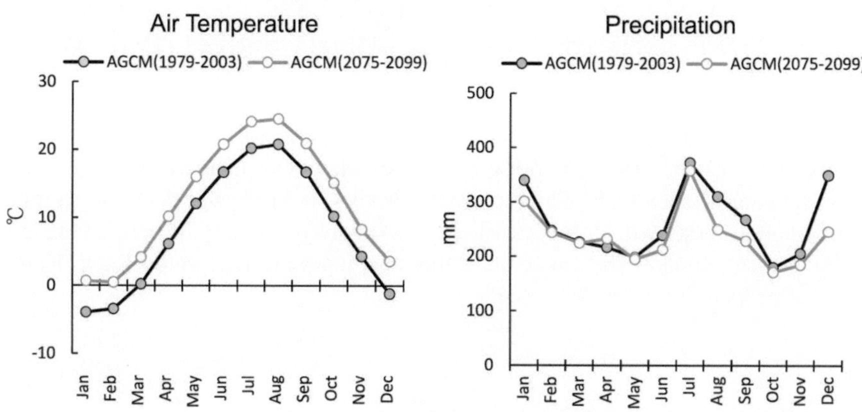

Fig. 5.12 Comparison of average monthly air temperature and precipitation projected by MRI-AGCM3.2S (RCP8.5 scenario)

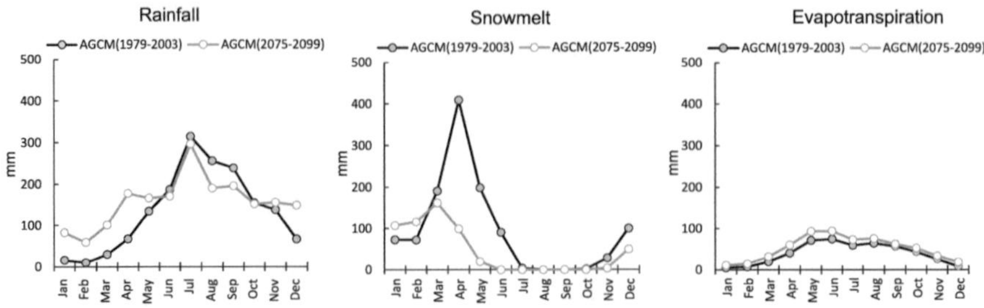

Fig. 5.13 Comparison of monthly rainfall, snowmelt, and evapotranspiration projected by MRI-AGCM3.2S (RCP8.5 scenario)

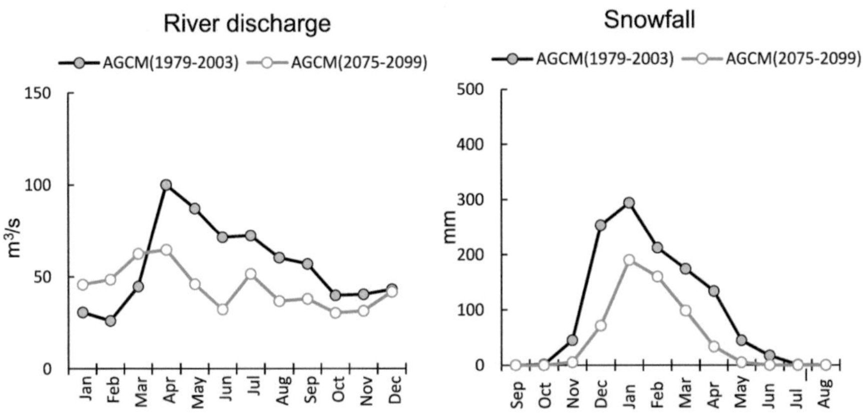

Fig. 5.14 Comparison of monthly river discharge and snowfall projected by MRI-AGCM3.2S (RCP8.5 scenario)

balance will change significantly with a rise in temperature. Rainfall in the winter (December to February) and spring (March to May) will increase. However, the peak amount of snowmelt, which currently occurs in April, will decrease significantly and the peak will shift backward approximately 1 month (March) compared to the present climate. The impact of evapotranspiration change in the Tedori River basin appears to be small.

4 Climate Change Impacts on Water Budget

In the future, river discharge in this region is likely to change significantly due to an increase in rainfall in the winter to spring, and a decrease in the snowmelt freshet in the early spring. In the case of the Tedori River basin, the monthly river discharge from January to March will initially increase and then decrease from April to November (Fig. 5.14, left). These changes will lead to uncertainty in seasonal river discharge and its uses. In the colder regions of Japan, paddy field irrigation depends heavily on river discharge in the late spring (April to May), and future decreases may thus have severe impacts on agricultural productivity in this region. Our data also indicate that the amount of snowfall and the period of snow cover will also decrease in the future (Fig. 5.6, right), which will impact local industries such as ski resorts, and will also affect public expenditures on snow removal from railways, airports, highways, and so forth.

The total annual snowfall will decrease to about 613 mm under the projected future climate conditions. This decrease corresponds to 495.9 million m^3, which is 2.3-fold larger than the current total storage capacity of the two reservoirs in the Tedori river basin. Snowpack accumulation during the winter acts as a natural water storage reservoir, which then supplies copious irrigation water in the spring. However, if the snowpack decreases due to climatic warming, that reservoir will be lost. It may become necessary to store rainwater during the winter season, either by adjusting the operation of the existing dams or by developing new water storage facilities.

5 Conclusion

To estimate the impacts of climate change on water resources in snow-dependent watersheds of Japan, this study focused on the Tedori River basin. Analysing the long-term observation, air temperature in this region has been rising continuously, and river discharge is already showing an increasing trend during the winter and a decreasing trend during the early spring. These data suggest that if warming trends continue, severe water shortages may occur during the cropping season in the near future. To evaluate these risks under future climate conditions, we conducted a hydrological model simulation using a super-high-resolution atmospheric general circulation model (MRI-AGCM3.2S). The results showed that, in the future, even when precipitation will not change dramatically, the river discharge will be significantly decreasing as a result of increasing air temperature (owing to increased rainfall and decreased snowfall in the winter, and, reduced snowmelt in early spring). These changes will be more serious in the cold northern mountainous regions, because the water resources recharge there is highly dependent on processes of snow-accumulation and snowmelt.

Acknowledgements This work was conducted under the framework of the “Precise Impact Assessments on Climate Change” of the Program for Risk Information on Climate Change (SOUSEI Program) supported by the Ministry of Education, Culture, Sports, Science, and Technology-Japan (MEXT).

References

Kojiri, T. (2006). Hydrological river basin assessment model (Hydro-BEAM). In V. P. Singh & D. K. Frevent (Eds.), *Watershed models*. Florida: Taylor & Francis, CRC Press.

Mizuta, R., Oouchi, K., Yoshimura, H., Noda, A., Katayama, K., Yukimoto, S., Hosaka, O. M., & Hosaka, M. (2000). Land surface processes in appendix to progress report on numerical weather prediction. *Japan Meteorological Agency, 46*, 48–66. (in Japanese).

Mizuta, R., Arakawa, O., Ose, T., Kusunoki, S., Endo, H., & Kitoh, A. (2014). Classification of CMIP5 future climate responses by the Tropical Sea surface temperature changes. *SOLA, 10*, 167–171.

Ohizumi, M., & Hosaka, M. (2000). Land surface processes: Appendix to a progress report on numerical weather prediction. *Japan Meteorological Agency, 46*, 48–66 (in Japanese).

Sato, Y., Kojiri, T., Michihiro, Y., Suzuki, Y., & Nakakita, E. (2013). Assessment of climate change impact on river discharge using the super-high resolution MRI-AGCM. *Hydrological Processes, 27*, 3264–3279.

Chapter 6
Ecohydrological-Based Forest Management in Semi-arid Climate

Antonio D. del Campo, M. González-Sanchis, A. Lidón, A. García-Prats, C. Lull, I. Bautista, G. Ruíz-Pérez, and F. Francés

1 Introduction

The role of forests on the provision and regulation of non-marketed ecosystem services is well known (Thorsen et al. 2014). This is especially important in areas like the Mediterranean, where protective forests play a major role against soil erosion and degradation, landscape quality and stabilization of the hydrological cycle. Socio-economic and cultural changes affecting rural society from the 1960s have produced a demographic decline, and with it, an abandonment of rural activities, leading to an expansion and densification of forest and scrub. Forest encroachment may decrease the streamflow from upper catchments (Gallart and Llorens 2004); this study reports a decrease in average annual flow of major Spanish rivers between 37 and 59%, partly explained by the densification of upstream forests, and increasing interception loss. Moreover, some Mediterranean basins (e.g. Segura and Júcar in Spain) present very serious problems of water scarcity, because of a combination of low/irregular rainfall and high rates of evapotranspiration, that has resulted in overuse of groundwater resources (Estrela et al. 2000a, b). These problems may even endanger urban water supply (approx. 15% of the total water supply in Spain). In addition, the Mediterranean region is already suffering some significant impacts of the climate change, such as longer dry seasons, or lower soil moisture content (Giorgi and Lionello 2008; García-Ruíz et al. 2011). All these issues have raised

A.D. del Campo (✉) • M. González-Sanchis • A. Lidón • A. García-Prats • C. Lull
I. Bautista • F. Francés
Research Institute of Water and Environmental Engineering, Universitat Politècnica de València, Valencia, Spain
e-mail: ancamga@dihma.upv.es

G. Ruíz-Pérez
Department of Crop Production Ecology, Swedish University of Agricultural Sciences (SLU), Uppsala, Sweden

© Capital Publishing Company, New Delhi, India 2017

J. Křeček et al. (eds.), *Ecosystem Services of Headwater Catchments*,
DOI 10.1007/978-3-319-57946-7_6

concern about the importance of forests and water interactions in the Mediterranean (Birot et al. 2011).

Nevertheless, in spite of increasing forest area, statistics show a decrease in forest management. In Spain, since 1990, the logging volume has dropped from 60 to 38% of annual growth (MMARM 2006). Current forest management practices in semi-arid low-productive forests are costly and completely unprofitable, so only public forests receive partial attention through fire-prevention silviculture or salvage logging after wildfires. The lack of both appropriate socio-economic structures and incentive policies has prevented the development of markets for forest services (Thorsen et al. 2014). As a result, low-productive forests usually end up unmanaged and abandoned, which leads to an increase of the rural disaffection with forests. Climate change is also a key factor posing additional strain onto unmanaged semi-arid forests, where an urgent adaptive management has been highlighted (Fitzgerald et al. 2013) in order to preserve their provision of goods and services. In contrast with other biomes, there exists a sharp consensus that semi-arid forests are negatively affected by changes in the growing season and on soil water regime (Lidner et al. 2014), resulting in loss of resilience and leading to reduced growth, higher susceptibility to drought and water scarcity, wildfires, pest and disease outbreaks, dieback, etc. Forestry is based on the sustainable management of natural resources (i.e. goods and services such as timber, pulp, water and soil conservation, wildlife, game and recreation). However, traditional forest planning and management has been more focused on productive functions (timber, pulp), leaving the rest of the resources almost unattended.

Typically, important decisions such as the rotation age, silvicultural system or type of regeneration are designed to maximize and sustain the yields of timber. This approach might not be profitable enough in the case of low-productive semi-arid forests, where the concept of multiple-use resource management, which pays attention not only to productive but also to protective and social functions, has gained increased attention during the last decades. Accordingly, these forests need an urgent change of the management approach that puts key goods and services (i.e. not timber) on the centre of the management decisions and includes as much resources as possible. Besides enhancing productive functions, forest management may also contribute to reduce the risk of forest fires, increase ecosystem resilience (Millar et al. 2007), increase water yield (Callegari et al. 2003; Molina and del Campo 2012; del Campo et al. 2014), improve tree growth and vigour (Mitchell et al. 1983), landscape value, etc. These goods and services are assumed to be modified (in some cases enhanced) when doing forest management, but in most cases they are not quantified. The tipping point with regards previous management approaches is that all these functions should be explicitly considered and quantified into forest planning and management. This paper aims to improve the understanding and quantification of forest and water relationships as affected by forest management in semi-arid environments.

2 Ecohydrology and Forestry

Ecohydrology is especially important in arid and semi-arid systems where the feedbacks between ecology and hydrology can be particularly tight. Forests have a strong interaction with water capable of controlling most of its functions. This interaction varies across hydro-climatic gradients (Asbjornsen et al. 2011), where different strategies are used. Water-limited environments, such as arid or semi-arid, show a more obvious water-plant relationship where plant growth is often controlled by stochastic pulses of water that directly affect plants' ability to adapt and survive. Several studies addressed specifically the effects of forest management on water yield (Bosch and Hewlett 1982). In semi-arid regions, adaptive management might be focused on forest and water relationships, so guidelines for this silviculture must address how the physical structure of forests can be shaped according to the water cycle components being modified (rainfall interception, throughfall, transpiration, soil moisture and deep infiltration) (Ungar et al. 2013). Sustainable forest management requires maintenance of the biological, chemical and physical properties and processes of soils. In natural forest ecosystems, soil organic carbon stock is usually high and in equilibrium but, theoretically, it can diminish following disturbance (e.g. after thinning treatments) by removing different amounts of biomass and timber (Jandl et al. 2007). However, some studies reveal that historic changes in forest management have not had a detectable effect on present forest soil organic carbon stocks in central Europe (Wäldchen et al. 2013). Similarly, microbial communities and nutrient cycling may be also affected—and enhanced—by forest management (Overby et al. 2015). Because of these potential effects, ecohydrological-based forest management must target and address both soil and biogeochemical issues.

Decreasing climate-related vulnerabilities of forests is one of the goals of adaptive forest management (Fitzgerald et al. 2013). In semi-arid climates, a generalized warming and drying trend has been reported to affect negatively the growth of forests (Sarris et al. 2007). The strong dependence of semi-arid forests to water makes them more sensitive to climate change, where variations in precipitation and temperature can markedly influence forest changes that usually lead to forest dieback and growth decline (García de la Serrana et al. 2015). In extreme cases, long, intense and continuous droughts have a serious impact on tree-growth leading to an increasing number of missing tree rings, water stress-related tree mortality and insect attacks (García de la Serrana et al. 2015). Studies on tree growth–climate relationships evidence that forest management makes stands less sensitive to water shortages in drought-prone sites (Sohn et al. 2013). Also, adverse meteorological conditions and dense forests stands increase the risk of forest fires. Climatic stressors such as extended periods of high temperatures or heat waves, low relative humidity and strong winds, will in all likelihood, alter the frequency, intensity, and the extent of fires (Planinsek et al. 2011). Changes in forest structure due to partial removal of the forest is a fire preventive silviculture, as its implementation breaks the fuel continuity and availability (structural effect) and modifies the microclimate and the vegetation status (short-term dynamic effect). Thus, reduction of fire risk through forest treatments should be quantified in order to provide a more comprehensive understanding of the effects of adaptive forest management on promoting enhanced resilience with regard to climate change (García-Prats et al. 2015).

3 Adressing an Ecohydrological-Based Forest Management

Integrating and targeting the above-mentioned issues into forest management will prove the added value of an adaptive silviculture in promoting tree and stand resilience, improve (or maintain) site productivity, enhance soil water content, reduce wildfire hazards, etc. Most of the functions, goods and services (including raw materials production) that forests provide in water-limited regions have an important ecohydrological background that makes forest management to be hydrologically oriented (Molina and del Campo 2012). However, this type of silviculture is underdeveloped in many aspects as compared to that traditionally oriented towards timber production (del Campo et al. 2014). It is necessary to develop guidelines for a more efficient implementation of ecohydrological-based forest management, which pursues the quantification and the manipulation of water cycle components in forests according to management objectives.

The ecohydrological-based forest management (proactive silviculture), as explained above, is an integrated approach considering water priorities in the forest planning and management. It aims to manipulate and optimize forest and water interactions through forest management by considering an ecosystem-based approach. This means that other ecosystem variables such as tree/stand growth and vigour, soil properties and biogeochemical cycles, tree-climate sensitivity and reduction of the risk of wildfire, must be also explicitly targeted and quantified into the management objectives (Fig. 6.1).

The desired future condition of the stand being ecohydrologically managed must be determined on a commitment between the effects of full site occupancy (best soil

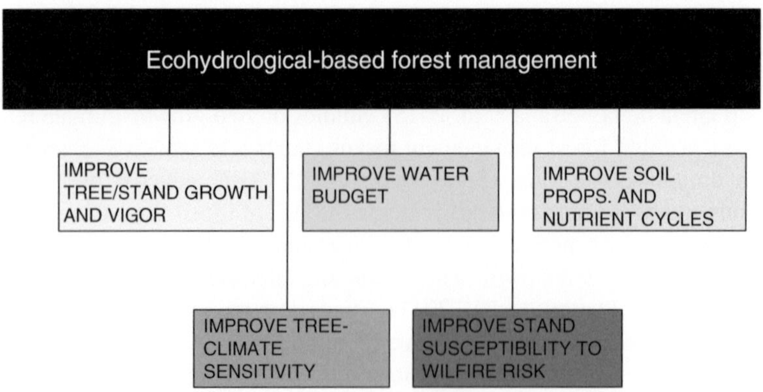

Fig. 6.1 Scheme of the effects of ecohydrological-based forest management

protection, maximum yield) versus more permeable forest structures allowing for lower interception losses (more blue water) and fuel loads and less competence among trees (enhanced resilience).

4 Best Management Practices in Semi-arid Forests of Eastern Spain

In order to address the effects of the ecohydrological-based forest management, the results of three experimental study sites are presented. The experimental studies represent three of the examples of abandonment of semi-arid forests: (i) artificial Aleppo pine plantation (del Campo et al. 2014), (ii) coppice *Quercus ilex* forest (González-Sanchis et al. 2015a) and (iii) post-fire natural Aleppo pine regeneration (del Campo et al. 2015) (see Fig. 6.2). The three forests are located within the Valencian region (E of Spain), and share low-productive value and water limitation. In all the three experiments, control and thinning plots are compared in terms of hydrological performance, biological and biochemical soil properties and nutrient cycles, tree growth and climate sensitivity and improved resilience to wildfire. Here, we summarize previously published results on these sites (Molina and del Campo 2012; del Campo et al. 2014; García-Prats et al. 2015; González-Sanchis et al. 2015a, b; Fernandes et al. 2015, 2016; Manrique-Alba et al. 2015; Ruíz-Pérez et al. 2016) and integrate them in order to provide an overview on the topic of this paper.

Climate in the Valencian region is typically semi-arid Mediterranean, characterized by a long summer arid period, during which temperature is high and rainfall is almost absent. Precipitations are scarce and torrential, highly variable in time and space and with a contrasted seasonal regime. The 10 days of heaviest rainfall in a year provide over 50% of the annual rainfall (Vicente-Serrano et al. 2004). Likewise, there is a recurrent drought with high temporal and spatial variability because of topographic controls and weather types (Estrela et al. 2000a, b), and this heterogeneity makes it difficult to establish an overall pattern (Vicente-Serrano et al. 2004).

The Forest extension in the Valencian Community is 1.3 million ha with the current trend of increasing forest area by 3300 ha per year, mostly through neglected

Fig. 6.2 Three experimental sites: artificial Aleppo pine plantation, coppice *Quercus ilex*, and, post-fire natural Aleppo pine regeneration

Table 6.1 Water balance of seven experimental plots during 4 years

		Water			Wildfire		Growth	Soil
Forest	Gr (mm)	Plot	ET(%)	$I_{>30cm}$ (%)	B/G	Fire risk imp (%)	ΔBiomass (%)	Nit (mgN/kg)
P. halepensis	1545 (25 months)	C	87	13	0.15		1.538	–
		T_L	74	26	0.34	5	4.049	–
		T_M	70	29	0.42	57	5.386	–
		T_H	58	42	0.72	35	16.16	–
		T_H98	68	32	0.48	47	3.91	–
Q. ilex	534 (12 months)	C	59	40	0.41	–	–	18.52
		T	31	68	2.2	–	–	27.64
	271 (12 months)	C	94	0	0.02	–	–	13.24
		T	98	0	0.02	–	–	24.32

C control plot (shaded, not thinned), *T* thinned, *L* low intensity thinning, *M* moderate intensity thinning, *H* high intensity thinning, *H98* high intensity thinning carried out 10 years ago, *Gr* gross rainfall, *ET* evapotranspiration, $I_{>30cm}$: deep infiltration, *B/G* ratio of blue (deep infiltration) to green (total evapotranspiration). Fire risk imp – improvement of the modified KBDI index (extracted from García-Prats et al. 2015); ΔBiomass – total biomass increment from 2009 to 2010 (extracted from del Campo et al. 2014), Nit – nitrification.

agricultural areas and their subsequent colonization by forest species. Forest woodlands now occupy about 54% of the forest land; they are scattered and with a low-productive value but with a high protective function. Forest abandonment has strong consequences that in the mid- and long-term will imply a higher investment than a punctual management. It is well known that forest abandonment increases forest density and expansion, fire hazard and the risk of drought stress. Thus, this lack of management is already costing money as both forest fires and the demand for water resources, have significantly increased over the last 20 years (Vicente-Serrano et al. 2004).

4.1 Water Budget

In our experiments, the hydrologic performance of the forests (at stand scale) after management (thinning) has affected most of the water cycle components such as rainfall interception, throughfall, transpiration, soil moisture and deep infiltration (del Campo et al. 2014). Each component has been registered, analyzed and quantified in all three experimental studies (Table 6.1). The results from all three studies indicate a general enhancement of the hydrological cycle by favouring deep infiltration after forest management. The general water balance indicates that the first main difference among treatments is found in the interception loss, which ranges from 12.4 to 33.5%. At the same time, there is a general dismissing of stand transpiration despite the fact that tree transpiration increases after management. Our results also indicate a little competence between transpiration and the deep infiltration terms for the throughfall water, as they present opposite seasonal patterns. This would mean that the differences in throughfall during wet and cool periods are relocated into deep infiltration water if the soil is wet enough (del Campo et al. 2014). Soil and understory evaporation yielded low differences among treatments (< 10% of rainfall), which would mean that, in spite of clearing vegetation, the evaporation was not severely increased. Nevertheless, in the case of the study with *Quercus ilex*, the positive effect of forest management appears not to be significant during the driest water year 2013–2014 (annual precipitation was only 271 mm, 58% of the average) (Table 6.1 and Fig. 6.4d). This behaviour could be due to the higher vapour pressure deficit registered during this water year, and the capability of *Quercus* spp. to extract water from deeper soil layers. In any case, it is important to point out that forest management could be less effective under dry situations and a more detailed analysis would be required.

The general effects of the ecohydrological forest management usually has led to a significant increase of the blue/green water ratio, which implies a significantly increasing of the water contribution from the forest to the catchment (Manrique-Alba et al. 2015; González-Sanchis et al. 2015b). The results obtained from these three studies agree with many others that also analyzed the effects of forest management on the water budget (Yaseef et al. 2010). Despite the fact the experimental results are extremely useful to analyze and understand the effects of forest management in the water budget, there is still the need of including the spatial and temporal variability in order to upscale and design a proper forest management strategy. In this sense, the use of modelling approaches becomes an important tool to include both variables. In this sense, Manrique-Alba et al. (2015) used a simple modelling approach to obtain the water interception pattern of a sub-catchment that includes the spatial variability in the water budget analysis. The authors developed a simple linear regression between throughfall and forest cover using experimental plots, and then extrapolated it by means of a LiDAR-derived forest structure model (Fig. 6.3 left). For other variables less predictable, a more complex non-linear modelling approach may be needed. In this sense, Fernandes et al. (2015) developed an artificial neural network to predict Aleppo pine' transpiration after forest management in the plantation site. This approach considered the temporal variability but not the spatial one. An alternative to include both, spatial and temporal variability, are the process-based models, which can be used to predict the hydrological, physiological and the bio-geochemical response of the forest under different scenarios. González-Sanchis et al. (2015b) used the BIOME-BGC model to analyze and optimize the hydrological balance of a whole Aleppo pine plantation (Fig. 6.3 right). Nevertheless, these process-based models are difficult to operate because of the high number of parameters that are required to estimate. As an alternative, there are less complex and parsimonious models, i.e. with a significantly lower number of parameters, but still capable to represent both spatial and temporal scales. Ruíz-Pérez et al. (2016) used the parsimonious and dynamic vegetation LUE-model and obtained similar results to those reported by González-Sanchis et al. (2015b) using a higher complexity model (BIOME-BGC).

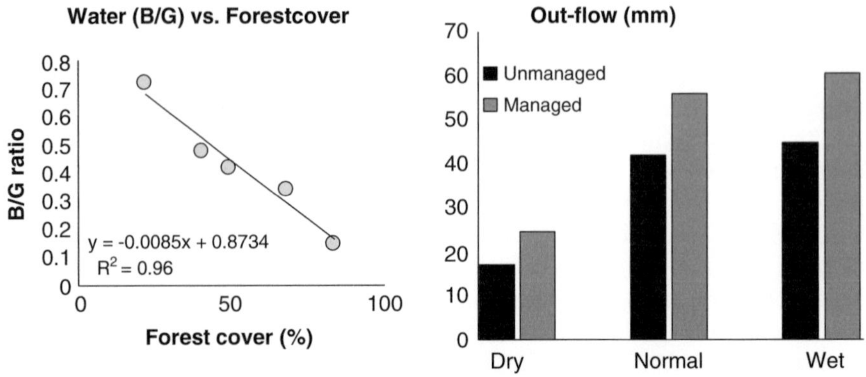

Fig. 6.3 Linear relationship between forest cover and blue/green water ratio B/G, and, out-flow of the managed and unmanaged Aleppo pine plantation during a dry, normal and wet year (González-Sanchis et al. 2015b)

Once the relationship between forest management and water budget is well established by means of the aforementioned mathematical models, it is possible to go one step beyond by using them in policy-making. A good example is that from Garcia-Prats et al. (2016) where a hydro-economic model is used with the aim to define the optimal forest-water management along with the planning horizon. The objective function maximizes the present net value of the net benefit of a stand, taking into account the value of the timber yield, the cost of the silvicultural operations and the value of the additional groundwater recharge produced by the forest management. In their study case, these authors report a groundwater recharge increase from 513 mm to 2435 mm between the no-management and the optimized scenario respectively.

4.2 *Soil Properties and Nutrient Cycles*

Soil properties can be modified with forest management. The alteration of the physical properties would mainly depend on the machinery used when managing, while the chemical and biological soil properties also depend on the silvicultural operations. Forest management usually increases soil exposition, but also modifies the water fluxes and the nutrients associated to them. According to our experimentations, forest management has increased nitrification process (Table 6.1 and Fig. 6.4c). This is likely due to the fact that the remaining trees can immediately compensate for the increased availability of inorganic N, but are rather expected to contribute to the increase in available ammonium and nitrate by reduced inorganic N uptake (Dannenmann et al. 2006). Contrarily, no significant effects of forest management have been observed in our plots in soil CO_2 emission and microbial activity (phosphatase and urease). Regarding the nutrient associated to the water fluxes, forest management appears to increase the potassium out coming via runoff, while the fluxes of carbon, nitrogen and phosphorus are not significantly affected with the management performed in our plots (Bautista et al. 2015).

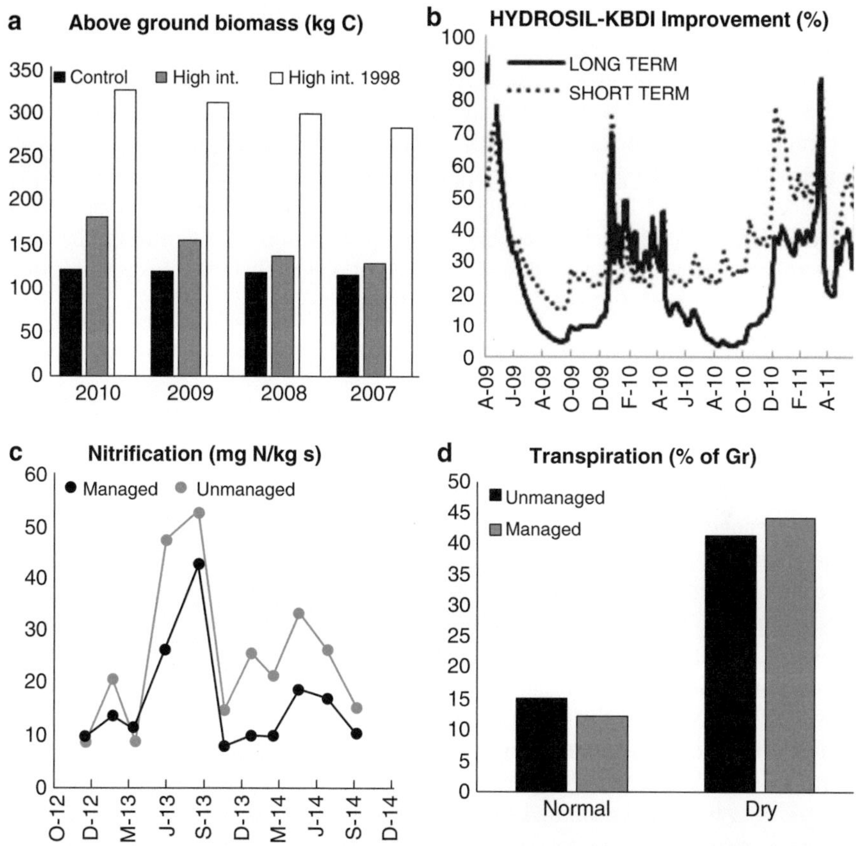

Fig. 6.4 Effects of forest management on aboveground biomass (**a**) fire hazard (**b**) in an Aleppo pine plantation and on nitrification (**c**) and transpiration (**d**) in a *Quercus ilex* coppice forest (a): Aboveground biomass evolution after two high intensity treatments, one in 2007 and the other in 1998, extracted from del Campo et al. (2014). (b): Improvement of fire hazard after the two high intensity thinning (long-term performed in 1998; short-term, 10 years after) by using a modified KBDI index, extracted from García-Prats et al. (2015). (c): Evolution of the nitrification process in managed and unmanaged *Quercus ilex* plots. (d): Transpiration of managed and unmanaged *Quercus ilex* plots during normal and dry water years.

4.3 Tree/Stand Growth and Vigour

Forest management dismisses the competence and increases the resources' availability, which involves a significant increasing of the plant sustainability (Mitchell et al. 1983). These effects are mainly observed by studying the primary and secondary growth (basal area, BA). From our experimental results, it can be stated that in general the forest treatment enhances both the primary and secondary growth as compared to the control plot (Fig. 6.4a). Regarding the Aleppo pine plantation, secondary growth was highly limited in the pre-thinning stages of the stand, with basal area showing increments around 4.1 cm^2/year, whereas thinned trees reached increments of 17.3 cm^2/year; the increase was proportional to the intensity of the thinning (del Campo et al. 2014). The thinning treatment in the post-fire Aleppo pine regeneration enhanced both the primary and secondary growth as compared to the control plot (del Campo et al. 2015). Length of the shoots per growth-cycle, needle length and needles per growth-cycle appear to be increased with the thinning. In this sense, the length of the cycle (each year flush is usually compound of several cycles), the length of the needles and the number of needles seem to be the most responsive variables. In the control, a reduction of polycyclism was observed, which is explained by adverse meteorological conditions.

4.4 Tree-Climate Sensitivity

Ecohydrological forest management might assist in adapting semi-arid forests to climate changes. However, to achieve this goal, a better understanding and quantification of tree-water relations and in particular of tree water use efficiency is required. Tree climate–growth relations, water use (WU) and water use efficiency (WUE) are central topics to be tackled when relating forest management and tree water (Sohn et al. 2013; Ungar et al. 2013). In our plots, Fernandes et al. (2016) reported the effects of thinning in the short and medium term by focusing on WUE and climate–growth relationships in the pine plantation. The results of WUE indicated that all thinning treatments improved the way trees used the soil water reservoir, but to different extents: WUE grew proportionally with decreasing forest cover. Likewise, these results have demonstrated that thinning makes trees less sensitive to water shortages, indicating that trees in the non-thinned plot need to rely more heavily on current year precipitation than those thinned (Fernandes et al. 2016). From this study it can be stated that the growth–climate relationship changed over the trees' lifetime especially after thinning, confirming that forest management increases the resilience of Aleppo pine trees to climate variations.

4.5 Fire Risk

As stated before, changes in forest structure due to partial removal of the forest canopy produce certain hydrological responses and consequently modify the water balance.

It is obvious that this hydrology-oriented silviculture is also a fire preventive silviculture, as its implementation breaks the fuel continuity (structural effect) and modifies the microclimate and the vegetation status (short term dynamic effect). In that sense, García-Prats et al. (2015) modified the Keetch-Byram drought index in order to make it sensitive to silvicultural operations (Table 6.1 and Fig. 6.4b). As can be seen in Fig. 6.4b, the average of wildfire risk reduction in the short term – expressed as KBDI units – was 22.5% and 26.4% for the two consecutive years after thinning. This reduction remained in the mid-term 12.5% and 6.7% respectively (since they were quantified for the same period in a plot thinned ten years before). This approach enabled to simulate the benefits achieved from thinning operations in our planted pine experiments in terms of a decreased forest fire risk and to link those to other benefits related to tree growth and water balance.

5 Conclusions

Semi-arid protective forests face the global change as crossroads with many potential impacts and threats such as green/blue water impairment, wildfires, pests, dieback, growth stagnation, etc. Whilst it is widely recognized the benefits of a timely proactive-adaptive silviculture, current challenges and needs of these ecosystems hit with the problem of lack of management and woodlands abandonment. To overcome this situation, non-marketed ecosystem services should be explicitly incorporated into forest management through new thinking and innovative approaches. Hydrology-oriented or ecohydrological-based silviculture is just an example that poses many opportunities to make silviculture more effective under water scarcity scenarios.

Acknowledgements This study is part of the research projects "CGL2011-28776-C02-02, HYDROSIL" and "CGL2014-58127-C3-2, SILWAMED," funded by the Spanish Ministry of Science and Innovation and FEDER funds, and "Determination of hydrologic and forest recovery factors in Mediterranean forests and their social perception," supported by the Ministry of Environment, Rural and Marine Affairs. The authors are grateful to the Valencia Regional Government (CMAAUV, Generalitat Valenciana) and the VAERSA staff for their support in allowing the use of the La Hunde and the Natural Park "La Sierra Calderona" experimental forests and for their assistance in carrying out the fieldwork.

References

Asbjornsen, H., Goldsmith, G. R., Alvarado-Barrientos, M. S., van Osch, F. P., Rietkerk, M., Chen, J., Gotsch, S., Tobón, C., Geissert, D. R., Gómez-Tagle, A., Vache, K., & Dawson, T. E. (2011). Ecohydrological advances and applications in plant-water relations research: A review. *Journal of Plant Ecology, 4*, 3–22.

Bautista, I., Pabón, C., Lull, C., González-Sanchís, M., Lidón, A., & del Campo, A. (2015). Efectos de la gestión forestal en los flujos de nutrientes asociados al ciclo hidrológico en un bosque mediterráneo de *Quercus ilex*. *Cuadernos de la Sociedad Española de Ciencias Forestales, 41*, 343–354.

Birot, Y., Gracia, C., & Palahi, M. (2011). *Water for forests and people in the Mediterranean region – A challenging balance*. Joensuu: European Forest Institute.

Bosch, J. M., & Hewlett, J. D. (1982). A review of watershed experiments to determine the effects of vegetation changes on water yield and evapotranspiration. *Journal of Hydrology, 55*, 3–23.

Callegari, G., Ferrari, E., Garfì, G., Iovino, F., & Veltri, A. (2003). Impact of thinning on the water balance of a catchment in a Mediterranean environment. *The Forestry Chronicle, 79*, 301–306.

Dannenmann, M., Gasche, R., Ledebuhr, A., & Papen, H. (2006). Effects of forest management on soil N cycling in beech forests stocking on calcareous soils. *Plant and Soil, 287*, 279–300.

del Campo, A., Fernandes, T., & Molina, A. (2014). Hydrology-oriented (adaptive) silviculture in a semiarid pine plantation: How much can be modified the water cycle through forest management? *European Journal of Forest Research, 133*, 879–894.

del Campo, A., González-Sanchis, M., Ruíz-Pérez, G., Llull, C., García-Prats, A., Lidón, A., & Bautista, I. (2015). *Hydrological response of pine saplings to forest management during the driest year on record.* Proceedings XIV. World Forestry Congress, Durban, South Africa.

Estrela, M. J., Peñarrocha, D., & Millán, M. (2000a). Multi-annual drought episodes in the mediterranean (Valencia region) from 1950–1996. A spatio-temporal analysis. *International Journal of Climatology, 20*, 1599–1618.

Estrela, M.T., Marcuello, C., & Dimas, M. (2000b). Las aguas continentales en los países mediterráneos de la Unión Europea. Centro de estudios Hidrográficos del CEDEX, Spain. http://hercules.cedex.es/Hidrologia/pub.htm. Accessed 16 Feb 2016.

Fernandes, T. J., Del Campo, A. G., Bartual, R., & González-Sanchis, M. (2015). Use of artificial neural network to estimate daily transpiration of *Pinus halepensis* mill. Stands thinned at different intensities. *IForest*. doi:10.3832/ifor1290-008.

Fernandes, T. J., Del Campo, A. D., Herrera, R., & Molina, A. J. (2016). Simultaneous assessment, through sap flow and stable isotopes, of water use efficiency (WUE) in thinned pines shows improvement in growth, tree-climate sensitivity and WUE, but not in WUEi. *Forest Ecology and Management, 61*, 298–308.

Fitzgerald, J., Jacobsen, J. B., Blennow, K., Thorsen, B. J., & Lindner, M. (2013). *Climate change in European forests: How to adapt*. Joensuu: European Forest Institute.

Gallart, F., & Llorens, P. (2004). Catchment management under Environmental change: Impact of land cover change on water resources. *Water International, 28*(3), 334–340.

García de la Serrana, R., Vilagrosa, A., & Alloza, J. A. (2015). Pine mortality in southeast Spain after an extreme dry and warm year: Interactions among drought stress, carbohydrates and bark beetle attack. *Trees*. doi:10.1007/s00468-015-1261-9.

García-Prats, A., del Campo, A., & Molina, A. J. (2015). Development of a Keetch and Byram-based drought index sensitive to forest management in Mediterranean conditions. *Agricultural and Forest Meteorology, 205*, 40–50.

García-Prats, A., del Campo, A., & Pulido-Velazquez, M. (2016). A hydroeconomic modeling framework for optimal integrated management of forest and water. *Water Resources Research, 52*, 8277–8294. doi:10.1002/ 2015WR018273.

García-Ruiz, J. M., López-Moreno, J. I., Vicente-Serrano, S. M., Lasanta-Martínez, T., & Beguería, S. (2011). Mediterranean water resources in a global change scenario. *Earth-Science Reviews, 105*(3), 121–139.

Giorgi, F., & Lionello, P. (2008). Climate change projections for the Mediterranean region. *Global and Planetary Change, 63*(2), 90–104.

González-Sanchis, M., del Campo, A., Lidón, A., Lull, C., Bautista, I., García-Prats, A., & Francés, A. (2015a). Incorporación de criterios eco-hidrológicos en la gestión forestal: Adaptación a la escasez de agua de una masa marginal de encina. *Cuadernos de la Sociedad Española de Ciencias Forestales, 41*, 211–218.

González-Sanchis, M., del Campo, A. D., Molina, A. J., & Fernandes, T. J. G. (2015b). Modelling adaptive forest management of a semi-arid Mediterranean Aleppo pine plantation. *Ecological Modelling, 308*, 34–44.

Jandl, R., Lindner, M., Vesterdal, L., Bauwens, B., Baritz, R., Hagedorn, F., Johnson, D. W., Minkkinen, K., & Byrne, K. A. (2007). How strongly can forest management influence soil carbon sequestration? *Geoderma, 137*, 253–268.

Linder, M., Fitzgerald, J., Zimmermann, N. E., Reyer, C., Delzon, S., van der Maaten, E., Schelhaas, M. J., Lasch, P., Eggers, J., van der Maaten-Theunissen, M., Suckow, F., Psomas, A., Poulter, B., & Hanewinkel, M. (2014). Climate change and European forests: What do we know, what are the uncertainties, and what are the implications for forest management? *Journal of Environmental Management, 146*, 69–83.

Manrique-Alba, A., del Campo, A. D., & Gonzalez-Sanchis, M. (2015). Light detection and ranging for implementing water-oriented Forest Management in a Semiarid sub-Catchment (Valencia, Spain). *Clean: Soil, Air, Water, 43*, 1488–1494.

Millar, C. I., Stephenson, N. L., & Stephens, S. L. (2007). Climate change and forests of the future: Managing in the face of uncertainty. *Ecological Applications, 17*, 2145–2151.

Mitchell, R. G., Waring, R. H., & Pitman, G. B. (1983). Thinning lodgepole pine increases tree vigor and resistance to mountain pine beetle. *Forest Science, 29*, 204–211.

MMARM. (2006). Anuario de Estadisticas Forestales 2006. http://www.mma.es/portal/secciones/biodiversidad.htm. Accessed 16 Feb 2016.

Molina, A., & del Campo, A. (2012). The effects of experimental thinning on throughfall and stemflow: A contribution towards hydrology-oriented silviculture in Aleppopine plantations. *Forest Ecology and Management, 269*, 206–213.

Overby, S. T., Owen, S. M., Hart, S. C., Neary, D. G., & Johnson, N. C. (2015). Soil microbial community resilience with tree thinning in a 40-year-old experimental ponderosa pine forest. *Applied Soil Ecology, 93*, 1–10.

Planinsek, S., Finer, L., del Campo, A., Alcazar, J., Vega-Garcia, C., Dimitrov, D., & Capuliak, J. (2011). Adjustment of forest management strategies to changing climate. In Bredemeier, M. et al. (Eds.), *Forest management and the water cycle: An ecosystem – Based approach* (Ecological Studies, 212, pp 313–333).

Ruíz-Pérez, G., González-Sanchis, M., del Campo, A., & Francés, F. (2016). Can a parsimonious model implemented with satellite data be used for modelling the vegetation dynamics and water cycle in water-controlled environments? *Ecological Modelling, 324*, 45–53.

Sarris, D., Christodoulakis, D., & Körner, C. (2007). Recent decline in precipitation and tree growth in the eastern Mediterranean. *Global Change Biology, 13*, 1187–1200.

Sohn, J. A., Gebhardt, T., Ammer, C., Bauhus, J., Häberle, K.-H., Matyssek, R., & Grams, T. E. E. (2013). Mitigation of drought by thinning: Short-term and long-term effects on growth and physiological performance of Norway spruce (*Picea abies*). *Forest Ecology and Management, 308*, 188–197.

Thorsen, B. J., Mavsar, R., Tyrväinen, L., Prokofieva, I., & Stenger, A. (Eds.). (2014). *The provision of Forest ecosystem services. Volumes I & II: What science can tell us* (Vol. 5). Joensuu: European Forest Institute.

Ungar, E. D., Rotenberg, E., Raz-Yaseef, N., Cohen, S., Yakir, D., & Schiller, G. (2013). Transpiration and annual water balance of Aleppo pine in a semiarid region: Implications for forest management. *Forest Ecology and Management, 298*, 39–51.

Vicente-Serrano, S. M., González-Hidalgo, J. C., de Martín, L., & Raventós, J. (2004). Drought patterns in the Mediterranean area: The Valencia region (eastern Spain). *Climate Research, 26*(1), 5–15.

Wäldchen, J., Schulze, E.-D., Schöning, I., Schrumpf, M., & Sierra, C. (2013). The influence of changes in forest management over the past 200 years on present soil organic carbon stocks. *Forest Ecology and Management, 289*, 243–254.

Yaseef, N., Yakir, D., Rotenberg, E., Schiller, G., & Cohen, S. (2010). Ecohydrology of a semiarid forest: Partitioning among water balance components and its implications for predicted precipitation changes. *Ecohydrology, 3*(2), 143–154.

Chapter 7
Ecosystem Services in African Headwaters

Jaroslav Balek

1 Introduction

Headwater regions are the key element in environmental services providing the human societies with several critical flows of values according to the nature capital of the African continent (MEA 2005). A sustainable management of land and water requires both the local knowledge (and local institutions rooted in the communities) and technological progress imported (Křeček and Haigh 2006). With the decentralisation, occurring in many African countries today, it is necessary to empower the local communities with appropriate responsibilities to give them a chance of an active participation in the stakeholder dialogue on management practices in a watershed scale.

Particularly, the African headwater wetlands ('dambos') play a significant role in the formation of hydrological regime of great rivers. Based on the long-term field investigations (Balek 1983; Balek and Perry 1973; Wood 2006), wetlands act as spongy reservoirs capable to absorb water during the rainy season and to release it slowly during the dry season. The aim of this study was to summarize the long-term investigation of wetlands performed in Zambia, and to compare the results with the research outcomes accomplished elsewhere in Africa.

2 Wetlands in Africa

The total area of African wetlands has been estimated to be 340,000 km^2. Among several types of classification, Balek (1977) recognised wetlands accordingly with their hydrologic properties (Table 7.1).

J. Balek (✉)
Independent Consultant, ENEX, Tábor, Czech Republic
e-mail: balekenex@volny.cz

© Capital Publishing Company, New Delhi, India 2017

J. Křeček et al. (eds.), *Ecosystem Services of Headwater Catchments*,
DOI 10.1007/978-3-319-57946-7_7

Table 7.1 Water balance estimates for selected African wetlands

Parameter	Unit	Bangweulu swamps	Kafue flats	Lukanga side wetlands	Dambo
Drainage area	km^2	102,000	58,290	19,490	143,000
Wetland area	km^2	15,875	2600	2600	15,000
Rainfall	mm/year	1190	1090	1250	1330
(drainage area)					
Rainfall	mm/year	1210	1110	970	1330
(wetland area)					
Evaporation from	mm/year	2340	2070	2070	1710
Open water surface					
Additionally	mm/year	1120–1260	196	252	–
Evaporated from					
Wetlands					
Evapotranspiration	mm/year	2000–2180	1005	1120	1075
From wetlands					
Loss in	% of inflow	60	4	7,8	–

Each of the major types of wetlands reflects special morphological features. The Bangweulu swamp is a combination of lakes mixed with deeply rooted wetland vegetation, papyrus and so called 'matete reeds'. As for the rainfall, some 1130 mm/year has bas been estimated for the Chambeshi river as an inflow and 1165 mm/year for the Luapula river as an outflow. The Chambeshi river is an ultimate source of the Upper Congo. Water balance calculations in the African swampy areas are complicated by the role of vegetation. For example, the evapotranspiration from papyrus can exceed the evaporation of an open water surface. Already Hurst (1954) claimed such a possibility; however, without the support of the detailed observations his conclusions were considered rather improbable. But, several laborious experiments, performed in the seventies (Balek 1983), confirmed his presumptions.

Owing to the wetland composition, about 60% of the total inflow is lost from Bangweulu by evapotranspiration. It can be said that the Bangweulu swamp functions similarly as an artificial reservoir, transforming the hydrological regime of the Chambeshi into a modified regime of the river Luapula.

Kafue flats represent periodically flooded lowlands along a shallow river bed. Perhaps similar function can be accounted to the renowned Zambezi flats; however, no sufficient data have been available for that extensive wetland. Generally, such type of the wetlands is seasonally saturated by the flooded river and slowly emptied during the dry season. Both inflow and outflow are dominant features. For a proper function a considerable amount of water is required and therefore the flats are usually found in the middle stretches of the river rather than close to the headwaters. As can be seen from Table 7.1, the evapotranspiration from the Kafue flats is 1005 mm/year. Excluding the swampy area it would be only 814 mm/year. Owing to much less developed channel system inside the wetlands and different vegetation, the inflow loss is much lower than at Bangweulu.

Lukanga, located in the centre of the Kafue basin and having a two-way connection with the main stream, is a representative of side-stream reservoirs. The water balance calculation is difficult considering a possibility that during the flood season there is a side spill of more than 500×10^6 m^3 of water in an average year. Considering very special morphologic features of such a type of wetland, the results cannot be widely generalised.

'Dambo' appears to be a negligible unit when its size is compared with the size of other wetlands. However, considering that total number of 'dambos' in Africa is estimated by 10^4–10^5, and that most of them are located in the upper parts of major river basins, they play a significant role in the African hydrology.

3 The Hydrologic Role of a 'Dambo'

Probably the first account of 'dambos' was given by Ackermann (1936) who described them as periodically inundated, stream-less and grass-covered depressions on the headwater end of a drainage system in the regions of dry forest or bush vegetation. According to Hindson (1961), dambo is the native name given to the seasonally waterlogged grass-covered, treeless areas bordering the drainage line. It is worth mentioning that such wetlands are called 'vlei' in South Africa and 'mbuga' in East Africa. They are very distinct units on the aerial photographs of the headwater of Zambezi, Upper Congo and Upper Nile.

Hindson (1961) reported that "'dambos' remain wet in dry seasons due largely to seepage which arises along the dambo fringe; this results from slow subsurface drainage from the upland area between 'dambos'". Such observation was very important considering that a delayed flow from dambos in the headwaters may be a beneficial factor for the stabilisation of the hydrological regime of great rivers particularly in Africa.

In the early seventies, a basic question related to the dambo behaviour was whether the geo-morphological conditions play decisive role or whether the dambo vegetation is a key factor. A more definite answer was obtained through laborious experiments in the Luano experimental catchments, established in the Zambian Copperbelt. The areas consisted of 89.3–95.1% *Brachystegia* woodland with other species such as *Julbernadia* and *Marquesia,* and 4.9–11.3% dambo with some sixty species of grass. Such composition was also representative for the Kafue/Zambezi headwaters. The area formed a part of the so-called Luano Forest Reserve (12° 34′S and 28° 01′E, altitude of 1300 m). Each of the catchments was less than 1.5 km^2 and located in the belt of 1270 mm mean annual rainfall.

During the first phase of the field monitoring, observations in four catchments included precipitation (40 rain gauges, 16 rainfall recorders), runoff (four discharge recorders with combined weirs and flumes), and groundwater level (66 level gauges). Additionally, soil moisture fluctuation was measured down to the depth of seven metres in 45 soil moisture profiles. Data of the standard meteorological station was used to calculate daily amounts of the Penman's potential evaporation. Results of the water balance were reported by Balek (1983).

Comparing 'dambos' and woodlands, two distinct groundwater regimes were identified. In the case of dambo, a rapid rise of the groundwater level was observed quite soon after the start of the rainy season. Then the groundwater level remained more or less constant because the dambo was fully saturated and all the rainfall water contributed exclusively to the surface runoff. On the contrary, the groundwater level in the woodland started to rise 2–3 months after the start of the rainy season and continued to rise up to the end of the rainy season.

After 3 years of observation, in the second phase of the experiment, a significant increase of floods during the rainy season was observed in two treated catchments by removal of the native trees. Soil moisture measurements were correlated with the root measurements performed by Maxwell (1972). The results clearly indicated that the trees taped the water for transpiration through a shallow root system till it was available in the upper soil layer, down to the depth of three metres. After the upper soil layer had been exhausted, a deep root system consumed water from deeper layers within the reach of the capillary zone above the groundwater level. Practically, all rainfall water, falling on the woodland, was transpired from the *Brachystegia*, *Julbernadia* and *Marquesia* species (Balek 1977).

4 Results of the Conceptual Approach

Based on a systematic observation and simulation of the hydrological processes, following results were obtained by the conceptual investigations:

- A high fluctuation in the hourly, daily, monthly and even annual precipitation within the total experimental area of 10 km^2 has been observed in that part of the tropics.
- The hydrological response of the catchment to the precipitation depended on the size of the dambo area within each particular catchment.
- The 'dambos' were recharged shortly after the rainy season had started, and released the water within a constant time interval after the last heavy storm. The length of the interval was dependent on the 'dambo' soil type and on the slope of it.
- From a higher percentage of 'dambo' area, a higher surface runoff occurred. Surface runoff formed the largest part of the total runoff in the catchments containing a dambo. The surface runoff was produced mainly as a result of the overstorage of the dambo aquifer.
- The duration of the surface runoff as compared with a non-swampy area of similar shape and size was delayed by the resistance of 'dambo' grasses towards the flow.
- The loss of evapotranspiration from the dambo grass, however dense it was, was lower than from the surrounding woodlands.
- Groundwater storage in the catchments was depleted more by evapotranspiration than by river base flow. The evapotranspiration by the woodland was approx. Three times higher than the evapotranspiration by the dambo grass. Therefore,

only a part of the woodland area, consisting of the mixture of trees and grass and therefore called a transitive region, contributed to the hedrograph formation. It seems very likely that the Brachystegia and relative species can transpire up to 90% of the potential evaporation and the ratio Et/Eo can exceed 1.0 during the rainy season.

- Based on the previous results it can be concluded that the small wetlands - dambos, located in the headwaters of great African rivers, play a unique role in the formation of their hydrological regime. This is because the thick grass cover of the dambos delays the surface flow, which has been produced by the over-stored dambo aquifer and thus contributes towards the stabilisation of their hydrological regimes.

5 Continued Challenge in the Investigation of Wetlands

During the second phase of this research, the *Brachystegia* forests, surrounding two experimental 'dambos', were harvested by clear-cutting. Successive events, the flood volumes and, particularly, volumes of fast (direct) runoff have increased significantly. These findings were interpreted as a confirmation of the traditional role of forests in the hydrological cycle. However, since then much attention has been paid to explain a complex role of 'dambos' in the context of watershed hydrology in Africa.

Bullock (1992a, b) reviewed and analysed major relevant papers, focused on various 'dambo' processes in other parts of Africa. Very few however, if any, have been based on direct observations or field experiments. In summary, the role of 'dambos' in the hydrological cycle considers:

1. Dambos appear to be an indiscriminatory factor in determining annual runoff from catchments at the regional scale.
2. There is no significant persistence in annual time series or reduction of the annual variability of flows with varying 'dambo' density.
3. There is a base flow factor in determining base flow and dry season flows (at least "at the national scale in Zimbabwe").
4. There is no evidence to suggest that 'dambos' can significantly contribute to the maintenance of base flow or dry-season flow regimes.
5. There is no significant influence exerted by 'dambos' in reducing or increasing either the magnitude or variability of annual maximum instantaneous flood discharges either at the national scale or within particular soil classes.

Furthermore, Bullock (1992a), after reviewing data and conclusions from past studies, stated that 'dambos' are considered to play an important role in the hydrological cycle "because of their extensive distribution in headwater regions"; and concluded: "that the mechanisms and processes by which 'dambos' determine the movement of water are still poorly understood".

Other attempts such as of von den Heyden (2006) were based on weak data and as such the conclusions about the negative role of 'dambos' could not be taken seri-

Table 7.2 Monthly and annual evapotranspiration (actual E_t and potential E_o) during the year with annual precipitation P_a = 1620 mm

Month	Evapotranspiration			
	Dambo		Woodland	
	E_t mm	E_t/E_o	E_t mm	E_t/E_o
October	6.58	0.03	49.05	0.24
November	80.72	0.53	85.39	0.56
December	86.36	0.66	129.04	0.99
January	87.50	0.59	153.16	1.03
February	67.54	0.55	163.75	1.32
March	64.14	0.47	193.66	1.42
April	29.26	0.21	192.72	1.39
May	20.40	0.17	119.82	0.99
June	9.17	0.09	82.47	0.83
July	9.63	0.09	72.49	0.67
August	6.65	0.05	68.09	0.52
September	5.84	0.04	56.23	0.36
Year	473.79	0.29	1356.87	0.83

ously. The author can not agree with the statement that the mechanisms and processes of dambos have been poorly understood, at least not in the Luano catchments. It is improbable that in the near future similar experiments will be repeated elsewhere.

By the results of continuing research in the Luano experimental area (Molnár 1974–1996), with reforestation of cleared units, and findings of Bullock (1992b), it is possible to conclude that in the watersheds with absence of 'dambos' (covered entirely by forests), much of the precipitation has been lost by evapotranspiration instead of supporting the prolonged runoff. The example of the Luano project is shown in Table 7.2.

6 Conclusions

It is evident that 'dambos' create conditions which are not favourable for the growth of woodlands. Shallow soil profiles underlining the 'dambo' areas, together with groundwaters stored in the non-capillary pores up to the soil surface, contribute to the development of thick grasses only there. A 'dambo' spot tends to support the grassy aquatic environment capable to extend surface runoff process. Thus, precipitation, morphology and vegetation play a complex and unique role in the hydrological cycle of 'dambos', and, generally, they play a positive role in the hydrologic regime of major African rivers.

Fortunately, the latest development in central and southern Africa indicates that in the near future more attention will be given to the protection of 'dambos', at least in the Limpopo and Zambezi basins. Hopefully, experiments, once conducted in remote parts of the Central African Plateau, will encourage future systematic and experimental research, focused on the hydrology and effective land use of similar wetlands in other parts of the world.

References

Ackermann, E. (1936). Dambos in Nordrhodesia. *Wissenschaftliche Veröffentlichungen, 4*, 149–157.

Balek, J. (1977). *Hydrology and water resources in tropical Africa*. Amsterdam: Elsevier.

Balek, J. (1983). *Hydrology and water resources in tropical regions*. Amsterdam: Elsevier.

Balek, J., & Perry, J. (1973). Hydrology of seasonally inundated African headwater swamps. *Journal of Hydrology, 19*, 227–249.

Bullock, A. (1992a). Dambo hydrology in southern Africa – Review and reassessment. *Journal of Hydrology, 134*, 373–396.

Bullock, A. (1992b). The role of dambos in determining river flow regimes in Zimbabwe. *Journal of Hydrology, 134*, 349–372.

Hindson, J. R. E. (1961). The water resources of Central Africa. *The Geographical Journal, 111*, 222–234.

Hurst, H. E. (1954). *Le Nil*. Paris: Payot. (in French).

Křeček, J., & Haigh, M. J. (2006). *Environmental role of wetlands in headwaters, NATO Science Series, 63*. Dordrecht: Springer.

Maxwell, D. (1972). Root range investigations. National Council for Scientific Research, Report 26, Lusaka.

MEA. (2005). *Ecosystems and human well-being: Synthesis, Millenium ecosystem assessment*. Washington, DC: Island Press.

Molnár, L. (1974-1996). Personal communication. Lusaka and Bratislava.

von der Heyden, C. J. (2006). *Financing freshwater Management in South Africa*. Chihuahua: HSBC – WWF Knowledge Exchange.

Wood, A. (2006). Headwater wetlands in eastern and southern Africa. In J. Křeček & M. J. Haigh (Eds.), *Environmental role of wetlands in headwaters, NATO Science Series, 63*. Dordrecht: Springer.

Part II
Enhancing Environmental Services in Headwaters

Chapter 8
Protective Stands: Lessons from the Past and Today's Challenges

Jérôme Lievois, Xavier Gauquelin, Anthony Dubois, and Alison Evans

The assertion of the mountain forest's protective function against natural hazards has been subject to debate for a long time. Its protective role has faced both socio-economic stakes in the concerned areas as well as scientific and methodological deficiencies. It is only recently that the protective role has effectively and objectively entered territorial planning documents and it remains incomplete. Its integration has been made possible thanks to the numerous works during the last 30 years to identify, quantify and strengthen the protective function.

Review of a fruitful collaboration between researchers and field workers, enriched by the cross-border cooperation (Fig. 8.1).

1 French Mountain Forests

In France, forests located in areas 600 metres above sea level are defined as "mountain forest". They cover a surface area of 4.4 million ha, i.e. 27% of French mainland forest[1] (Table 8.1).

These forests are governed by different regulations, which in various ways interact with the eventual protective functions against natural hazards. A summary of this governance is shown in Table 8.1.

[1] Sources: IFN.

J. Lievois (✉) • A. Evans
National Forest Office, Haute Savoie RTM Unit, Haute Savoie, France
e-mail: jerome.lievois@onf.fr

X. Gauquelin
National Forest Office – DFRN, Laon, France

A. Dubois
ONF International Grenoble, Grenoble, France

© Capital Publishing Company, New Delhi, India 2017
J. Křeček et al. (eds.), *Ecosystem Services of Headwater Catchments*,
DOI 10.1007/978-3-319-57946-7_8

Fig. 8.1 Sapet state-owned forest (Drôme): Comparison of a ravine between 1904 and 1995 (*Photos:* ONF RTM archive)

Table 8.1 The different statuses of French mainland mountain forest

	Public forest	State-owned RTM forest	Private forest	Protection forest
Status	Public		Private	Public or private
Applicable regulation and reference documents	Governed by public Forest regulations. Compulsory *Forest management plan* (definition of goals and planning for 10–20 years, approved by the ONF[a])		*Simple Management Plan* for forests of more than 25 ha	Special management scheme. The grading of a forest as "protection forest" is subject to a decree of the council of state
Estimation of the protective function against natural hazards	Diagnosis of hazards and stakes and definition of the forest's function. A number of these forests contain a protective function	Natural or artificial woodland and of which the silviculture aims to control natural hazards (*See Table* 8.2)	Private mountain forest with no identified characteristics or function.	Grading motivated by reasons of natural hazard control (but also because of ecologic functions or concerning the well-being of populations)

(continued)

Table 8.1 (continued)

	Public forest	State-owned RTM forest	Private forest	Protection forest
Status	Public		Private	Public or private
Level of restrictions	Management operations shall take the protective function into account if identified	Hazard control oriented forest management. Harvesting is possible, but rarely profitable	No particular restrictions	The most binding legal instrument
Surface area[b]	+/– 1,230,000 ha	411,000 ha with 63% forest cover (257,000 ha)	+/– 2,900,000 ha	49,952 ha (exclusively "mountain" motive)

[a]Office national des forêts – French National Forestry Office
[b]Sources: IFN; ONF RTM; French Ministry of Agriculture, Food and Forestry

2 Questions Concerning the Protective Function

In regard of every situation's specificity, the knowledge of the protective function of these forests remains substantially heterogeneous (Sonnier 1991; Gauquelin and Courbaud 2006; ONF 2010, 2012; FAO 2014; Vésino and Marco 2014, ONF – DFRN 2015). Considering this lack of knowledge, a fear of reduced hazard control capacity due to the aging of tree stands emerged in the 1990's, particularly for artificial forest stands established more than a century ago within the framework of the RTM policy (See Box) (Fig. 8.2).

The main principles for hazard control by forests are however well known since several decades and are given in Table 8.2. But we needed more information:

- How to characterise these forest stands?
- Where are they situated?
- What kind of silviculture should be applied to them?
- With what kind of financing?
- Within what regulatory framework?

The "RTM" Policy in France

The French policy of Mountain Terrain Regeneration was initiated in 1860 and effectively launched by the Forestry and Waters administration in 1882.

Within the framework of repeated natural disasters of great magnitude in the middle of the nineteenth century, a series of laws were promulgated to initiate wide-ranging programmes of State terrain acquisitions, reforestation campaigns and torrential flood correction works in French mountain areas.

(continued)

Some figures: (*ONF RTM*)

- 390,000 ha acquired by the State
- Close to 200,000 ha reforested
- 93,000 works of torrential flood correction
- 75,000 linear metres of avalanche barriers (excluding walls)

The implementation of the RTM policy was transferred to the French National Forestry Office (ONF) when it was created in 1966. Today, the ONF is in charge of RTM forest management, maintenance of works and the prevention and management of natural hazards.

Fig. 8.2 Protection forest above a departmental road (Haute Savoie) (*Photo:* ONF RTM – 07/04/2010)

Table 8.2 Principles of controlling various types of hazards through forest stands with protective functions

Type of gravitational hazard	Dendrological characteristics for a protective function
Rock fall	Stem density combined with basal area disperse the energy of rocks in motion
Avalanche	Crown coverage during winter (evergreen conifers) reduces the probability of avalanche
Torrential floods and surface erosion	The sole presence of a plant cover with a forest soil contributes to less surface erosion and to increase the response time between a given rainfall and the associated torrential flood output by proportionally reducing the peak flow
Landslides	The interaction between forest and landslides appears as complex and often uncertain

3 Implemented Actions

Different projects carried out since the 1990's have contributed to clarify these interrogations:

1990–2000: Strengthening the scientific foundations

A research programme operated by the Irstea[2] and the ONF has studied the structure of subalpine spruce stands in order to determine stability factors faced with natural hazards in mountains. It resulted in a typology to grade these forest stands according to their vertical stratification criteria, which guarantee physical stability.[3]

2003–2006: Elaborating tools for the manager

The Interreg project "Sustainable management of mountain forests" (Italy, Switzerland and France) allowed the elaboration and popularisation of technical tools for optimising mountain forest silviculture notably in regard of natural hazards. The main goal was to identify and to propose silvicultural techniques to maintain or improve the stability of forest mountain stands neither hindering harvests, nor natural stand renewal. Discussions and exchanges between managers and researchers have been made easier by numerous cross-border workshops. In France, Guidelines for Silviculture in Mountainous areas (GSM) were first published for the Northern Alps (2006) then for the Southern Alps (2012).

2007–2011: Inventorying and estimating the needs

At the same time, the ONF RTM initiated an inventory[4] of actual surface areas where the forest presents a protective function vis-a-vis hazards beyond identified challenges. This inventory showed that 139,431 ha of state-owned RTM forest should be subject to specific silviculture actions to renew or maintain its protective function.

2008–2013: Testing the operationality of tools, understanding and estimating costs

The Interreg IV "Protection Forests" evaluated the characteristics of forests with protective functions (FFP) on five territories in the Northern Alps, carried out 19 pilot work projects and elaborated a silvicultural intervention planning for all the identified FFP. This was a follow-up of previous actions:

- Develop cartography methods for FFP at the level of small regions (intercommunalities) in order to identify priority intervention zones;
- Implement techniques recommended by Swiss, Italian and French reference documents (GSM) on pilot projects;

[2]The National Research Institute of Science and Technology for Environment and Agriculture (formerly Cemagref).

[3]Renaud, J.P. et al., CEMAGREF - ONF, 1994.

[4]Phase 1 of the Programme for the Renewal of Protection Forest Stands (2007–2011).

- Having knowledge of intervention costs (a rough order of magnitude at 70€/ha/year was in this way determined) and
- Improve forest management strategies in alpine territories.

2014: Capitalise

This thematic was the subject of a workshop[5] between European practitioners within the framework of the FAO Working Party on the Management of Mountain Watersheds (WP MMW).

To ONF and since 2012, these works have, amongst others, resulted in a method to determine the present challenges concerning FFP, to structure the methods of forest management planning and to deploy silvicultural techniques adapted to different types of gravitational hazards. Reference documents have been integrated in the forest management planning manual for this purpose.

4 Tools Made Available

In the past, the forest management planning manual was the only regulatory tool available to foresters in order to estimate the protective function of a state-owned parcel in regard of silvicultural operations to be implemented. This came very often down to: either grading as silviculture for production, or an "off scope" grading for the implementation of a minimalistic silviculture, also called protective silviculture.

Two types of tools have been elaborated in order to move beyond this practice judged too binary:

- The **GSM**[6]: These mountain forestry guidelines are highly operational and directed towards all the players within forest management and for all forests. They provide techniques of analysis and suggest multifunctional silvicultures adapted to each station, goal or natural constraint while avoiding incompatible positions between protection and production. These guides are based on research works carried out by the Irstea in France, the WSL[7] in Switzerland and the University of Turin in Italy.
- The **Hazard Control Index** (*IMA in French*): This index is used to grade the protective function of forest stands in regard of one or several natural hazards in a mountainous area. It is based on the combination of dendrometic data such as the number of stems per hectare and basal area and expresses the hazard control degree on a scale from 0 to 5. (Fig. 8.3)

The Hazard Control Index was firstly mapped within the framework of the Interreg project *Protection Forests* in five areas in the Northern French Alps, then at the request of intercommunity bodies to clarify their forest management issues (Fig. 8.4).

[5] Hazard zone mapping and the integration of the forest protection function. Good practices and recommendations for national/local policies (Sixt, Haute Savoie, 2014). Report available on the FAO web site (http://www.fao.org/forestry/89331/en/).

[6] Co-written by the Irstea, the ONF and the French Regional Centres of Forest Property.

[7] Swiss Federal Institute for Forest, Snow and Landscape Research (Switzerland).

ROCKFALLS (blocks of less than 1 m^3) Hazard Control Index			Stem diameter predominantly below 17.5 cm		Stem diameter predominantly above 17.5 cm	
			Stem density (N/ha) diameters of 7.5 to 17.5 cm		Stem density (N/ha) diameter of 17.5 cm and +	
			N < 2 000	N > 2 000	N < 350	N > 350
AREAS OF TRANSIT AND DEPOSIT	Wooded strip > 200 m (planimetric)	G < 25 m^2/ha	1	3	1	3
		G > 25 m^2/ha			4	5
	Wooded strip < 200 m (planimetric)	G < 25 m^2/ha	0	2	0	1
		G > 25 m^2/ha			2	3

Fig. 8.3 Example of a table in the forest management planning manual used in defining the hazard control index in relation to rockfalls. (where G is forest stand basal area (m^2/ha))

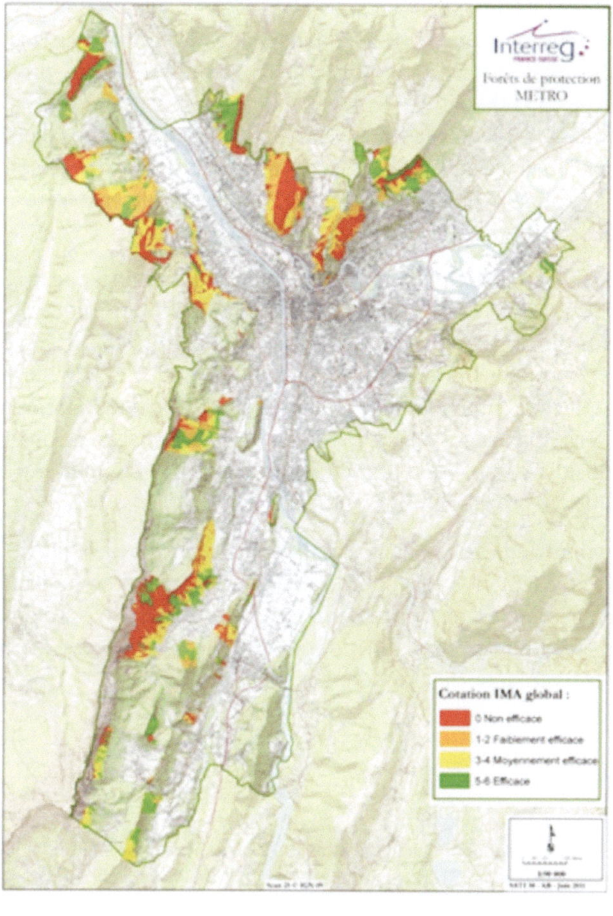

Fig. 8.4 Cartography of the Hazard Control Index for aggregate forest stands with an identified protective function above the Grenoble agglomeration

5 The Operational Strategies Showing Positive Results

- The succession of research and development programmes running for 25 years is a good example of advancement strategies. Tools based on research works have been implemented and adapted to every stage of integration of the protection forest, while involving all the players in the mountain forestry sector.
- The ONF has launched a programme of renewal of protection forest stands in state-owned RTM forests. Silvicultural works integrated in this programme will provide for a progressive and natural regeneration of forest stands. The annual financial requirements for these forests in light of performed pilot projects are estimated at three million euros.
- The Ministry of Agriculture wishes to give priority to state-owned RTM forests involved in controlling important hazards concerning well identified socio-economic stakes (12% of state-owned forested RTM surface areas) and those already equipped with torrential flood correction works, for which forest stands are a support (14% of state-owned forested RTM surface areas). The purpose of the second priority is to reduce the maintenance cost of these structures. In this context, the total surface area concerned by the renewal of protection forest stands reaches 65,000 ha.
- The civil society expects the forest's protective function to be taken into account in the management of all mountain forests and not only for state-owned RTM forests. The implementation of technical routes, in an incentive or regulatory manner, is underway.

  – The forest's protective function is being studied and mapped by ONF officers since 2012 within the framework of the elaboration of *Aménagement forestiers* (Forest management planning) of all public forests.
  – Within the framework of integrated management of natural hazards, the forest's protective function may be included in Natural Hazards Prevention Plans. The European Agricultural Fund for Rural Development (EAFRD) may subsidise up to 80% of the cost of protective silvicultures proposed by any public or private player.

- All these concerted actions have supplied mountain forestry players with tools, legal texts and financial means to assess, quantify and strengthen the forest's protective function in mountainous areas. This results in similar expectations regarding other non-market functions such as public access to forests, water quality preservation or the preservation of biodiversity.

References

FAO. (2014). Hazard zone mapping and the integration of the forest protection function. Report of the workshop 1. Working Group II of the EFC WPMMW. French forest code (Version of 01/07/2012).

Gauquelin, X., & Courbaud, B. (2006). Guide des sylvicultures de montagne – Alpes du Nord françaises. CEMAGREF – CRPF Rhône Alpes – ONF - ISBN 2–84207–306-1.

ONF. (2010). National Guidelines for Management and Management Planning (ONAG) of public forests.

ONF. (2012). Détermination de la maitrise des aléas par les peuplements en forêt à fonction de protection. Fiche du Manuel d'Aménagement 9200–12-GUI-EAM-018-VERSION A.

ONF – DFRN. (2015). Mountain watersheds and ecosystem services: Balancing multiple demands of forest management in head-watersheds. French national report for the 30th session of EFC WP MMW – FAO.

Renaud, J. P., et al. (1994). Softwood forests with a protection function in the French Northern Alps: General typology of structures, descriptive method, analysis of population stability and silvicultural interventions. CEMAGREF; ONF.

Sonnier, J. (1991). Analyse du rôle de protection des forêts domaniales de montagne. *Rev. For. Fr. XLIII* – 2-1991, 131–146.

Vésino, R. N., & Marco, O. (2014). Renouvellement des Peuplements de Protection (RPP) Phase 1: La cartographie aléa/enjeu et sa prise en compte dans les aménagements forestiers. DERN – DTN RTM – ONF.

Chapter 9
Headwaters Management Constraints Within the Kinneret Ecosystem Services

Moshe Gophen

1 Introduction

The Lake Kinneret and its drainage basin, mostly the Hula Valley, are two ecosystems with a very high level of the national importance. This status was given to them because they form a major source of drinking water supply and being the only national site for inland aquatic recreation, commercial fishing, and international and national tourism. During the last 80 years, the external constrained pressure on those ecosystems was significantly enhanced. The research and development of these systems were intensively carried out. Presently, we are located at a point where we can look backward and be ready for actual decision making and missing information can be pointed out, all aimed at formulating the innovated design of appropriate design for future activities. The aim of this paper is to give an overview of the environmental problems faced in the past and recent water resource management of the Lake Kinneret (Fig. 9.1), to be implemented by the future proposed construction to meet the optimum watershed management design.

2 Background

2.1 The Kinneret Lake

The Lake Kinneret is the only natural freshwater reservoir in Israel. The lake-catchment area is located in the Syrian-African rift valley (northern Israel, subtropical climate conditions). Total annual water balance of Israel is formed by

M. Gophen (✉)
MIGAL Scientific Research Institute, POB 831, Kiryat Shmone 11016, Israel
e-mail: gophen@migal.org.il

© Capital Publishing Company, New Delhi, India 2017

J. Křeček et al. (eds.), *Ecosystem Services of Headwater Catchments*,
DOI 10.1007/978-3-319-57946-7_9

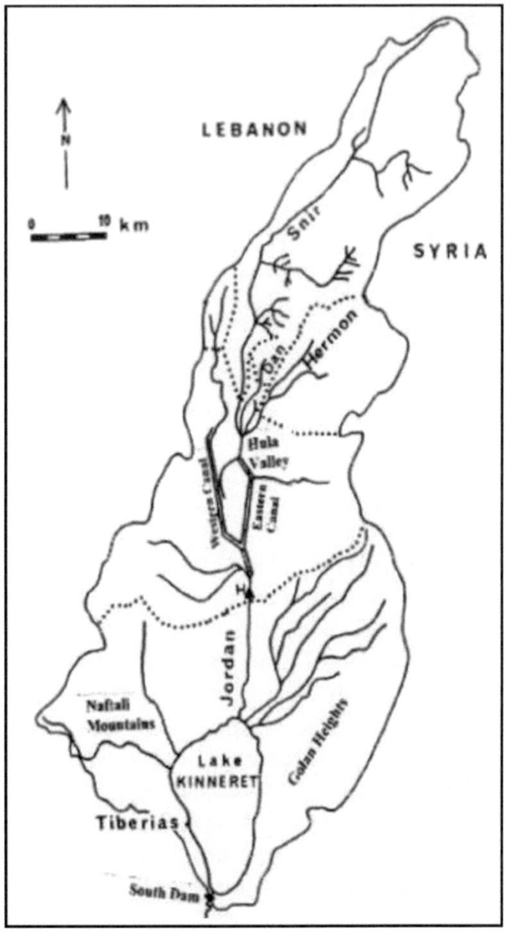

Fig. 9.1 The Lake Kinneret system, northern Israel

precipitation volume of 8 billion m^3 (between >1000 mm in the north, and <100 mm in the southern desert), approx. 70% is evapotranspiration, 5% runoff, and 25% deep infiltration. Above 95% of the Israeli natural water resources are exploited, and 85% of waste waters are also utilized. The national annual water supply is approx. 2.11 billion m^3, of which 0.55 billion m^3 are due to the Kinneret–River Jordan system, and 0.65 billion m^3 are produced by desalinization of Mediterranean waters. Before 2005, the average annual input to the water budget of the Lake Kinneret was 723 million m^3/year; 462 (River Jordan), 110 (Golan Heights rivers), 70 (Eastern Galilee), 68 (direct precipitation) and 13 (Diversion of Jarmuch). The output considers 262 (10^6 m^3/year of evaporation), 323 (National water carrier), 76 (outflow to the South Jordan River) and 62 (local consumers). The Lake Kinneret water level fluctuates between 208.80 and 214.87 m below sea level. The upper water level is limited due to previously constructed housing which might be damaged by higher levels. The determination of the lower limit is due to the limnological concept based on the consequent prediction of water quality.

Both rainfall and the national water demand were priorities of the NWC strategy, formulating the management design propositions. The optimal management was therefore aimed at maximizing close dam to get the highest storage capacity. Before the dam construction, nutrient-rich winter floods crossed the lake in the upper layers as a result of their higher temperature, than that of the epilimnetic water and left through the open outlet. A 'close dam' management might be, therefore, supportive for nutrient retaining during the winter, and settling later to the hypolimnion, though pumping is done from the epilimnion. The 'close dam' management in comparison with an 'open dam' conditions, might therefore, enhance the nutrient accumulation. Moreover, winter demands and pumping are lower than in the summer, predicting surplus nutrients retention. Water retention time, strongly affecting the nutrient dynamics, is a significant parameter of the lake's trophic status.

During the early 1950s, the Lake Kinneret was proposed to serve as the main supplier of the national domestic and agricultural demands. The lake outflow has been controlled by the lake management policy integrating both hydrological and environmental constraints. The National Water Carrier (NWC) was initiated in the 1950s and started to operate on the 10th June 1964. The NWC system regulated the withdrawal of water according to the water resource recharge and water quality. In the Kinneret Lake, water salinity highly fluctuated; before 1960, the content of chlorides was 300–330 ppm, during the early 1960s 400 ppm, and during early 1980's 200–210 ppm. Later, this content gradually elevated to 280–300 ppm found in 2002. As a result of the water supply from the Lake Kinneret, the NWC system transferred salts to the Israeli soils and aquifers in the southern part of the country. About nine million tons of dissolved elements (chloride, carbonate, sodium, magnesium, potassium, calcium, sulfate) from the Lake Kinneret were conveyed as well as deteriorated soils and aquifers in the southern part of Israel. The lake Kinneret is also exploited for fishing: approx. 180 licensed fishermen which remove commercially, an average of 1600 tons of fish (94 kg/ha) per annum. The zooplanktivorous fish, lavnun (bleak, Mirogrex, Acanthobrama) comprised more than 50% of the lake stock biomass. Among eight commercial species out of 24 natives recorded, were tilapias (averaged in 326 tons per year) and the exotic mugil (averaged in 160 tons per year) are the most important in the commercial landings.

However, the Kinneret ecosystem has undergone significant changes during the last 70 years. Some of the changes are natural and others are anthropogenic. In the drainage basin: (1) Increasing population up to above 200,000; (2) Sewage removal and fishpond restriction; (3) New agricultural technologies (crop types, irrigation); and (4) Development of eco-tourism. Prominent changes in the lake ecosystem among others, are affected by: (1) Salt removal; (2) High (6.3 m) amplitude of water level fluctuations; (3) Fluctuations in the annual inflow ($> 10^9$ m^3, and <260 10^6 m^3 per annum) resulting from variations in annual precipitation ($>$ 1060 mm, and $<$333 mm); (4) Plankton biomass and composition changes; (5) Fish stocks; (6) Epilimnetic temperatures; and (7) Nutrient concentrations. The Lake Kinneret water quality is of national concern and therefore, the lake depends on the nation and the nation is dependent on the lake. The Kinneret lake is a warm monomictic water reservoir, stratified from May through mid December (anoxic hypolimnion), and being totally mixed during mid December through April. The seasonal distribution of hydrological, chemical and biological parameters represents their subtropical climatic conditions: high levels in winter and low in the epilimnion during the summer.

2.2 *The Drainage Basin Management*

The Kinneret drainage basin area is 273,000 ha, including the Hula Valley (10% of the basin area; elevation of 60–180 m). The length of the drainage basin is 61 km, the mean slope 5%, and the highest elevation 2814 m (north). The long-term average annual volume of runoff of three major rivers of this area, Hatzbani, Banyas and Dan, is 130, 115 and 260 million m^3, respectively. These rivers converge into the Jordan River (annual discharge of 500 million m^3). The Jordan River flows downstream from the Hula Valley along 15 km south into the Lake Kinneret.

Until 1957, the Hula Valley was covered by the shallow swampy Lake Hula (mean depth of 1.5 m; area of 1300 ha), and of peat bottom wetland (4500 ha). During 1952–1957, the old Lake Hula and the surrounding wetlands were drained, and the area of 5900 ha was converted to agriculture (Table 9.1). About 10% of the area was cultivated inappropriately, causing serious soil deterioration, accompanied by the threat of nutrient flux downstream to the Lake Kinneret. Consequently, in 1990–1997, a reclamation project (the Hula Project) was implemented there. This project included drainage network revision, construction of a shallow swampy lake (area of 110 ha), renovation of the irrigation systems, and an implementation of a new generation of the agricultural development (stressing the continuous vegetation cover of the soil: 'green covering').

During the last 40 years, the drained area was successfully re-cultivated, but still facing several serious problems. For example, the bare peat soil was causing several

Table 9.1 Land use (percentage of the land area) in the Hula Valley, 1949–2010 (drainage in 1952–1957 and reclamation in 1990–1997)

Land use	1949	1958	1976	1986	2010
Water	24	–	–	2	2
Swamps	54	7	7	3	7
Flooded	22	–	–	–	–
Field crops	–	59	79	58	68
Fishponds	–	8.5	8	5	2
Uncultivated	–	17	–	14	5
Orchards	–	–	3	8	9
Others	–	8.5	3	10	7

heavy dust storms. Fortunately, the nutrient flux from the Hula Valley did not endanger water quality of the lake Kinneret. As a result of the inappropriate watershed management, drainage canals were blocked, soil structure was deteriorated and dehydrated, groundwater table declined, outbreaks of underground fires occurred more frequently, and rodent population outbreaks caused severe damage to crops, particularly, 10% of the drained area, located at the lowest latitude of the Hula Valley, was extremely damaged. Therefore, in 1990–1994, reclamation activities (the Hula Project) were implemented. The main aim was to reduce the nutrient flux into the Lake Kinneret while continuing the economical utilization of the land by conversion of 110 ha from agriculture to ecotourism. This conversion included also the construction of another wetland lake (the Agmon Lake of 0.4 m average depth and 110 ha surface area) serving mainly in the nutrient removal and recreation (eco-tourism). Additionally, a plastic barrier (4 mm thick plastic sheet installed to the depth of 4.5 m, crossing the valley along 2.8 km from east to west) near the Agmon Lake served to prevent a ground flux of nutrients to the Lake Kinneret.

3 Regional Climate Change

The long-term record (1940–2015) of rainfall represents a decline reflecting the regional climate changes. As a result of natural modifications and the previously described anthropogenic interventions, the ecological conditions in the Lake Kinneret ecosystem has undergone several changes, for example, air, and consequently, lake water temperatures were lowered until the mid-1980s, and later increased again. The air temperature increase, together with the water level decline, leads to the enhancement of heat capacity of Lake Kinneret epilimnion. Shilo et al. (2015) suggested reduction of wind velocities resulted enhancement of evaporation from the lake. In 1970–2010, the regional precipitation decreased with the exceptions of heavy rainfalls more or less every 10 years. A positive correlation was found between air temperature and land use in the Hula Valley (Table 9.1), affected particularly by changes in albedo (Gophen et al. 2014). Changes in the air temperature in the Hula Valley during the period of 1946–2008 indicate a decline from 1946 till the mid-1980s, and rising later (Table 9.2).

The air temperature before the drainage (1946–1958) was higher (by 0.8 °C) than during the post-drainage period (1959–1990). With the renewed cultivation

Table 9.2 Mean air temperature in the Hula Valley

Period	Number of years	Mean air temperature (°C)
1946–1958	13	20.2
1959–1982	24	19.4
1983–1990	8	18.9
1991–2008	18	19.8

policy (1990–1997), the air temperature was elevated by 0.9 °C, until 2008. Prior to the drainage of the Hula Valley (the Hula lake and surrounding wetlands), the surface was covered by water (100%) and aquatic plants (ca 50%). After the drainage, in 1958, water covered only 15% of that area, while the rest was covered seasonally by field crops. After the implementation of the Hula Project, in 1997, water covered only 11% of the surface, and the rest by field crops. The Hula Project brought the green cover almost for the whole year.

The draught has been monitored there by the Standard Precipitation Index (PSI, based on a long-term rainfall records, McKee et al. 1993, 1995; Gutman 1999). The PSI values were evaluated from the long-term record of precipitation observed in the Kfar Giladi Station since 1930 (Givati 2016). In the period of 1930–1980, the PSI index indicates 17 years of 'near normal conditions', 13 years of 'moderate drought', and 1 year 'severe drought'. In the later period, 1981–2014, a progress in aridization is evident: 1 year of 'extreme drought', six of 'moderate drought' and 7 years of 'near normal conditions'. Anyway, the SPI method can detect only trends in the climate. In 1970–2000, the hydrological survey of the Kinneret basin has indicated a long-term decline of hydrological parameters (Givati 2016). In the northern part of the catchment annual precipitation decreased from 954 to 708 mm. The annual discharge in lake inflows has decreased from 279 to 216 million m^3 (Dan river) and from 73 to 59 million m^3 (Banias river). In total, water available in the Kinneret lake (total inflow minus evaporation) was reduced by 150 million m^3 per year. Gophen (2014) presented also increasing temperatures of the epilimnetic water during the winter. Shilo et al. (2015) indicated the drop in the lake evaporation (ca 5%) as a result of reduced wind energy over the lake (ca 80%), while daily temperatures of the air and water surface increased by 2–4 °C.

4 Groundwaters in the Hula Valley

The groundwater table in the northern part of the valley is shallower than in the south, with smaller seasonal fluctuations. Higher levels of nitrogen and phosphorus in groundwaters of the southern valley are documented by Gophen et al. (2014, 2016). Increased contents of nutrients in the south were affected by three main factors: (1) Underground water migration; (2) Eroded substances from the southern mineral soil; and (3) Intensive evapotranspiration. Due to the lower level of organic content in the southern soils and in spite of possible underground accumulation, no significant difference between southern and northern underground reservoirs was indicated for concentrations of nitrate. Unlike nitrates, the concentration of reduced N form of ammonium in the shallow aquifer (depth 0–5 m) was higher in the south. It is suggested that nutrient concentrations (dissolved and suspended) in groundwaters of the south are higher than in the north, as a result of groundwater flow through preferential pathways and bypassing the plastic barrier underneath and/or aside. The flow direction from north to south is controlled by the hydraulic gradient (a result of both the dip slope and more intensive evapotranspiration in the south). In deep layers of the Hula Valley (depth of 10–100 m), lignite has been formed. In the early 1970s, a project on thermoelectricity production by lignite combustion was studied. As part of a feasibility study, investigation was carried out about the chemical composition of the interstitial waters within the lignite matter. These waters were pumped from several deep boreholes (Serruya and Edelstein 1979).

These results supported the identification of two distinct underground water aquifers in the Hula Valley: (1) The upper (shallow) aquifer confined between surface and 10–15 m, where cracks were formed as a consequence of the entire valley drainage and peat soil dryness enabling water migration through those "preferential pathways", tunnel, cracks or other types of underground free spaces; and (2) The deeper storage confined within depths of 10–100 m. Vertical or horizontal water migration or exchange with the upper stock is not known and has never been studied. In the upper reservoir, an intensive water migration through preferential pathways was documented. The chemical composition of the 'lignite water' is partly known, and no evidence was found to approve mass exchange or migration within the deeper stock or intermediate contact phase, with the upper stock. The geochemical composition and condition of the upper and lower aquifers are clearly different (particularly, different contents of ammonia, nitrate and total phosphorus were identified). The geochemical conditions within the deep lignite water layer are anaerobic, and in the upper underground reservoir aerobic. Therefore, in the upper shallow stock, NH_4 is lower and NO_3 is higher. The migration of outsourcing waters like irrigation, rain or drainage are confined within the shallow underground aquifer. Significant differences of total phosphorus and total nitrogen contents between surface waters (open canals) and the underground shallow water were documented by Gophen (2014).

5 Conclusions

Ecosystems of the Lake Kinneret and its drainage area are crucial for the country life. Particularly, the Hula Valley (20 thousands ha) is an important part of the Lake Kinneret basin (273 thousands ha). Moreover, the Hula Valley supports the agricultural land use, farming and ecotourism. This area went through processes of soil deterioration which resulted in a soil consolidation and dust storms. According to results of Levin and Shoham (1983), Ben-Gai (2002) or Rybakov et al. (2003), drying of peat soils contributed to the formation of preferential water pathways, and increasing loss of water in the Hula Valley. The observed decline of groundwater table enhanced the oxygen penetration into deeper peat layers, and, thus, also underground fires were accelerated.

The anthropogenic pressure on this system is extremely high; the Lake Kinneret is the only natural freshwater lake which has been used for drinking water supply, fishery and recreation. The management of such a system requires a sophisticated and regularly updated design. Namely, the management of the Hula Valley (aimed at directing crops and design of the drainage system) might support the reduction of pollutant fluxes into the Lake Kinneret.

It is evident that a design of effective 'ecosystem services' in such a complex system (Lake Kinneret – Hula Valley – upper catchments) needs long-term and detailed investigations and learning from the past management. The main future objectives are seen, namely, in:

- The management of the Lake Kinneret and its drainage basin system has to meet principles of human welfare and ecological (limnological) principles.
- The optimum solution results in a complicated management design, formulated as the 'golden path' between all parameters involved.
- The follow-up study should concentrate on the migration of water and nutrients in ground- and soil-water environments of the Hula Valley, and the genesis of pollution fluxes to the Lake Kinneret.
- Impacts of water scarcity (global warming and aridization) on the planned agricultural development and water quality in the Lake Kinneret.

Acknowledgement This research project was funded by the European Union's Seventh Framework Programme FP7- REGPOT-2012-2013-1 (Agreement n. 316157).

References

Ben-Gai, Y. (2002). Near-surface geophysical studies in a drainage wetland – the Hula Basin, Upper Jordan Valley. GII Rep. No. 835/216/02, GLOWA UCJR Project.

Givati, A. (2016). The impact of climate changes on the Northern water resources. Hydrological Service, Water Authority, State of Israel (Personal Communication).

Gophen, M. (2014). Land-use, albedo and air temperature changes in the Hula Valley (Israel) during 1946–2008. *Open Journal of Modern Hydrology, 4*, 101–111.

Gophen, M., Meron, M., Orlov-Levin, V., & Tsipris, Y. (2014). Seasonal and spatial distribution of N & P substances in the Hula Valley (Israel) subterranean. *Open Journal of Modern Hydrology, 4*, 121–131.

Gutman, N. B. (1999). Accepting the standardized precipitation index: Calculation algorithm. *Journal of American Water Resources Association, 35*, 311–322.

Levin, I., & Shoham, D. (1983). Ground surface elevation loss in the reclaimed Hula Swamp in Israel during the period 1958–1980. In *Proceedings of the 2nd International Symposium Peat in Agriculture and Horticulture*, Bet Dagan.

Mckee, T. B., Doesken, N. J., & Kleist, J. (1993). The Relationship of drought frequency and duration to time scale. In *Proceedings of the 8th Conference on Applied Climatology*, Aneheim, 17–22 January 1993. Boston American Meteorological Society.

Mckee, T. B., Doesken, N. J., & Kleist, J. (1995). Drought monitoring with multiple time scale. In *Proceedings of the Ninth Conference on Applied Climatology*, Dallas, 15–20 January 1995. Boston American Meteorological Society.

Rybakov, M., Fleischer, L., & Ten Brink, U. (2003). The Hula valley subsurface inferred from gravity data. *Israel Journal of Earth Sciences, 52*, 113–122.
Serruya, C. and Edelstein, M. (1979, December). Impact of the Hula Lignite in Mining on Lake Kinneret Water Quality. Feasibility Study, Hula Lignite Project, Kinneret Limnological Laboratory, Ministry of Energy and Infrastructure.
Shilo, E., Ziv, B., Shamir, E., & Rimmer, A. (2015). Evaporation from Lake Kinneret, Israel, during hot summer days. *Journal of Hydrology, 528*, 264–275.

Chapter 10
Mitigation of Natural Hazards in Mountain Watersheds of Japan

Hideaki Marui

1 Introduction

In Japan, the disasters associated with seismic and volcanic activities occur frequently (Ministry of Land, Infrastructure and Transport 2002). In recent decades, more intensive rainfalls (associated with the global climate change) caused serious sediment disasters. Inhabitants living in the orogenic zone had to live with such disasters and endeavoured to develop technologies to mitigate their impacts (Hokuriku Regional Development Bureau 2007). Nowadays, standardized techniques for torrent control and for landslide stabilization are available. As a result, sediment discharge is strongly reduced in many torrents and a lot of landslide slopes are stabilized at least concerning smaller scale or middle scale landslides. However, a completely different solution is needed for the mitigation of catastrophic disasters caused by large scale landslides, because structural measures are not able to sufficiently resist against large scale landslides (National Institute for Land and Infrastructure Management 2012).

In the last 20 years, researchers and engineers have made significant efforts to clarify necessary targeting items concerning large scale landslides. Although certain progress is recognized, some essential uncertain items are still remaining. It is urgently needed to clarify such uncertain items in order to arrange more sophisticated and effective way to mitigate future disasters caused by large scale landslides. In this article, background conditions and actual state of large scale landslide occurrences are introduced and specific items to be clarified are pointed out.

H. Marui (✉)
Reseach Center for Natural Hazards and Disaster Recovery, Niigata University, Ikarashi-Ninocho 8050, Nishi-Ku, Niigata 950-2181, Japan
e-mail: maruihi@cc.niigata-u.ac.jp

© Capital Publishing Company, New Delhi, India 2017

J. Křeček et al. (eds.), *Ecosystem Services of Headwater Catchments*,
DOI 10.1007/978-3-319-57946-7_10

2 Large Scale Landslides

The Ministry of Land, Infrastructure and Transport, which is responsible for mitigation of sediment disasters, carried out nationwide evaluation of danger degree of large scale landslides. At first, "Estimated frequency map of large scale landslides" was formulated in 2010, as shown in Fig. 10.1a. In order to estimate frequency of large scale landslides, 120 examples since last 150 years were collected. The frequency was estimated based on analysis of the 120 examples with consideration of quaternary uplift value and geological formation. The frequency is classified into four categories from extremely high to extremely low. The areas of very high frequency are concentrated in the central part of the main island, namely in the Japanese Alpine Region. Areas of very low frequency are distributed, for example, in the Kanto Plain including Tokyo metropolitan area.

As a next step, more detailed "Density map of scars of large scale landslides" was formulated in 2012, as shown in Fig. 10.1b. In this map, the whole territory of Japan was divided into grids of 5 km meshes. The relative density of scars of large scale landslides was indicated concerning individual grids of 5 km meshes and is classified into four categories from extremely high to no landslide. The areas of very high density of scars of large scale landslides are remarkably concentrated in the Japanese Alpine Region.

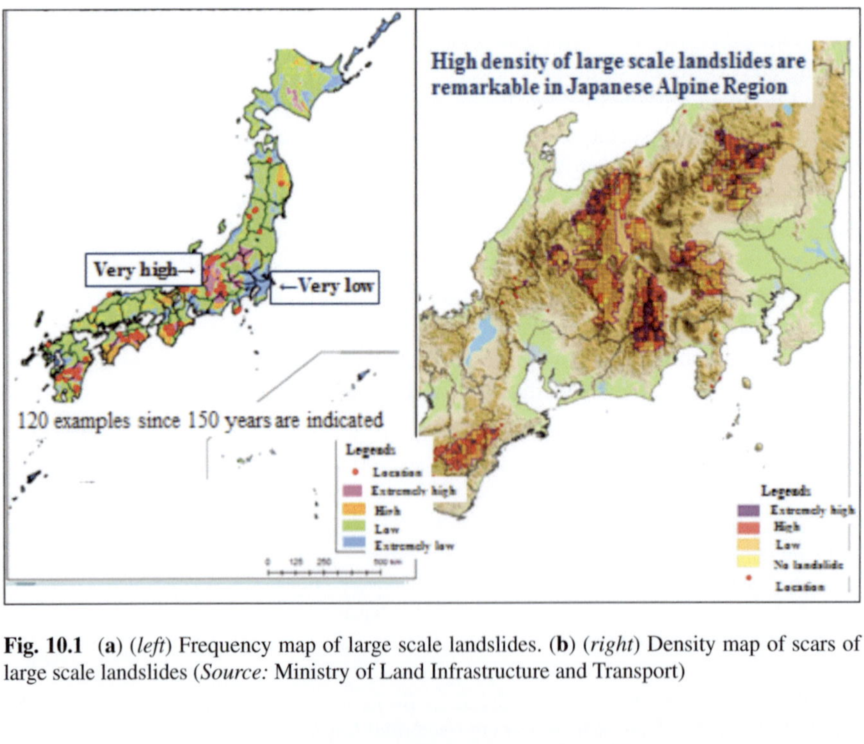

Fig. 10.1 (**a**) (*left*) Frequency map of large scale landslides. (**b**) (*right*) Density map of scars of large scale landslides (*Source:* Ministry of Land Infrastructure and Transport)

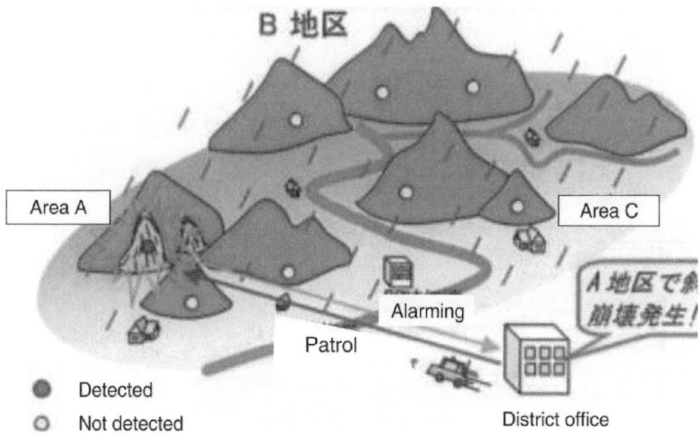

Fig. 10.2 Network system of detecting sensors

Further detailed investigations were carried out concerning the areas of very high density of the scars. As a result of the detailed investigations, the danger degree of large scale land-slides was evaluated concerning individual torrent watersheds with approximately 1 km^2 in the target areas.

Practical management for mitigation of disasters caused by large scale landslides is to be planned and operated by the responsible agency, namely by Regional Development Bureau under the Ministry of Land, Infrastructure and Transport. The agency is responsible for the emergency operation in case of occurrence of large scale landslides. Dissemination of necessary information like danger degree of the concerned torrent watershed to local inhabitants should be also arranged beforehand. For this purpose network systems of detecting sensors using seimograph are installed to catch occurrence of large scale landslides (Fig. 10.2). Furthermore, aerial laser profilings are carried out before and after the landslide occurrence. Surface displacement after disaster can be analysed in detail by comparison between both profiles before and after.

3 Recent Research Results on Large-Scale Landslides

There are certain difficult items on “definition and classification”, “accuracy of prediction” and “feasibility of counter measures” in order to plan sophisticated and effective mitigation measures against large scale landslides. At first, various scales and types are included in the categories of “large scale landslides”. It is basic prerequisite for practical mitigation measure to clearly identify scale, type and characteristics of each individual large scale landslide. At the next point, it is considered that the evaluation of “relative danger degree” is possible and therefore areas with

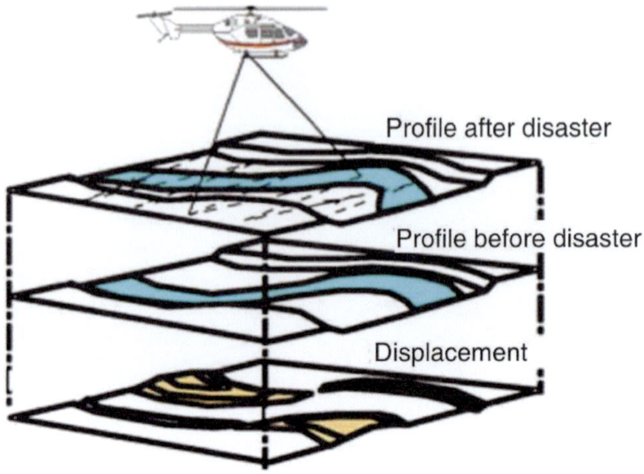

Fig. 10.3 Aerial laser profiling before and after disaster

"high danger degree" can be predicted with certain accuracy. However, the prediction of an individual large scale landslide event and especially its occurrence time is considered to be very difficult to assess. Further, applicability and limitation of each individual countermeasure should be clearly evaluated. It is to be recognized that structural measures are effective only for possible scale of landslides (Fig. 10.3).

The principal triggering factors of large scale landslides are increase of groundwater level or increase of pore water pressure caused by intensive rainfalls and/or snowmelts, and strong shaking of the ground surface caused by earthquakes. An example of a large scale landslide caused by snowmelt accompanied by intensive rainfall is shown in Fig. 10.4. Another example of a large scale landslide caused by earthquake is shown in Fig. 10.5. It is to be pointed out that such a scale of landslide with volume of several millions m^3 can be caused only by strong earthquake and not by rainfall.

Recently researches on large scale landslides are intensively carried out from various interdisciplinary viewpoints. Geological and geomorphological approach to detect dangerous slopes on large scale landslides was developed by various researchers (Chigira 2014). National Institute for Land and Infrastructure Management has pointed out several specific topographic features which are directly related with occurrence of large scale landslides, such as "double ridge", "linear depression" and "arc-shaped crack" (Fig. 10.6).

Prediction method of occurrence of large scale landslides by monitoring of electric conductivity of torrent water is currently being developed (Jitozono 2014). Longitudinal distribution of electric conductivity of torrent water is measured to detect areas with high concentration of groundwater, which indicates high possibility of large scale landslide occurrence. An example of such measurement of electric conductivity of torrent water is shown in Fig. 10.7. Measured values of the longitudinal distribution of electric conductivity along the torrent channel in

Fig. 10.4 Hachimantai landslide caused by snowmelt (Length: 700 m, Width: 400 m, Volume: 70 million m^3)

Fig. 10.5 Aratosawa landslide caused by earthquake (Length: 1.3 km, Width: 1 km, Volume: 6 million m^3)

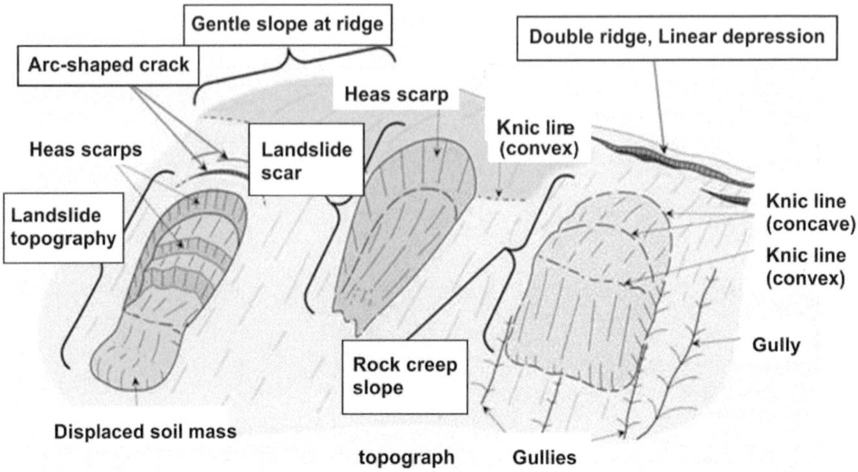

Fig. 10.6 Specific topographic features related with large scale landslides (*Source:* National Institute for Land and Infrastructure Management)

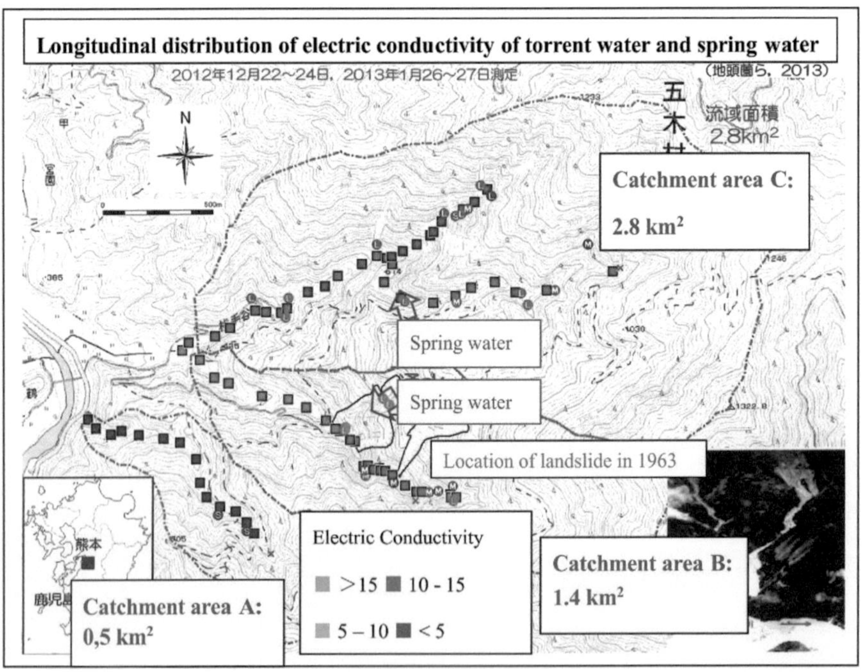

Fig. 10.7 Prediction of large scale landslides by monitoring of electric conductivity of torrent water

three catchment areas are indicated in the figure. In the catchment area A, measured values at all points show very low electric conductivity. In the catchment area B, in the middle part of which a landslide occurred in 1963, measured values show high and very high electric conductivity. In the catchment area C, measured values show different intensity of electric conductivity from medium to high according to the location of torrent channel.

Prediction method of occurrence time of large scale landslides using effective rainfall amount is also currently being developed (Kosugi 2013). If the rainfall amount of the selected target catchment area is continuously measured, effective rainfall amount at any time can be calculated. Key issues for arrangement of practical alarming and evacuation are to select appropriate indicator of rainfall amount and to formulate appropriate critical line as threshold line. Generally, accumulated rainfall amount is used as a long-term indicator to evaluate effects of antecedent rainfall and 1 h rainfall is used as short-term indicator to evaluate effects of rainfall intensity. A prediction method of dangerous situation using snake curve is shown in Fig. 10.8. In the diagram, horizontal axis shows long-term rainfall indicator and

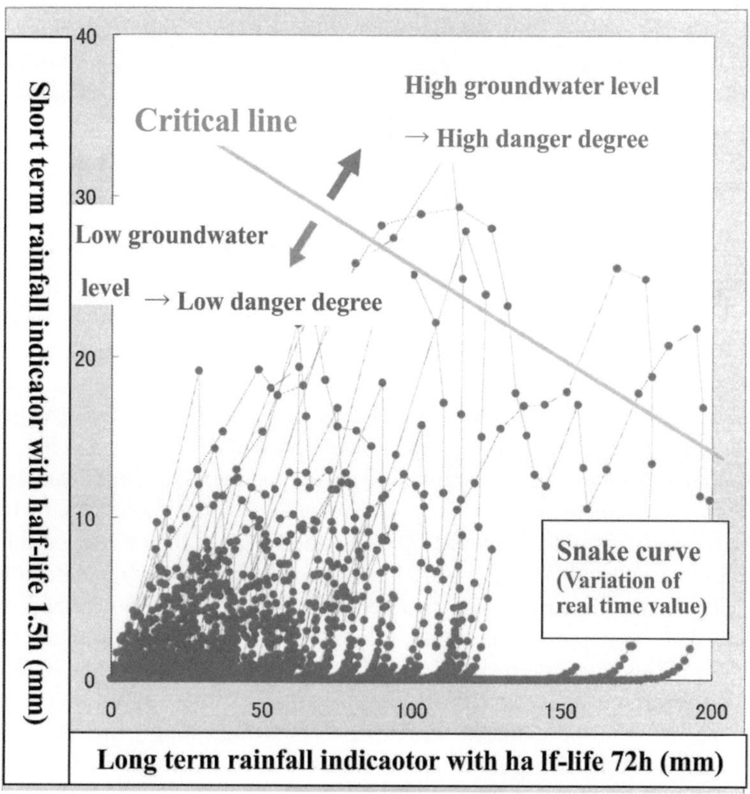

Fig. 10.8 Prediction method of occrrence time of large scale landslide using effective ranfall (Kosugi 2013)

vertical axis shows short-term rainfall indicator. On the basis of past occurrence of large scale landslides in the neighbouring area, a critical line as threshold line can be introduced. Lower area of the critical line should be safe area and upper area should be dangerous area. Snake line shows variation of real time value of both long-term and short-term indicators. If the snake line exceeds the critical line, it means dangerous situation.

Although a certain progess as mentioned above are recognized as a result of recent intensive research activities, some essential uncertain items are still remainig. It is urgently needed to improve accuracy of prediction concerning the occurrence of individual event of large scale landslides, especially on occurrence time.

4 Comprehensive Watershed Management

Landslide mitigation measures should be conducted in order to prevent or reduce the movement of the sliding soil mass so that the resulting damages can be minimized. However, in general effectivity of structural countermeasures both in the source area and downstream area is limited against large scale landslides. Non-structural measures focussing on arrangement of evacuation and appropriate land-use are mostly recommended. Two representive examples of comprehensive watershed management are described.

4.1 Holistic System of Torrent Control Works in Tateyama

The Tateyama Caldera is located in the most northern part of the Japanese Alpine Region and is headwater catchment area of the Joganji River (Fig. 10.9). The length and the total catchment area of the Joganji River are 56 km and 386 km^2 respectively. Joganji River has extremely steep longitudinal profile comparing with large rivers in Europe (Fig. 10.10). Overview of Tateyama Caldera is shown in Fig. 10.11. The combination of weathered materials on slopes and intensive rainfall amount can bring high landslide frequencies and also removal of sediment materials. In 1858, a large scale landslide with displaced soil volume of 1.27×10^8 m^3 triggered by the Hietsu Earthquake with M 7.1 occurred and formed a landslide dam. After 14 days and 59 days from the earthquake, two debris flows caused by the collapse of the landslide dam subsequently brought serious damage to the downstream. These destructive debris flows destroyed 163 villages and killed 1800 people. Until now several sizes of landslides and debris flows triggered by intensive rainfall have frequently occurred in the calderea area. Due to landslides and debris flows in the caldera, many people living in the downstream area of the Joganji River suffered from sediment related disasters.

Holistic countermeasures were intensively implemented in the upstream and also in the midstream area by state government since 1926 (Fig. 10.12). A fundamental

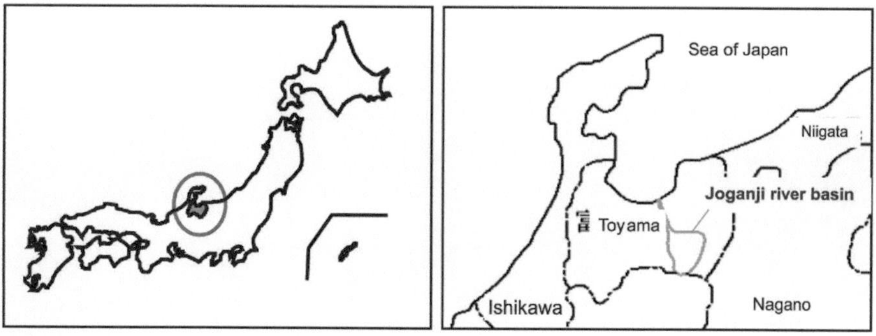

Fig. 10.9 Location of the Joganji river basin

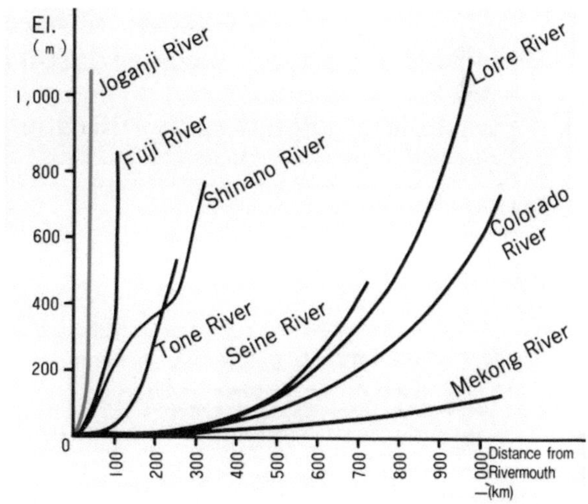

Fig. 10.10 Longitudinal profile of Joganji River

high check dam (Shiraiwa dam) was constructed at the exit of the calderea area in order to keep the basic level against erosion and to deposit sediment materials as much as possible in the caldera area (Fig. 10.13). Further more in several torrential gullies in caldera a series of steps-like check dams were constructed for consolidation of torrent gully bed (Fig. 10.14). After 70 years this gully is completely covered by vegetation. A large scale check dam was constructed at the exit of the middle stream. This dam with very large sediment trap capacity of 5 millions m^3 aimed to retain sediment materials to protect Toyama Plain in downstream area.

This holistic system of torrent control works was conducted by Dr. Akagi with utilization of modern torrent control technology developed in European Alpine Region in early stage of the twentieth century in combination with Japanese traditional technology of hillside works. After that, since 80 years, a lot of control works are added and also maintenamce works are continuously carried out to keep the expected functions of the old facilities.

Fig. 10.11 Overview of Tateyama Caldera

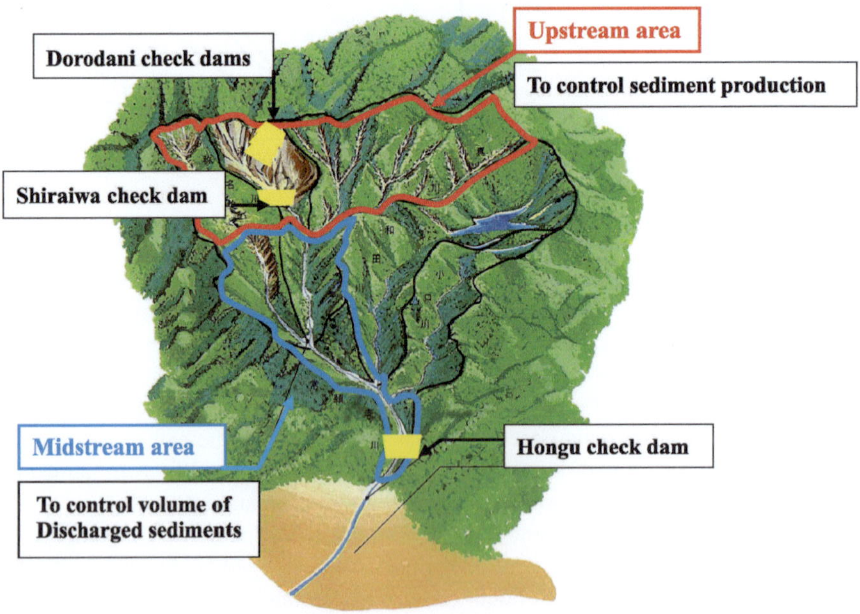

Fig. 10.12 Schematic diagram of holistic system of torrent control works in Tateyamara Caldera

Fig. 10.13 Shiraiwa check dam

4.2 Comprehensive Watershed Management Against Jinnosukedani Landslide

Jinnosukedani landslide, which is located in the headwater catchment area of the Tedori River, is one of the most representative landslides in Japan. The length and the total catchment area of the Tedori River are 72 km and 809 km^2 respectively (Fig. 10.15). Tedori River has very dangerous torrential characteristics and repeatedly flooded. Especially the downstream area of the Tedori River has experienced disastrous flood after the landslide occurrence triggered by heavy rainfall and snowmelt in the headwater catchment in 1934. More than 100 people were killed by this flood. Abundant water amount in the watershed has been utilized for electric power generation, industrial use and irrigation for agriculture.

Tedorigawa dam was constructed between 1969 and 1979 as a multipurpose dam, which is utilized for flood control, industrial use and hydroelectric power generation. It is a rockfill type dam (Fig. 10.16) with the height of 153 m, the length of

Fig. 10.14 Steps-like check dams in a steep gully (Dorodani). (**a**) Construction of dams in 1933. (**b**) Recent situation in 2007

420 m and the maximum volume of stored water about 10 million m^3. The total volume of the stored water amount is 231 million m^3 and 190 million m^3.

Concerning the countermeasures against Jinnosuke landslide, structural measures have been carried out only in the active sliding blocks (1.7 million m^3). However, very large potential sliding block (34 million m^3) is showing certain continuous movement. It is not feasible to stabilize such dimesion of potential block by installation of structural measures. Furthermore, extremely large area (500 ha) is designated as operational area for mitigation of landslide disaster as shown in Fig. 10.17. In this case, practically structural measures are only applicable for the active sliding blocks. Continuous monitoring should be carried out to detect the movement of the potential sliding block. If some dangerous movement over threshold value will be detected, appropriate alarming and evacuation should be operated. Concerning the unpredictable critical event of a extremely large scale landslide, displaced soil mass should be eventually retained in the reservoir area of the Tedorigawa dam. For this case of Jinnnosukedani landslide, comprehensive watershed management including torrent control works should be only possible way of disaster mitigation.

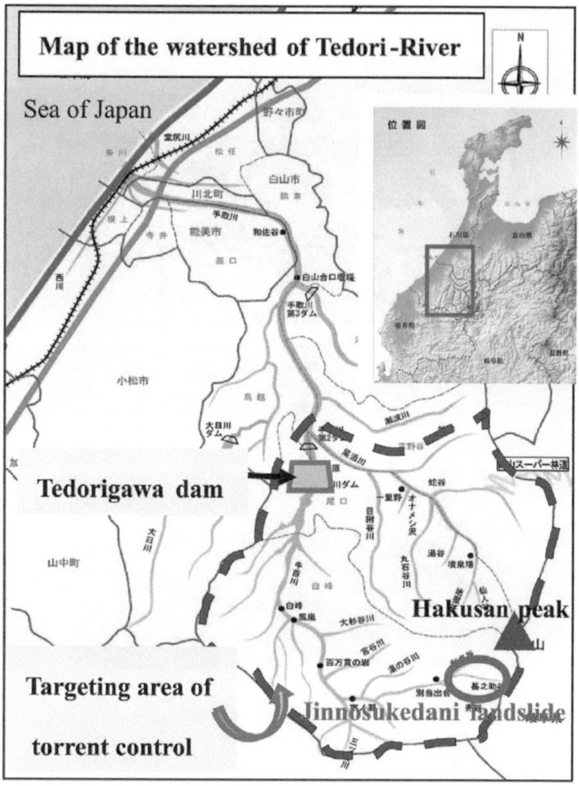

Fig. 10.15 Overview of watershed of Tedori River

Fig. 10.16 Overview of Tedorigawa dam

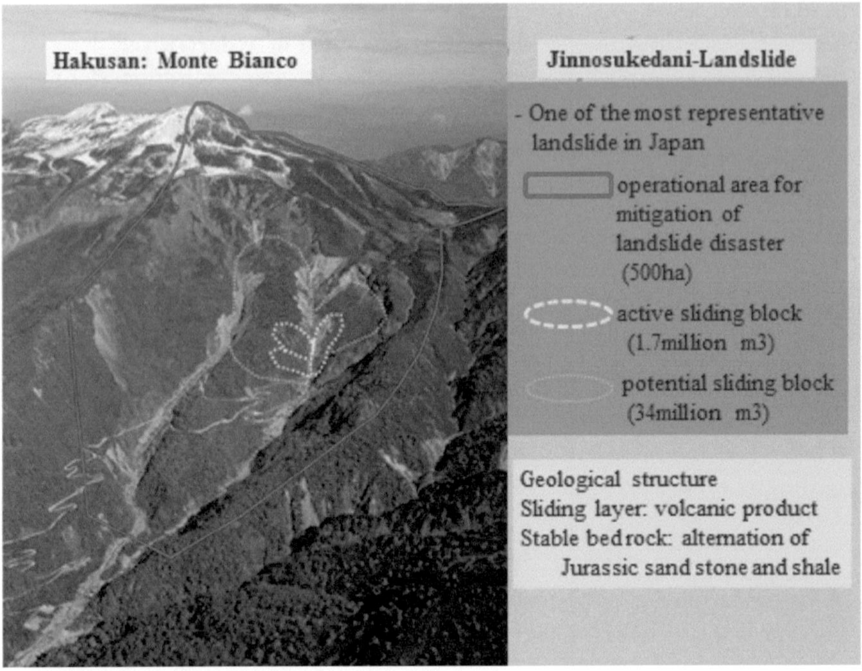

Fig. 10.17 Overview of Jinnosuke landslide

5 Conclusions

Mitigation of large scale landslide disasters is still very difficult task, although a certain basic knowledge is available. Main difficulties are lying on the complexity of the targeting phenomena, which have wide range of scales and types, and on the insufficient accuracy of prediction method and also on the unclear feasibility of countermeasures.

Practical disaster management should be planned on the basis of the clear identification of the target phenomena. It is urgently needed to develop appropriate prediction method with sufficiently higher accuracy in selection of endangered areas and also in occurrence time using hydrological approach.

Due consideration should be paid on the limitation of effectivity of structural measures against large scale landslides. With recognition of such limitation, practical non-structural measures like "arrangement of early warning and evacuation system" and "management of appropriate land-use" should be operated.

Concerning the unpredictable large scale landslide, most probably comprehensive watershed management should be only possible solution. For this purpose preparedness with "formulation of emergency disaster management plan" should be an essential item.

References

Chigira, M. (2014). Geological and topographic detection of dangerous slopes on large scale landslides. Abstract for the 63th annual meeting of SABO Society (in Japanese).

Hokuriku Regional Development Bureau. (2007). *Report on Jinnosukedani landslide*. www.hrr.mlit.go.jp/kanazawa/hakusansabo/H18_4-7_jinnosukepdf

Jitozono, T. (2014). Site prediction of potential deep-seated landslides by using electric conductivity of mountain stream water. *Journal of the Japan Society of Erosion Control Engineering, 66*(6), 56–59. (in Japanese with English abstract).

Kosugi, K. (2013). Prediction of occurrence time of large scale landslides using effective rainfall. Report for the seminar on deep-seated landslides, 2 (in Japanese).

Ministry of Land, Infrastructure and Transport. (2002). *Density map of scars of large scale landslides*. www.mlit.go.jo/common/000223656.pdf.

National Institute for Land and Infrastructure Management. (2012). *Actual state and management of deep-seated landslides*. www.nilim.go.jp/lab/bbg/kouenkai/kouenkai2012/pdf

Chapter 11
Environmental Management in the Headwater Catchments of Kiliki River, Nagaland, North East India

Mohan S. Rawat

1 Introduction

The Brahmaputra is one of the world's largest rivers with a drainage area of 580,000 km^2 (50.2% in China, 33.6% India, 8.1% Bangladesh and 7.8% Bhutan). The watershed has been degraded by land use changes, mainly due to increased shifting cultivation, which affects about 64% of the basin, as well as by deforestation and road construction (Keitzar 2009). These changes have accelerated soil erosion, flooding and landslide activity and caused the general deterioration of natural resources (Valdiya 1985, 1987; Rawat and Rawat 1994). The problem is greater because these Himalayan headwaters are critical for ensuring the food, water and energy security for large downstream regions.

The river catchments of North East Hill (NEH) region are dominated by shifting cultivation, locally called jhum, a long-rotation forest-fallow farming system that provides a livelihood for a large number of ethnic communities. The system is common across Bangladesh, Bhutan, Northeast India, Nepal and Myanmar and has a very long history (ICIMOD 2006). This traditional method of cultivation has proved sustainable over millennia. However, it relies upon long periods of forest fallowing to allow the land and soil to recover from its slash, burn and plant mode of agricultural exploitation (Haigh 1990). Unfortunately, as population pressures increase, fallow periods have decreased to the point where it has become the cause of land and resource degradation, sometimes to the point of crisis (Rawat 2013). The problem has been compounded because recent environmentally-oriented watershed management practices have focused on specific problems rather than a generalized sustainable approach. The aim of this paper is to explore the planning strategies for environmental management in six selected headwater catchments with contrasting

M.S. Rawat (✉)
Department of Geography, School of Sciences, Nagaland University,
Lumami, 798627 Zunheboto, Nagaland, India
e-mail: msrckr@rediffmail.com

© Capital Publishing Company, New Delhi, India 2017

J. Křeček et al. (eds.), *Ecosystem Services of Headwater Catchments*,
DOI 10.1007/978-3-319-57946-7_11

land use and to propose micro (grass root) planning measures for better environmental management and sustainable development.

2 Regional Impacts of Shifting Cultivation

Jhum, shifting cultivation, is a traditional practice and still predominant in the studied region. As a result of population expansion, demand for food and fuel has increased and land availability for agriculture has been reduced. Therefore, in recent years, the cycle of 10–15 years has decreased to 3–5 years, which allows insufficient fallowing time for the soils and vegetation to recover. Consequently, the indiscriminate felling of trees on hill slopes has brought higher erosion of soil and increasing rates of the siltation of streams and water reservoirs, leading to unprecedented floods. The amount of soil erosion mainly due to the shifting cultivation is estimated as ca 88.3 million tonnes annually (Singh and Singh 1981; Toky and Ramakrishnan 1981). Table 11.1 describes the annual cropping cycle of a typical jhum system and highlights those points where the soil is most vulnerable in this local shifting cultivation system.

Table 11.2 shows the general land use statistics for the North East Hill region, which is considered a 'mega-biodiversity' area. Unfortunately, deforestation linked to shifting agriculture is reducing the region's forest areas at an annual rate of 0.7%. The estimated amount of dry/wet leaves, wooden material, and forest litter burnt in shifting cultivation comes to ca 25.61×10^6 t; which results in an additional loading to the environment of ca 10.88×10^6 t of pollutants annually (Table 11.3).

It has long been argued that settled agriculture is the solution to the region's excessive use of shifting agriculture. Unfortunately, this is not popular with the cultivators because it demands more labour input, which successive governments have attempted to compensate by providing schools, housing, electricity, etc. (Haigh 1990). However, the shift has major environmental benefits. For example, in shifting cultivation areas, the amount of suspended sediment in runoff can vary from 1.3 to 20.3 mg/l. Mean annual soil losses varied from 11.2 to 97.2 t/km^2 for the new land uses compared with 3621 t/km^2 for the shifting cultivation system (Table 11.4). Long-term study confirmed that 93–99% of rainwater can be retained in-situ, compared to only 66% by shifting cultivations.

Table 11.1 Soil erosion calendar of shifting cultivation system in the North East Hill region of India

Month	Agricultural operationt	Erosion problem	Soil erosion (t/ha)	
			Minimum	Maximum
January to April	Selection of plot, forest cutting, burning and cleaning of hill slopes and sowing begins	Displacement of loose soil materials to down hills and rolling down of earthworm casting, soil erosion as above and wash due to rains	0.0	22.4
May	Sowing/weeding	Heavy soil wash, faint drilling at foot hills on silt deposits	0.2	61.9
June	Weeding	Heavy wash of soil aggregates	0.2	45.4
July	Weeding/harvesting begins	Heavy wash of soil aggregates, crop root exposed, farm soil visible	1.8	21.9
August	Harvesting and occasional weeding	Soil wash continues	1.0	29.6
September	Harvesting	Moss appears, soil erosion slows down	0.1	13.8
October	Harvesting	Soil erosion appreciably reduced	0.0	2.7
November	Harvesting	No erosion, moss turns blackish	0.0	0.0
December	Harvesting/threshing/carry harvest back to home	No erosion	0.0	0.0
Year	Cropping with zero tillage on steep slope	Heavy soil wash	3.3	201.4

Table 11.2 Land use classification in states of North East region of India (area in thousand hectares)

State	Geo-graphical area	Forest area	Area not available for cultivation	Other unscientific land	Waste land	Percent of geographical area	Net sown area
Arunachal Pradesh	8174	5154	77	41	8934	16	149
Assam	7844	3071	2455	535	2439	25	140
Manipur	2233	1515	1445	24	5594	25	140
Meghalaya	2243	851	226	646	4745	21	202
Mizoram	2108	1593	211	81	4456	21	65
Nagaland	1658	862	28	224	2053	20	190
Tripura	1049	631	131	40	3224	31	270

Source: Anonymous (2015)

Table 11.3 Nutrient released through ash and blow-off under cultivation cycles in lower elevations in the North East Hill region

Nutrients (kg/ha/year)	Cultivation cycle					
	30 years		10 years		5 years	
	Release	Blow off	Release	Blow off	Release	Blow off
P	313	147	262	156	151	43
K	1739	817	2070	1229	262	156
Ca	956	449	193	115	116	33
Mg	209	98	152	90	114	32

Source: Toky and Ramakrishnan (1981)

Table 11.4 Soil erosion in headwaters under different land use practices in the North East Hill region

Land use practice	Soil loss (t/ha/year)	Author
'Bun' system for raising tubes crops	40–50	Singh and Singh (1981)
Pineapple along the hill slope	24–63	Ghosh (1976)

3 Study Methods

3.1 The Study Area

The study area is located in the central part of Nagaland (the Brahmaputra river basin of the North East India) and includes both urban and rural landscapes with different land use. The hilly terrain of the Kiliki river basin (25°59′33″–26°06′43″N and 94°25′55″–94°32′52″E) in Zunheboto district of Nagaland (Fig. 11.1), covers an area of 80 km^2. The elevation ranges between 510 and 2066 m (Rawat and

Fig. 11.1 The Kiliki River basin in Nagaland (North East Hill region of India)

Furkumzuk 2007). The length of the main Kiliki river is 17.94 km. In this catchment, about 32% is covered by dense natural forests. The main land uses are shifting cultivation and fallow land (55%), terrace cultivation (2%), and 11% plantations and settlements. Six micro watersheds have been selected for detailed environmental study that focuses on environmental status and the resolution of development problems.

3.2 Data Collection

Data collection included long-term meteorological and hydrological observations supported by regularly organized field surveys to assess the physical environment and natural resources. Socio-economic studies were conducted through questionnaires to identify the current socio-economic and cultural conditions within the local communities and stakeholders involved. Additional information on the forest cover, land use, geology and geomorphology were obtained through the detailed studies of Survey of India Topographical Sheets while satellite data were used for detailed resource mapping and land use/land cover identification. The supporting research materials used were: (1) (1) Survey of India Topographical Map Nos. 83 J/8, 83 J/12, 83 K/5 and 83 K/9 with scale of 1:50,000, (2) IRS P6 LISS III and IRS P6 PAN imageries, (3) Arc Info GIS Software of the Arc GIS Master Lab Kit and (4) Field Instruments-Altimeter, Pedometer, Brunton Compass, Geology Hammer, Pigmy Current Meter; Measuring Tape and Abney Level etc.

4 Sample Development Recommendations

The Kiliki river basin undergoes rapid geomorphologic change during the monsoon season; landslides are an acute problem. Socioeconomic and questionnaire studies confirmed that the conservation and management of water and forest resources in the study area is a big issue for local inhabitants. Facing increasing population pressure, shorter cultivation cycles have resulted in drastic declines in the crop yield, loss of forest wealth, soil fertility, and biodiversity. Local people are concerned to reverse the decline in their economic well-being and environmental quality.

The Kiliki basin study area was divided into six headwater catchments for planning purposes. Each unit reflected different land uses, environmental characteristics and development problems: The six were: (1) Shichi, (2) Azukikihi-Lomopula, (3) Asukhomi, (4) Kiliki Hu- Lokobo, (5) Upper Shepoki, and (6) a miscellaneous land (Fig. 11.2). This article illustrates the process by detailing the planning strategies proposed for the first two of these catchments.

The Shichi (1) watershed (20°02′23″–26°03′53″N and 94°30′10″–94°32′04″E, area of 5.02 km^2) is situated in the middle part of the Kiliki river basin. The main village (about 10 households) is located in the elevation of 1500 m. The cultivation area is 55 ha, forested land 147 ha, scrubs and fallow 336 ha, settlement 0.2 ha and 9 ha of a terrace cultivation.

The proposed management strategy is shown in Fig. 11.2. The functionality of headwater drainage network requires, first of all, that springs should be perennial. Therefore, the traditional concept of spring sanctuaries, protected forests on the higher slopes, should be reconsidered. Rain and surface water harvesting practices should be adopted in water scarcity areas. The Shichi stream water could be utilized for minor irrigation (by agriculture and horticultural development near villages).

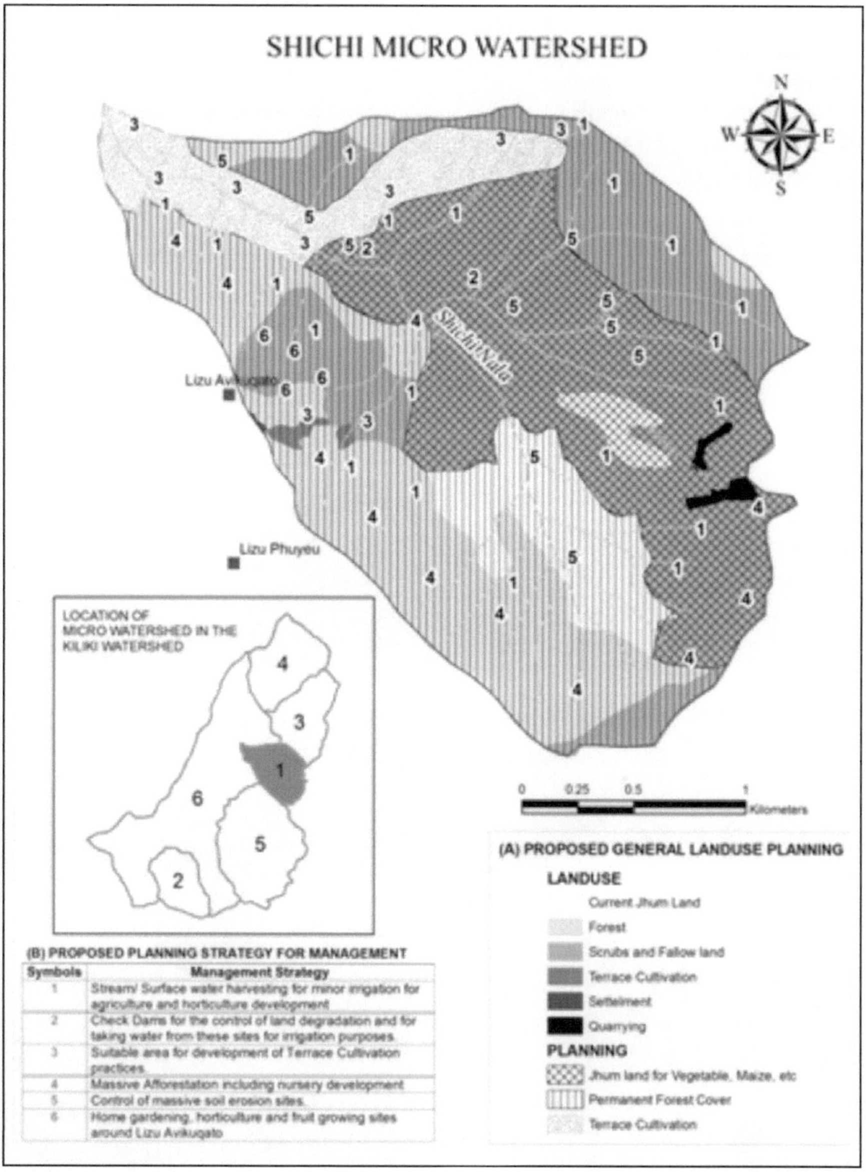

Fig. 11.2 The proposed micro level planning strategy in the Shichi micro watershed

Downstream, check dams are proposed to control the overland flow. Around the outlet of this headwater catchment, there is land suitable for the further development of terrace cultivation. On the eastern side of Lizu Avikuquato village some gently sloping land is recommended for future terrace cultivation, while close to Lizu Avikuquato village, home gardening, horticulture and fruit growing areas might be still developed. However, around the Zunheboto road, massive forestation is required. A network of roads is needed to access distant areas for development activities.

The Azukikihi-Lomupula (2) watershed (25°59′36″–26°01′20″N and 94°27′32″–94°28′57″E, area of 6 km²) is situated in the altitude of 650–1800 m. Three villages are located by the watershed divide (Fig. 11.3). All the villages suffer economic

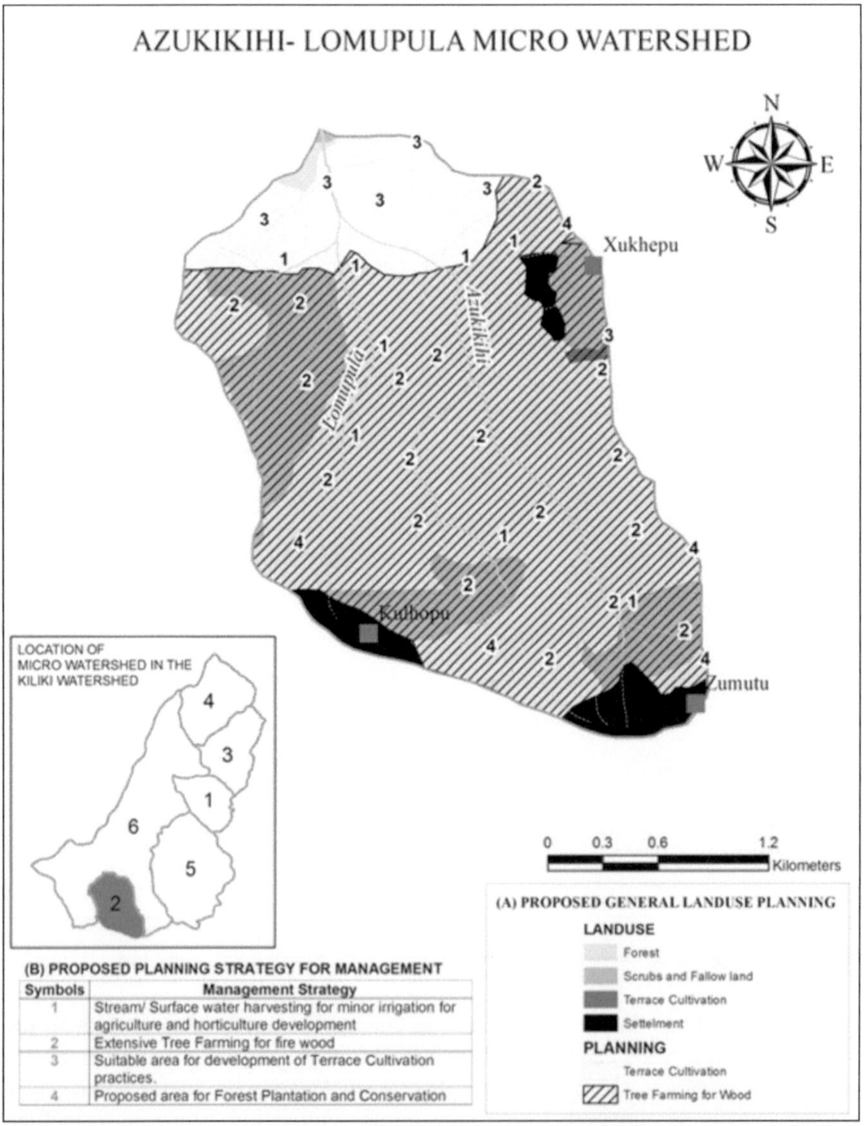

Fig. 11.3 The proposed micro level planning strategy in the Azukikihi-Lomupula watershed

backwardness, resource scarcity and the emigration of people to urban areas. The land area includes 403 ha of forests, 156 ha of scrubs and fallow, 36 ha of settlement areas and 7 ha of terrace cultivation. In this catchment, there is an urgent need of forest conservation: on the gentle hill slopes, nurseries may be developed for the forestation by native species. The proposed strategy considers two categories: (A) general land use planning, and (B) a planning strategy for management of environment and development activities. Surface water harvesting from the nearby streams and rooftop rain water harvesting systems are required for domestic and minor irrigation purposes. In and around the villages 'spring sanctuaries' will be reintroduced for the regular and proper functioning of water sources. Several check dams along the streams are proposed to control runoff. Terrace cultivation may be encouraged in the valley areas of both the small rivers alongside the introduction of minor irrigation facilities. Tree farming will be developed on the steep slopes for fuel, fodder and timber purposes. An urgent requirement was felt for the improvement of the basic school education, and to introduce understanding of the watershed concept.

Similar micro scale planning measures were proposed for watersheds (3), (4), (5) and miscellaneous land (6). In each case, the systematic planning strategy concentrates on achieving sustainable development and evoking active citizenship.

In each case, an 'Environment Management Committee' was constituted for the supervision of the stakeholder dialogue and to consider the logistics needed for the implementation of land development and the monitoring of 'environmentally sound planning' measures. The 'Environment Management Committee' was constituted to respect the priority of effective resources management and environmental hazard mitigation. In addition an 'Environmental Conservation Fund' would support environmental conservation and the restoration of ecological stability with the goal of enhancing local self-reliance.

5 Conclusions

This study shows how recent cultivation activities, which have caused drastic decline in both the crop yield and the environment of Nagaland (East India), may be tackled. The problem is that the environment has been degraded by shifting cultivation, the slash and burn system on steep slopes, deforestation, and also by road construction. These phenomena have accelerated soil erosion, landslides, the deterioration of natural resources and the consequent loss of the well being of inhabitants. Action is urgently needed to halt the further environmental degradation and the micro watershed planning approach described here, which is based on ecosystem services considerations, seems to be a good way to secure a better future for the citizens of these isolated mountain headwaters.

The proposed concept of micro watershed planning includes combination of the traditional local knowledge applied through citizen participation with modern technology skills. Of course, actions at all levels are required to develop and promote the programmes, policies and approaches that integrate environment, ecology, economy and social components of Nagaland for its overall sustainable development.

Participatory micro-watershed planning, which is guided by the involvement of rural people, illustrates how small-scale changes, locally agreed, can help restore and sustain the ecosystem services and the environmental services of hilly headwater catchments and might become an integral part of the food, water and energy nexus. This study illustrates some practice-oriented examples of such effective planning in small headwater catchments. It shows how decision making and planning procedures have to integrate and unite environmental and socio-economic approaches with the local stakeholder wishes and control.

Acknowledgement This study was supported by the G.B. Pant Institute of Himalayan Environment and Development, Kosi-Katarmal, Almora, Govt. of India (Project Sanction No. GBPI/IERP/04-05/21/860), and the Integrated Eco Development Programme for the Himalayan Region. The author also wishes to thank Dr. C. Furkumuk and Dr. P. Toiho Sema, Department of Geography, Nagaland University, Lumami for their assistance.

References

Anonymous. (2015). *Basic statistics of North Eastern region 2015*. Shillong: Government of India, North Eastern Council, Secretariat (Evaluation and Monitoring Sector).

Ghosh, S. P. (1976). *Development of suitable planting technique of pineapple in hill slope*. Annual Report of ICAR Research complex for NEH region, Shillong.

Haigh, M. J. (1990). Shifting agriculture (jhum) and environmental devastation: The search for a solution. In N. K. Sah, S. D. Bhatt, & R. K. Pande (Eds.), *Himalaya: Environment, resources & development*. Almora: Shree Almora Book Depot.

ICIMOD. (2006). *Debating shifting cultivation in the Eastern Himalayas: Farmer's innovations as lessons for policy* (Compiled by Kerkhoff, E. and Sharma, E.). Kathmandu: International Centre for Integrated Mountain Development (ICIMOD).

Keitzar, S. (2009). *A new approach to hill agricultural technology with special reference to Nagaland*. Nagaland: Department of Agriculture, Government of Nagaland.

Rawat, M. S. (2013). Sustainable development in Nagaland through integrated watershed management: A case study from Kiliki River Basin. *International Journal of Development Studies and Research, 2*(1), 60–68.

Rawat, M. S., & Furkumzuk, C. (2007). Environmental degradation due to the expansion of urban land use in headwater areas in Nagaland. In *Proceedings of national seminar on urban expansion and environmental degradation in the North East India*. New Delhi/Imphal: University Grants Commission/Department of Geography, Manipur University.

Rawat, J. S., & Rawat, M. S. (1994). Accelerated erosion and denudation in the Nana Kosi Watershed, Central Himalaya, India. Part I: Sediment Load. *Mountain Research and Development, 14*(1), 25–38.

Singh, A., & Singh, M. D. (1981). *Soil erosion hazards in North-Eastern Hill region, Research Bull. No. 10*. Umiam: ICAR Research Complex for NEH region, Division of Agricultural Engineering.

Toky, O. P., & Ramakrishnan, P. S. (1981). Runoff and infiltration losses related to shifting agriculture (Jhum) in northeastern India. *Environmental Conservation, 8*, 313–321.
Valdiya, K. S. (1985). Accelerated erosion and landslide-prone zones in the Central Himalayan region. In J. S. Singh (Ed.), *Environmental regeneration in Himalaya: Concepts and strategies*. Nainital: Central Himalayan Environment Association.
Valdiya, K. S. (1987). *Environmental geology: Indian context*. New Delhi: Tata McGraw Hill.

Chapter 12
Agro-Environmental Sustainability of the Yuanyang Rice Terraces in Yunnan Province, China

Michael A. Fullen, Zhu Youyong, Wu Bozhi, Li Chengyun, Li Yong Mei, An Tong Xin, and Gilles Colinet

1 Introduction

Better water comes from better trees, better paddies develop with better water, and better offspring prosper with better paddies (Ancient song of the Hani minority people).

The Yuanyang terraces (22°49′–23°19′N, 102°27′–103°13′E, Fig. 12.1) are extremely spectacular for photographers and tourists. Thus, they have long been a magnet for photographers and tourists, evidenced by many excellent web-based photo albums. However, the remoteness from host investigative teams (Yuanyang is 326 km south of Kunming, the capital of Yunnan Province) and the former political tensions on the nearby Chinese-Vietnamese border, meant the area has received scant scientific study.

2 The 'Magic Mountain Sculptors'

The Hani minority are believed to have originated as the nomadic Qiang tribe, which migrated from the Qinghai-Tibetan Plateau to Yunnan Province in ~300 AD, where they developed their complex rice dominated terraces agricultural system (Wang 1999). During the Ming Dynasty (1368–1644), an Emperor awarded the

M.A. Fullen (✉)
Faculty of Science and Engineering, The University of Wolverhampton,
Wolverhampton WV1 1LY, UK
e-mail: m.fullen@wlv.ac.uk

Z. Youyong • W. Bozhi • L. Chengyun • L. Yong Mei • A.T. Xin
Yunnan Agricultural University, Kunming, Yunnan Province, People's Republic of China

G. Colinet
Gembloux Agricultural University, Gembloux, Belgium

© Capital Publishing Company, New Delhi, India 2017

J. Křeček et al. (eds.), *Ecosystem Services of Headwater Catchments*,
DOI 10.1007/978-3-319-57946-7_12

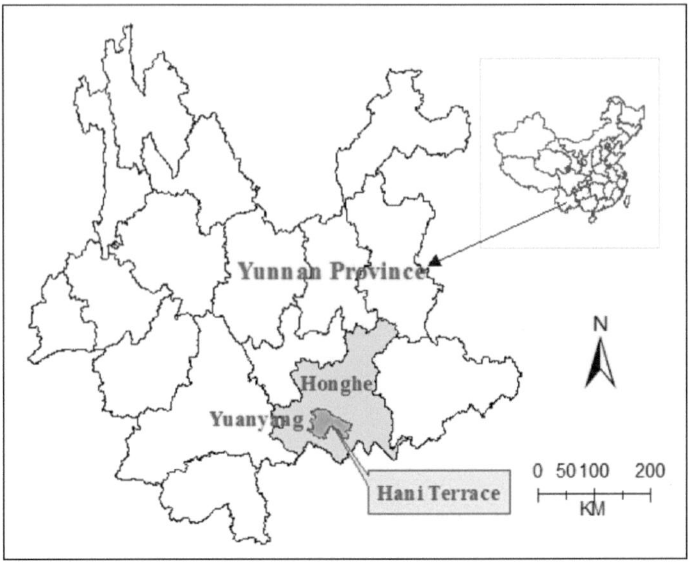

Fig. 12.1 Location of Yunnan Province, Honghe Prefecture (county), Yuanyang Township and the Hani Terraces

Hani people the title of "*Magic Mountain Sculptors*" for "*building their ladders to heaven*." The Hani language belongs to the Tibeto-Burmese linguistic family, which adds to their distinctness from the Han Chinese majority. Generally, terrace construction commenced on lower slopes and each generation adds another terrace level to the system. Thus the oldest, as yet undated, terraces probably lie at the base of the terrace system. The research team intends to extract organic remains, especially wood, from beneath the oldest terrace. Carbon-14 dating of such buried organic material should indicate the inception of terrace building (Fig. 12.2).

The environment of the Yunnan uplands is suitable for rice cultivation with an average annual sunshine of 1670 h and temperature of 15.4 °C (Wang 1999). The area possesses considerable water resources, with ample annual precipitation. Average annual precipitation is ~1400 mm, ranging between 1200 and 1500 mm on lower slopes to 1500–2000 mm on upper slopes. Most precipitation falls as summer monsoon rains between May and October. The meso-scale climate effectively retains and recycles water within the Yuanyang basin, a subcatchment of the Red River (Hong He) basin. The valley is noted for frequent fogs. Furthermore, differential late afternoon solar heating of mountain sides induces slight pressure differences, which cause upslope (anabatic) winds to transport moist airstreams from the valley base upslope. The monsoonal precipitation falls on the Ailao Mountain Range, which rises from ~1400 m above sea-level to 2940 m.

Lithology plays an important role, with upslope rocks consisting mainly of impervious metamorphic gneiss and mica-schists, which encourage runoff into the mid-slope spring lines. Furthermore, structural geology influences hydrogeological

Fig. 12.2 Rice terraces of Yuanyang

behaviour. The main valley follows a syncline structure and so the hydrological catchment is larger than the physiographic catchment, with groundwater permeating along bedding planes into the main catchment from adjacent catchments. Furthermore, deep (~50 m) pockets of weathered granite act as important water storage features and springs often originate at their bases. Thus, the mid- to lower slopes possess a multitude of springs, which are the main water source for the paddy fields (Fig. 12.3).

Soil survey revealed that inherent soil fertility is surprisingly low. Soil pH is typically low at pH 5–6 and cation exchange capacity (CEC) is also low, with typical values of 4–7 cmolc kg^{-1} (Colinet et al. 2011). Crop productivity is thus maintained by efficient recycling of nutrients within the system (Colinet et al. 2011). Topsoils (Ap horizons) are generally clay loams and the depth of the minero-organic topsoil seems to increase in the lower, and thus older, terraces. Typically, the depths of Ap horizons are ~0.6 m in the upper terraces, ~1.0 m in the mid-level terraces and ~1.2 m in the lower terraces. This leads to the research question whether the depth of Ap horizons can be used as an indirect surrogate measure of terrace age.

The total Hani population is 1.44 million, of which 1.42 million live in Yunnan Province (2000 Census). About 12,500 Hani people live in Vietnam. In Yunnan, most Hani live in the Ailao Mountain Range, concentrating in southern Yunnan between the Red River and Mekong River. The Hani population in the seven counties of Yuanyang, Mejian, Honghe, Yuanjiang, Jinping, Luchun and Jiangcheng accounts for 76% of the total Hani population and >50% of the local population. There are two towns and 970 villages in the 2190 km^2 Yuanyang Prefecture. The

Fig. 12.3 Weathered granite intrusion used for extracting building sand. These 'pockets' play a crucial role in water storage and supply, 22 October 2009

total population was 371,489 (late 2004), of which the Hani minority population was 326,100 (87.8%) and the agricultural population was 351,543 (94.6%).

The seven major nationalities in Yuanyang County live at different altitudes, largely in accordance with their traditions and customs. Generally, the people living on the relatively flat valley bottoms (elevation from the lowest point of Xiaohekou village at 144–600 m) are mostly Dai (Tai, that is ethnically Thai) people. At the bottom of the valley side (altitude 600–1000 m) are Zhuang people. Yi people mainly live on the lower mid-section of the valley side (altitude 1000–1400 m). The Hani people mainly live at altitudes of 1400–2000 m. Higher elevations (>2000 m) are the habitat of Miao and Yao people. Normally, the Han majority live in towns or alongside the main roads.

The Hani people have a unique cultural system that reveres the land. Even the name of their main village 'Qingkou' (Catchment Gate) is connotative of their landscape ecology. The site selection of Hani villages also indicates their environmental understanding, as Hani villages are usually located on the middle and sunny mountain slopes. At the middle part of the mountain, temperatures are mild with less possibility of disease and pest damage than at the lower warmer and more humid basal slopes and less chance of animal attack (i.e. by bears) than the higher part. Normally, a Hani village consists of 20–50 households and the distance between two adjacent villages is ~2 km. With such small villages and their close proximity, people can efficiently manage their lands without long walks to the fields.

The Hani religion embraces polytheism and the worship of ancestors and nature. They pay particular devotion to the 'forest god,' which is perceived as the source of life-giving water. Deforestation is considered a religious violation and the Hani people actively teach their children to respect the forest. This concept significantly contributes to forest conservation and ecosystem stability. Consequently, the existing forest area in Yuanyang Prefecture is ~28,000 ha, that is 26.7% of land cover.

The Hani have developed complex agricultural systems. Rice is the dominant crop, but it is maintained with few additions of artificial fertilizers or pesticides. Complex fertigation systems provide most nutrients, especially nitrogen (N). Mixes of manure from cattle and pigs and 'night soil' are placed adjacent to channels. When needed, irrigation waters are diverted into paddy fields, inputting a N flush, especially in the early spring growing season. The Hani have the philosophy of dividing the rice crop into three. The upper third 'is for the people' (i.e. the crop for domestic consumption), the middle third is 'for the animals' (i.e. straw for animal consumption) and the lower third is 'for the soil' (i.e. retained as a stubble, returning nutrients to the paddy soil system and improving soil structure). Rice straw is also used for thatching the distinctive Hani 'mushroom shaped' houses (Fig. 12.4).

Increased crop genetic diversity can extend the cultivation period of a crop variety and the Yuanyang rice cultivars have considerable genetic diversity (Zhu Yongyong et al. 2000). Genetic diversity and careful agronomic management means rice crop yields are high, typically 4–6 t ha^{-1}, compared with the usual 2–3 t ha^{-1} in traditional cropping systems. Minimal artificial fertilizers and virtually no pesticides are used. The traditional rice varieties used by the Hani require less fertilizer inputs than modern varieties and are noted for their resistance to disease (Zhu Yongyong et al. 2000).

Fig. 12.4 The thatched 'mushroom houses' of Qingkou village, 12 November 2010

The red rice of Yuanyang is a local speciality type of brown rice. Red rice is highly nutritious, and has 18 types of amino-acid. The red rice has strong adaptability, low fertilizer requirements and is generally pest-resistant. The thick stems and compact plants means red rice can resist lodging, blast and white leaf disease. In 2013, the total output of red rice from the Yuanyang Terraces was ~9400 t.

Ancillary food production includes ducks, fish (crucian carp (*Carassius carassius*), eels and mudfish), frogs and snails. Pond weed is utilized for pig-feed (Fig. 12.5). These measures provide valuable supplements to the diet and increase the nutrient loads of the perennially wet paddy fields. The nutrient budgets and aquatic ecology of this integrated water management system are under further study. Indeed, the Yuanyang Terraces are the only human created, upland wetlands registered under the Ramsar Convention (Yang Fusheng 2010).

Land use patterns promote soil and water conservation. Hilltops are afforested, with ~15 identified tree species, predominantly Yunnan pine (*Pinus yunnanensis*). This land use is very effective in conserving soil and water and releases high quality water from the upper to lower slopes. In addition, a diverse range of mushrooms are harvested from the forest and bee culture provides both honey and a pollination mechanism for local crops and wild vegetables.

Fig. 12.5 Work in progress on the Yuanyang Terraces (12 November 2010)

3 The Yuanyang Terraces and UNESCO World Heritage Status

The special characteristics of the Yuanyang Terraces made the area most suitable for UNESCO World Heritage Status. Plans were initiated in October 2000 to propose that the Yuanyang Terraces become a UNESCO World Heritage Site. UNESCO convened an international conference in Mengzi City, Honghe Prefecture, on 11–15 November 2010 (The People's Government of Honghe Prefecture 2010). This formed the platform for the final proposal for World Heritage Status. On 22 June 2013, the Yuanyang Terraces was awarded UNESCO World Heritage Status and became China's 45th World Heritage Site.

The Honghe Conference provided a blueprint for the special status of the Yuanyang Terraces, considering both opportunities and challenges. The management plan emphasizes *"protection, comprehensive management and reasonable, sustainable development,"* recognising the special attributes of the "*river-forest-villages-terraces four dimensional ecological system*" (Yang Fusheng 2010). In terms of agriculture, the proposed strategy is for the terraces to focus on the production of high-value organic cash crops. The distinctive 'red rice' is of high-quality, nutritious and possesses a distinctive enjoyable flavour. The rice and rice products, especially red rice noodles, attract a high market price (approximately double the normal price). Combined with the production of high-value organic honey and fruit, these products should yield high and sustainable agricultural incomes.

Relationships between agriculture and tourism pose special challenges. The rapid and unplanned development of commercial tourism could seriously damage the terrace system. Attracting people away from agricultural work to financially more lucrative tourist services would mean terraces were not properly maintained and so would fall into disrepair. Eventually, the very landscape tourists come to see could be undermined. Europe has already experienced such problems, especially in coastal Spain.

The plan for the Yuanyang tourist sector is to develop green and sustainable tourism. Therefore, the development of tourism aims to adopt sustainable policies, which cater for the current tourists, who number about 1.2 million per year. Without careful planning, these tourists could 'swamp' the 29 villages in the Yuanyang terraces, which are home to ~23,247 villagers living in 5036 families. Local Government policies encourage the development of "*ecological, green, eco-tourism*." The promotion of tourism includes the phased and controlled development of the hotel sector and for hotel architecture to harmonize with the local environment. Tourists also stay at terrace farmhouses, thus generating income for householders and tourists gain first-hand experience of Hani culture. Instead of facilitating car driving in the area, tourist minibuses are encouraged, along with walking and bicycles, allowing tourist access without congestion and minimizing air pollution. Tourist services also focus on the consumption of local food and thus stimulate the local agricultural economy.

The relative isolation of the Yuanyang Terraces may well be helpful in the future prospects for the local environment. The terraces remain difficult to access, especially since there is no local airport. Therefore, the terraces do not attract mass tourism, but tend to attract tourists concerned with the environment and conservation, such as photographers and environmentalists. The successful harmonization of high-value organic agriculture and eco-tourism provides both opportunities and challenges to the sustainable development of the Yuanyang Terrace system. The intelligent implementation of these policies should conserve this World Heritage, landscape and biodiversity resource for the benefit of the current and future generations.

4 Global Lessons for Sustainability

The combination of all these identified cultural and agro-environmental factors appear to have allowed the Yuanyang rice terrace system to remain in a sustainable state. The research team postulate that these lessons will have some applicability for agro-environmental sustainability on the global scale. Identified lessons relate to resource optimization, landscape multifunctionality and cultural attitudes. However, our understanding is far from complete. Thus, a joint Chinese-European team has established the 'Yuanyang Project' to explore, study and understand the factors which promote sustainability. The team is approaching these questions as inquisitive students, trying to understand the 'secrets' of the Hani people. Understanding and then presenting these findings should make a positive contribution to the broader development and dissemination of sustainable agricultural technology.

The ancient ballads (poetic songs) of the Hani people are highly instructional. Until recently, the Hani had no written language. These ballads were collated by Zhang Hongzhen (2010), based on meetings with the Mopi (elders of the Hani). The collated ballads consist of 121 verses and cover events throughout the four seasons. Below is a selected verse from each season.

Winter

I cut down water gourd trees,
And turned over straw ash for the need of growing buckwheat.
The leaves of water gourd trees and the straw ash are good fertilizer,
So the buckwheat surely grows well
With plump seeds.
Do remember to cultivate the field in the coming year,
No matter what a good harvest of buckwheat it is.

Spring

In the right condition raindrops beat down the withered leaves,
Trinkling the edge of the leaves like grease.
The raindrops fall on the muddy road.
Without raindrops in winter, the stream may dry up,
And is revived in March with the coming of raindrops.

The grass of Eyuecao buds gradually on the edge of the village,
Woods and grass look green by the pool.
And Wushanoao grass returns to life again.
Horses nicker in the stall,
And the fat and husky oxes grow well.

Guricao grass by the fields bud with tender leaves,
Offering succulence for the pigs,
The boar grows fatter and stronger day by day.

Summer

Hani's happy Kuzhazha Festival falls in June,
It is time for the Hani to play swing;
The swing is set by the Quifang house on the verge of the village,
And quality bamboo is used in making the swing,
Attached to the swing,
Four strong ropes are put firmly into the earth,
And the ropes are fastened together at the top,
With a pedal board being pegged at the bottom,
And a bridge-like beam being set between.

The swing is played high,
With one's arms extending like birds;
With the first swing,
Bad plants are swung out,
And good plants are swung in;
With the second swing,
Evil persons are swung out,
And pure persons are swung in;
With the third swing,
The lazy domestic animals are swung out,
And diligent ones are swung in;
Playing the swing symbolizes the bumper harvest,
And safety for every household,
As well as the prosperity of domestic animals.

Autumn

We harvest ample foods to treat hundreds of Dai people,
And hundreds of Pula people;
The men and women are engaged in endless labour,
On which their value depend.

The autumn is almost gone.
The cicada goes back to earth to rest,
And Dayue returns to earth to sleep;
They live in the section between the loess and black-land,
Where the thin roots in earth are used as clothes and quilts.
They have to drink dew which has fallen into the earth,
And be taken care of by Zheboranyi.

The three months of autumn have come to an end,
A round of 13 days and nights have come to an end,
And a year of 12 months has come to an end,
The three months of winter is just around the corner.

Thirty-three more agriculturally-orientated ballads have been serialized in the 'Newsletter of the European Society for Soil Conservation' (ESSC), in 2012–2016.

References[1]

Colinet, G., Koulos, K., Wu Bozhi, Li Yongmei, LaCroix, D., Su Youbo, Chapelle, J., Fullen, M. A., Hocking, T., & Bock, L. (2011). Agro-pedological assessment of the traditional Yuanyang rice terraces of Yunnan Province, China. *Journal of Resources and Ecology, 2*(3), 289–299.

The People's Government of Honghe Prefecture. (2010). *Proceedings of the first terraced landscapes conference*. Honghe, P.R. China. (In Chinese and English).

Wang, Q. (1999). *On the culture of the terraced fields*. Kunming: Yunnan University Press. (In Chinese).

Yang Fusheng. (2010). Honghe experiences: Protection mechanism of Honghe Hani terraces. In *Proceedings of the first terraced landscapes conference*. Honghe, P.R. China. (In Chinese and English).

Zhang Hongzhen. (2010). *Interpretations of the Hani Seasonal Production Ballads*. Kunming: Yunnan Press Group Company and Yunnan Art Press Company. ISBN:978-7-5489-0034-4. (In Chinese and English).

Zhu Yongyong, Chen Hairu, Wang Yunyue, et al. (2000). Genetic diversity and disease control in rice. *Nature, 406*, 718–722.

[1] Note: Excellent web-based photo albums are available on the Yuanyang Terraces:

https://www.google.co.uk/search?q=yuanyang+rice+terraces&tbm=isch&tbo=u&source=univ&sa=X&ei=GVkPU8OCG9LwhQfv3oGoCQ&ved=0CCoQsAQ&biw=1280&bih=828 (accessed 26/03/16).

http://www.trekearth.com/gallery/Asia/China/South/Yunnan/Yuanyang/(accessed 26/03/16).

http://natureproducts.net/Ecotourism/rice_terraces.html (accessed 26/03/16).

Chapter 13
Environmental Security and Headwater Control in Brazil

Ladislav Palán and Petr Punčochář

1 Introduction

The forest percentage is an important catchment characteristic often used in hydrological studies. With the rising area of forests, generally, a decreasing risk of flood damages is considered (FAO 2006). Well managed forests are also supposed to mitigate impacts of the expected global climate change on increasing frequency and extent of floods (FAO 2006; IPCC 2007). The accelerated runoff with catchment deforestation has been documented worldwide but, Botzen (2013) still calls for a better understanding of hydrological processes within a catchment to design more effective ecosystem services in headwater regions. It is evident that a poor watershed management leads to higher monetary demands as well as the loss in insurance, not only from a flood damage perspective, but also due to the water quality decline (Křeček and Palán 2015).

The aim of the study is to analyse the effect of forests on flood flows in selected headwater catchments of Brazil. The outcomes will be used by watershed planning, as well as by the assessment of environmental security or insurance analyses.

L. Palán (✉)
Department of Hydrology, Czech Technical University, Prague, Czech Republic
e-mail: ladislav.palan@gmail.com

P. Punčochář
Aon Benfield – Impact Forecasting, Singapore, Singapore

© Capital Publishing Company, New Delhi, India 2017

J. Křeček et al. (eds.), *Ecosystem Services of Headwater Catchments*,
DOI 10.1007/978-3-319-57946-7_13

2 Material and Methods

2.1 Motivation and Background

Aon Benfield (2016a, b) designed a catastrophe model to analyse financial implications of natural catastrophes around the world. The forecast experts have developed software providing clients with better information on the risk of nature catastrophe occurrence (hurricanes, tornadoes, earthquakes, floods, wildfires), consequent damages on the property, casualty and crop insurers and reinsurers.

Globally, since the 1950s, mere ten biggest floods have caused damages of approx. 31,638 million USD and have killed 2061 lives (Aon Benfield 2016a, b). In Brazil, flood events belong to the most damaging natural phenomena of rising frequency; since 2009, five biggest floods were recorded there. Thus, a more effective control of floods (including a better design of ecosystem services in headwater catchments) is in the focus of landscape engineers, decision makers and insurers. The analysis of flood events has addressed, particularly, the type and genesis of floods and probability of their occurrence.

2.2 Analysis of Available Hydrological Data

The modelled area included the state of Sao Paulo (70% of total population of Brazil) and additional 53 units across the country (Fig. 13.1). There are two key achievements based on the hydrological data processing: (1) the probability of extreme flows, and (2) synthetically generated peak-flows.

In this study, selection of gauging stations included three main criteria: (1) relatively high economic and insurance hazards, (2) discharge records longer than 25 years and data quality, and (3) similar morphology, catchment area and land-use. The time-series of flow records were tested by standard methods (Shaw 1991); and the CUSUM analysis (cumulative sum control chart) was performed. More than 3000 stream recording stations (operated by the Brazilian Water Authority) were taken into account, but, only 265 of them passed the test of required data homogeneity and consistency, and were accepted by further analyses.

CEH (1999) provides the two main approaches to flood frequency estimation: the statistical analysis of peak flows, and the rainfall-runoff method. The statistical analysis is recommended as the first choice method for long records of gauged floods. According to the extreme value theory, several probabilistic distributions were applied to identify the flood frequency (Fig. 13.2): two-parameter log normal (LN2), three-parameter log-normal (LN3), Gumbel (GUM), two parameter gamma (GAM 2), three parameter gamma (GAM 3), log Pearson type III (LP3), generalized extreme values (GEV), Weibull (WEI), generalized Pearson (GP) and generalized logistic distribution (GL). The method of L-moments (based on

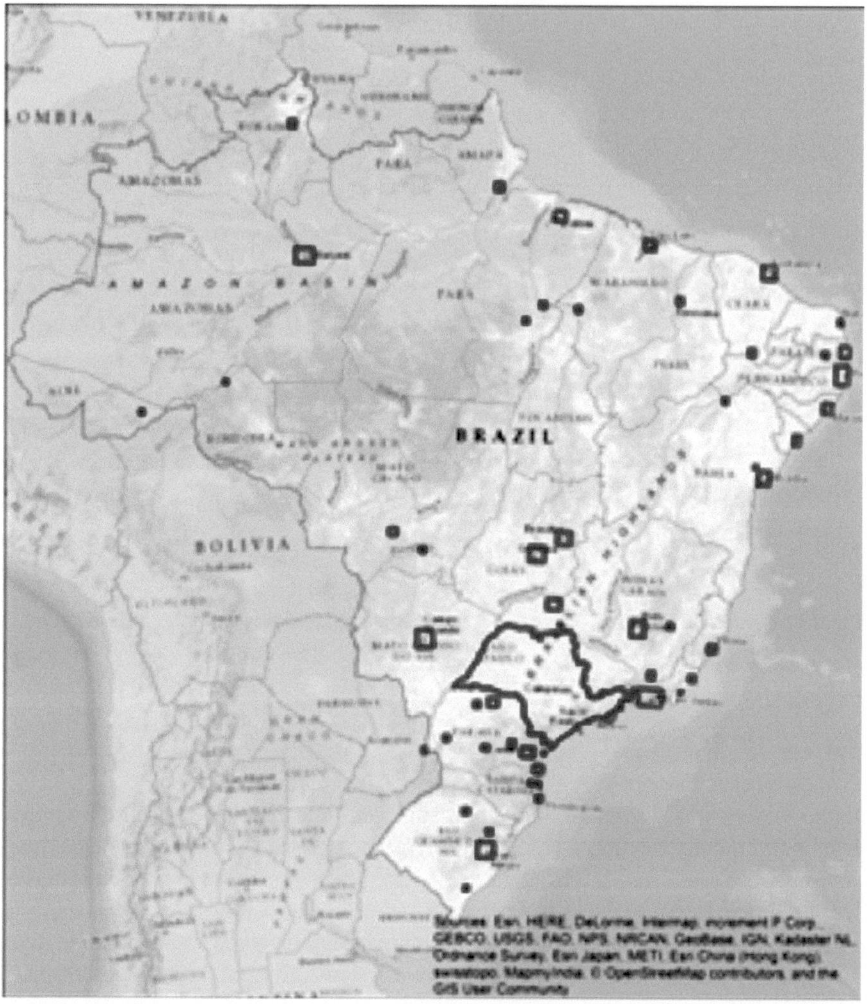

Fig. 13.1 The area of interest

probability-weighted moments) (Hosking1990) was employed to estimate the parameters of theoretical distributions. The advantage of this method is seen in possibility of a wider application and lower sensitivity to the outlier data in analysed records (Rowinski et al. 2002). Based on the most appropriate distribution, the design flows of return periods 2–10,000 years were identified in the gauging stations, and, for the nearby un-gauged profiles, design flows were inferred from catchment similarities (Shaw 1991).

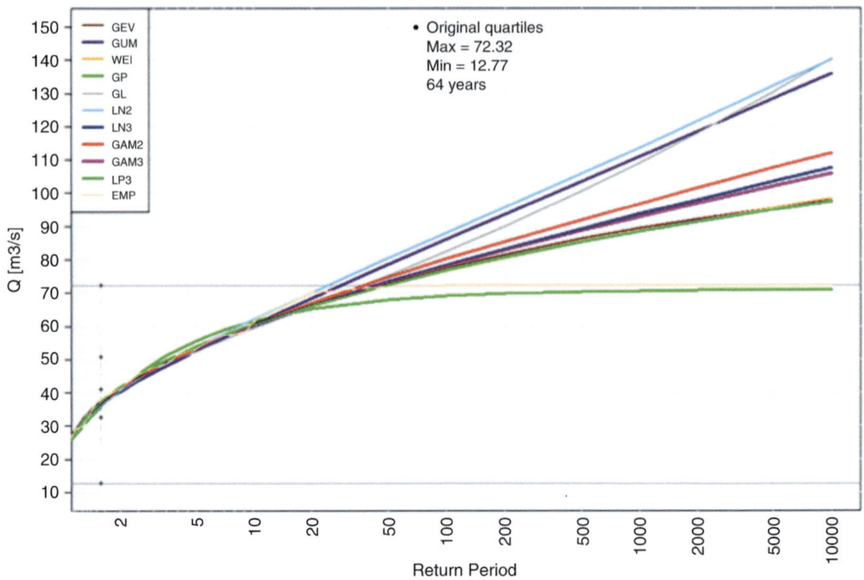

Fig. 13.2 Estimated flood frequency curves for the station Luiz Alves (Rio Luiz Alves river)

2.3 Role of Forests in Flood Control

In the studied area, 13 headwater catchments were selected with respect to required length and quality of runoff observation. The mean observation record of gauging stations is 52 years, varying from 29 to 74 years; their basins are located in four different climate regions (Fig. 13.3).

Respecting the climate differences, studied catchments were divided in four groups A–D (Table 13.1): including dry winter (Aw, group A), humid subtropical oceanic climate, without dry season and with hot summer (Cfa, group B), humid subtropical oceanic climate, without dry season with temperate summer (Cfb, group C), humid subtropical climate with dry and hot winter (Cwa, group D) (Alvares et al. 2013).

This study was concerned only on the role of forest percentage in rainfall-runoff responses, in a catchment scale. Unfortunately, any detailed data of other important factors (e.g. forest species, age, stocking, canopy density, or an occurrence of additional fog/cloud precipitation) were not available.

In the investigated catchments, the areal extend of forests was identified by the satellite images ESRI compared with the historical Landsat imagery: the forest percentage varies from 0 to 90%. The role of forests in flood control has been reported particularly by higher frequency (lower extremity) rainfalls because of the limited retention capacity in watersheds. FAO (2006) reported the approximated drop of 1.5–2.0% in catchment water yield with each 10% of rising forest area (mature

Fig. 13.3 The Köppen climate in respected basins

Table 13.1 Basic characteristics of studied catchments

Group A–D	Catchment river name	Area [km^2]	Observation [years]	Elevation [m]	Köppen class
A	Rio das Antas	223	29	905	Aw
A	Rio Rodeador	110	56	1050	Aw
A	Corrego Cabeca de Veado	32.3	68	1020	Aw
B	Rio Jaragua	281	58	35	Cfa
B	Ribeirao Garcia	131	32	30	Cfa
B	Rio Luiz Alves	104	61	120	Cfa
C	Rio Miringuava	272	32	870	Cfb
C	Rio Barigui	262	53	870	Cfb
C	Rio São Lourenço	236	55	670	Cfb
D	Ribeirao Jequitiba	238	36	670	Cwa
D	Rio Bagagem	93	60	950	Cwa
D	Rio Bagagem	868	74	760	Cwa
D	Rio Pouso Alegre	195	61	750	Cwa

evergreen stands). On the other hand, Versini et al. (2013) highlighted impacts of deforestation on the growth of impervious areas and more rapid hydrological responses to rainstorms. Particularly, rising peak-flows (more frequent flood event and greater flood volumes) cause a higher economic (and also insurance) loss. Therefore, there are urgent calls for the flood prevention and active flood protection. Mills (2009) already presented a concept where the insurance sector should be a significant supporter of preventive measures to reduce the risk of floods, including the ecosystem services in watersheds.

In Brazil, the rising insurance loss (registered with more frequent floods) initiated the development of a more effective methodology of the loss estimated (including the flood assessment model for Brazil). Since these data are based on confidential and proprietary information of Brazilian insurance companies they cannot be shown to the public. Therefore, for the aim of this study, only a simple flat portfolio was applied in two areas of interest. The area of a lower risk includes catchments in municipal agglomerations, and the second one contains catchments of a higher flood risk. The property analysis focused on the nearby municipalities downhill to evaluate a potential insurance loss. Analysis included the total loss prior to the application of any insurance conditions (such as limits and deductibles) and by specific values of the average annual loss (AAL).

3 Results and Discussion

Results of this study confirm the hypothesis that higher discharge might be expected with decreasing forested area in all climate types considered. The analyzed discharge was specified according to the catchment area, and the role of forest percentage on specific peak-flows was compared. In catchments of the group A (Fig. 13.4, left), the forested areas do not differ much (5–15%); however; it is evident that with decreasing forestation (approx. three times lower forest percentage), peak-flows increased by 41% for the return period of 2 years, still propagating to 45% for the return period of 20 years. In catchments of the group B (Fig. 13.4, right), of a relatively high forested area (60–90%), forest percentage decreasing by 30% leads to rising specific peak flows by 98% (from 0.24 to 0.47 $m^3.s^{-1}.km^{-2}$) for the return period of 2 years, and by 117% (from 0.47 to 1.03 $m^3.s^{-1}.km^{-2}$) for the return period of 20 years.

In groups C and D, significant differences in the design flows are shown in Fig. 13.5. In the group C, there were found the highest differences: with dropping forest percentage from 70 to 20%, the specific discharge might increase by 380% for the return periods of 2 years, and by 400% for the return period of 20 years. Similarly, in the group D, changes in the specific maximum discharge are 160% for the return period of 2 years, and 290% for the return period of 20 years. These results confirm the hypothesis of more frequent peak-flows and flood damages occurring in the forest-free catchments. However, in Brazil, the disastrous floods have been caused mainly by extreme rainstorms falling by a high initial saturation

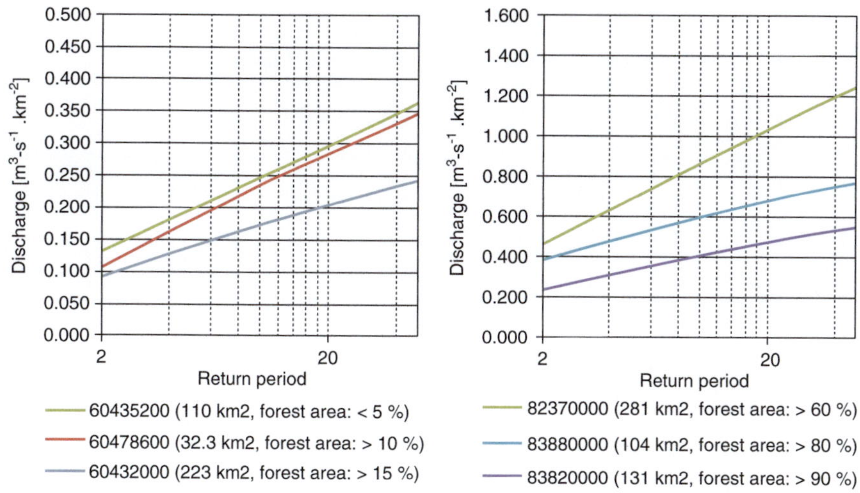

Fig. 13.4 Estimated flood frequency curves for basins in A (*left*) and B (*right*) groups

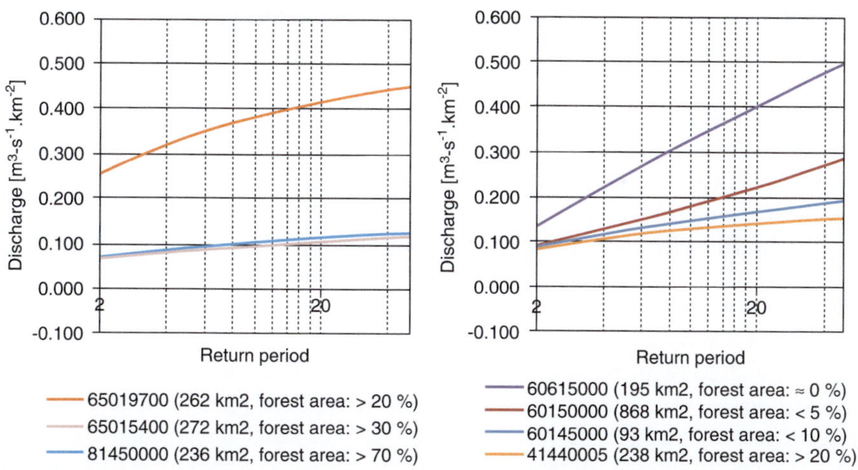

Fig. 13.5 Estimated flood frequency curves for basins of C (*left*) and D (*right*) groups

(Klijn and Schweckendiek 2013), where the retention potential is over-helmed by rainfall intensities. It is evident that by certain circumstances the protective role of forests can be driven down. For example, in some alpine catchments, Punčochář et al. (2012) reported the highest interception storage (from 10 to 15 mm) by mature spruce forests, but this retention potential is rather limited by the extreme rainfalls.

Results of this study indicated the impact of land-use in a watershed (increasing impervious or lower-retention spots) on the runoff acceleration, causing more frequent floods. This situation may result in some additional investments to the techni-

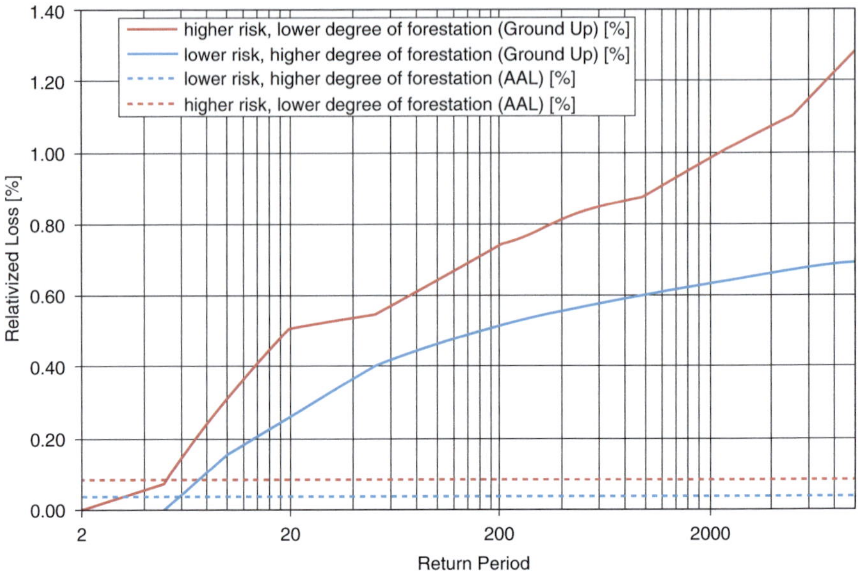

Fig. 13.6 Average annual loss (AAL) in the investigated area of Brazil

cal flood protection in municipalities, as well as in increasing insurance loss. According to the expected global climate change (more frequent hydrological extremes) (IPCC 2007), generally, the risk of natural disasters and consequent damages will rise up (Fig. 13.6).

The modelled losses in municipalities affected with a high risk of flooding coming from catchments of low share of forest are generally higher (~1.4–2.1 times), while the AAL are even almost three times higher. The highest impact of forested area on flood control might be expected by lower return periods; by the return periods of 2 and 20 years, the specific loss is 0.26 and 0.51%, respectively. According the ALL value, watersheds with higher forested areas can produce up to 2/3 lower insured loss. Therefore, the investments to well managed forest stands in catchments, especially in the neighbourhood of urban areas, can lead to: (1) lower economic and insured loss, (2) lower insurance costs in areas with a relatively low flood risk, which could stimulate development in safer areas and restrain development in high-risk spots (Botzen 2013) and (3) lower investments in technical flood defenses.

The analyzed impact of forested land on flood risk in Brazil is affected by numerous uncertainties including the climate change forecast. These uncertainties come from the ambiguity of forest areas and land-use determination, and also from the quality of available hydrological information. Particularly, in the land-use classification of selected catchments, the possible forest degradation or growing extend of impervious surfaces were not investigated. Although the studied catchments of four respected groups are relatively close in term of the comparative analysis (elevation, climate, geology, morphology, etc.) there are still significant differences.

In higher elevations, fog and cloud precipitation (not registered by a standard rain gauge network) can saturate the canopy and significantly reduce the retention capacity (Křeček and Palán 2015). Also, differences in morphology and geological background could affect changes in the observed runoff regime. Therefore, the peak-flows maxima considered by this study could be affected by several other critical phenomena.

However, the results of this study can provide the stakeholders in focused regions of Brazil with important information on the future flood risk development. Definitely, much more attention must be paid to a better understanding of the role of vegetation cover (extend, seasonality and quality) in the mitigation of the flood genesis. The risk of future flood intensification with expected climate change scenarios seems to be an opportunity to involve the insurance business in the sector of watershed ecosystem services. The insurance industry may contribute to limit the damage caused by natural disaster by acting as a price signal of risk and promoting the undertaking of mitigation measures.

4 Conclusions

Flood damages represent the most important risk in headwater catchments of Brazil. In this study, ecosystem services in watersheds considered only the forest extend approximated by the forest percentage. In general, watersheds of a relatively low forest percentage are prone to a higher risk of flood damages:

1. The obtained results confirmed the significant protective role of forests in the flood control.
2. Deforestation and growing impervious surfaces tend to more frequent and higher peak-flows.
3. Flood peak-flows affect directly both the economy and insurance loss.

The challenge of future flood intensification needs to involve the insurance business in the sector of watershed ecosystem services. The average annual loss (AAL) can be used in the approach of cost benefit analysis to assess the effect of ecosystem services in flood risk.

References

Aon Benfield. (2016a). *Impact forecasting*. http://www.aon.com/impactforecasting/impact-forecasting.jsp. Accessed 5 Feb 2016.

Aon Benfield. (2016b). *Catastrophe insight*. http://catastropheinsight.aonbenfield.com. Accessed 5 Feb 2016.

Alvares, C. A., Stape, J. L., Sentelhas, P. C., de Moraes Gonçalves, J. L., & Sparovek, G. (2013). Koppen climate classification map for Brazil. *Meteorologische Zeitschrift, 22*(6), 711–728.

Botzen, W. (2013). *Managing extreme climate change risks through insurance*. Cambridge/New York: Cambridge University Press.
CEH. (1999). *Flood estimation handbook*. Walingford: Centre for Ecology and Hydrology.
FAO. (2006). *The new generation of watershed management programmes and projects, FAO Forestry Paper 150*. Food and Agriculture Organization of the United Nations: Rome.
Hosking, J. R. M. (1990). L-moments: Analysis and estimation of distributions using linear combinations of order statistics. *Journal of the Royal Statistic Society, B, 52*, 105–124.
IPCC. (2007). *Climate change 2007: The physical science basis. Fourth assessment report of the intergovernmental panel on climate change*. New York: Cambridge University Press.
Klijn, F., & Schweckendiek, T. (2013). *Flood risk management: Research for policy and practice*. Boca Raton: CRC Press.
Křeček, J., & Palán, L. (2015). Effects of fog precipitation on water resources and drinking water treatment in the Jizera Mountains, Czech Republic. *Civil Engineering Journal, 10*, 1–7.
Mills, E. (2009). A global review of insurance industry responses to climate change. *The Geneva Papers on Risk and Insurance Issues and Practice, 34*, 323–359.
Punčochář, P., Křeček, J., & van de Griend, A. (2012). Interception storage in a small alpine catchment. In J. Křeček et al. (Eds.), *Management of mountain watersheds*. Dordrecht: Springer.
Rowinski, P. M., Strupczewski, W. G., & Singh, V. P. (2002). A note on the applicability of log-Gumbel and log-logistic probability distributions in hydrological analyses. *Hydrological Sciences Journal, 47*, 107–122.
Shaw, E. M. (1991). *Hydrology in practice* (2nd ed.). London: Chapman & Hill.
Versini, P. A., Velasco, M., Cabello, A., & Sempere-Torres, D. (2013). Hydrological impact of forest fires and climate change in a Mediterranean basin. *Natural Hazards, 66*, 609–628.

Part III
Environmental Services in the Changing World

Chapter 14
Maintaining Environmental Services in Mountain Watersheds

Hans Schreier

1 Introduction

Mountain watersheds are considered the water towers for humanity (Viviroli and Weingartner 2004; Viviroli et al. 2007) and generally provide high quality water at critical times of the year to the urban, industrial and agricultural activities that primarily take place in the lowlands. At the same time these watersheds are fragile, more vulnerable, and more dynamic than lowland systems. The processes that impact streams in the headwaters are different from those in the lowland because the topography and geophysical setting is conducive to provide a direct pathway from the land to the stream. In the mountains climatic events and land use changes can convey soils directly into streams, while the processes in the lowlands are dominated by sediment transport, resuspension, bank erosion and seepage of contaminants. Climatic change has a magnifying impact on stream hydrology in mountains and this often results in more extensive flooding events in the lowland and can augment or decrease stream flows during drought periods. Similarly, water that originates from the headwaters is usually of high quality and most often degrades in the downstream direction as land use activities increase.

Climate models and field observations have shown that due to higher temperatures and more variable precipitation, the stream hydrology is changing in different ways depending on whether the watershed is rain dominated, snow dominated, has a glacial source, or a combination of all three. The watersheds at intermediate elevation ranges have already seen a major decrease in the accumulation of snow and this has resulted in seasonal advances in stream peak flow by 1–3 weeks (Hamlet et al. 2007). Earlier peak runoff often results in a decline in summer flow due to reduced soil water storage in snow and increased demand for water by human activities. Using the projections of many different global models there is good agreement that

H. Schreier (✉)
Faculty of Land & Food Systems, University of British Columbia, Vancouver, Canada
e-mail: hans.schreier@ubc.ca

© Capital Publishing Company, New Delhi, India 2017

J. Křeček et al. (eds.), *Ecosystem Services of Headwater Catchments*,
DOI 10.1007/978-3-319-57946-7_14

the temperatures are increasing in most mountain regions in the northern hemisphere and differences between modelled temperature projections are relatively small (Hamlet et al. 2005, 2007). There is also considerable evidence that mountain areas are warming up faster than lowland areas (Auer et al. 2007; Rangwala et al. 2013). However, projections of precipitation are highly uncertain in mountains and as shown in a case study in the Columbia Basin in North America some models project a 25% increase and others a 20% decrease in precipitation in the same watershed (Hamlet et al. 2007, 2013). It is becoming clear that the uncertainty and variability of precipitation is increasing due to the many microclimatic effects in mountain environments. Downscaling efforts to improve the modelling of precipitation is unlikely going to improve the predictive capability in these mountain environments. Dealing with these uncertainties means we will have to adapt and find new ways to manage and protect mountain watersheds.

In mountain watersheds, where the human impact is generally small, nature provides a wide range of essential environmental services that have been underappreciated and the value for such services has for a long time eluded satisfactory economic evaluations (Engel et al. 2008). In many mountain watersheds the use of water has been a human priority and as long as there was sufficient water available, little attention was given to the water needs to maintain all the environmental functions provided by nature. Many of the human water allocation rights were initiated during a period when water resources were thought to be plentiful and climatic conditions were relatively stable. There is now sufficient evidence to show that climatic variability is increasing and relying on historic streamflow records is no longer entirely valid because not only has the climate shifted but land use intensification and human water use have significantly increased. This has altered the hydrological regime in a complex manner. How do we partition and re-allocate and protect instream and off-stream water use in an equitable manner to meet the essential human and environmental needs under these changed climatic and land use conditions? The simple determination of what minimum flow requirements should be to maintain environmental services has proved to be extremely challenging (Postel and Richter 2003; Arthington et al. 2006; Pahl-Wostl et al. 2013).

2 The Role of Forests in Mountain Watershed

As shown in Fig. 14.1 forests moderate the hydrological cycle by evapotranspiring precipitation by 20–30%, allowing the intercepted rain to enter soil and groundwater, and depending on site conditions only 5–10% of the rainfall is converted into surface runoff that enters streams in forested watershed. These functions not only allow the water to be retained but trees within the riparian buffer zone shade the headwater streams and moderate water temperatures during the summer. They also provide large woody debris that improves the pool and riffle systems in the stream and enhances the habitat for the aquatic biota. Turbulent flow improves aeration which enhances the decomposition, increases water purification processes and

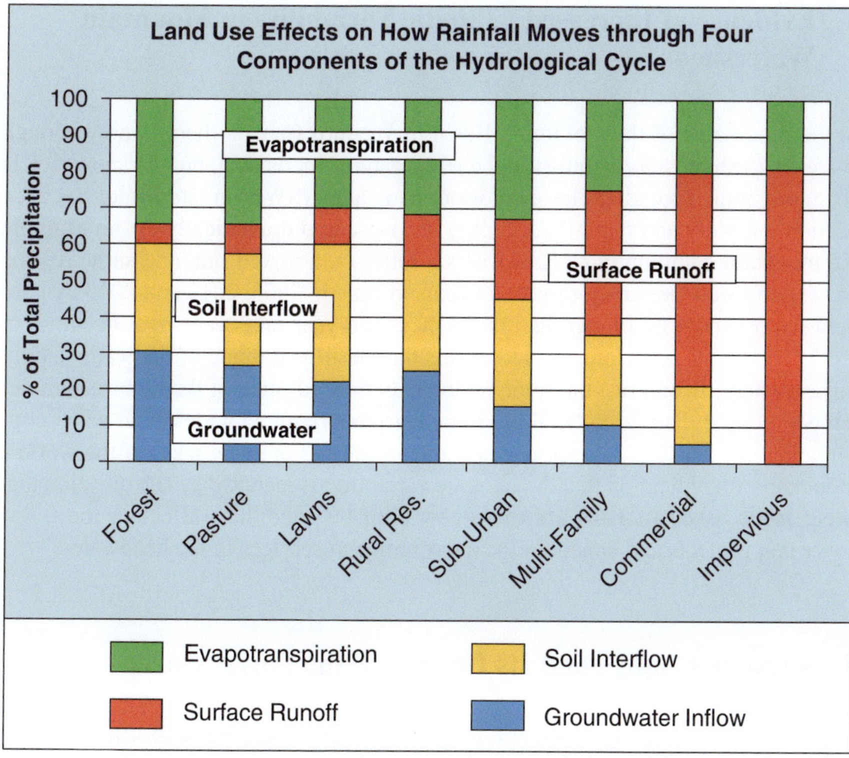

Fig. 14.1 Redistribution of water in the hydrological cycle by different land uses

nutrient cycling. Wetlands within the riparian zone store carbon, act as filters and moderate streamflow. There is general agreement that maintaining forests in headwaters are the most effective way to provide and enhance the wide range of environmental services that nature provides. In different parts of the world these functions are disturbed by increased global warming and pressures on the forest resources for economic exploitation. The conditions in the European Alps are somewhat different from those in other mountain areas in the world because societies in the Alpine region have learned from the effects of historic land use conversions from forest to agriculture. These forest conversion and conservation efforts have now resulted in significant increases in forest cover in many headwater areas. As a result, the risk to these protected forested area is now not from land use changes but primarily from increased climatic variability (Lindner et al. 2010), while in the North American context the combined effect of forest exploitation and climate change are of major concern.

3 Evidence of Increased Climatic Variability in Mountain Watersheds

In the mountains of the Columbia Basin in Western Canada there is now enough evidence to show that not only are the temperatures increasing, but the snowpack is becoming more sporadic. The snow water equivalent (SWE) in late winter has been declining rapidly and rainfall variability has increased dramatically. As an example the mountain community of Kaslo which relies on a mixed rain and snow regime has experienced three major rainfall events in June in 2005, 2012 and 2013 that are between 30 and 70% higher than the highest 100-year rain events that occurred in 1963. This led to widespread flood events that destroyed much of the water supply infrastructure. In contrast the same region experienced some of the highest summer temperatures and an extended drought in 2015 that required widespread water conservation efforts. These extreme events have occurred in many parts of the world's mountains, an indication that the climatic variability is increasing. The question that needs to be asked is: How are these new climatic conditions affecting the forest cover that protects and enhances the environmental services in the headwaters?

4 Climate Change and Its Impact on the Forest Cover

There is new evidence that higher temperatures and extended droughts increase the stress on trees, encourage pest and disease outbreaks and negatively impact tree physiology leading to increased tree mortality (Allen et al. 2015). These new climatic conditions resulted in a massive outbreak of pine beetle infestation in British Columbia, which affected 65% of the conifer dominated forest in the Province. Pine beetle infestations have periodically occurred over historic time but usually lasted only short time periods because of rapid dieback of the beetles whenever the average January temperatures reached −30 °C or lower levels. Since 1999 winter temperatures have increased dramatically and are now well above those experienced over historic times and this resulted in the greatest historic outbreak of pine beetles infestation that lasted from 1999 until 2014. It consumed 18 million hectares of forests, and affected 800 million m^3 of timber (Kurz et al. 2008a). It devastated 1/3 of the standing forest in British Columbia with Lodgepole pine being the most affected species. The problem could not be contained and the only reason for the decline was that the beetles have now exhausted the desired and available wood supply. This created a major dilemma. Leaving dead trees creates an enormous fire risk and if trees are harvested within 1–3 years after infestation the wood can still be used. Given the concern of potential devastating effects of wildfires on mountain communities, and the concern about excessive CO_2 emissions from large fires (Kurz et al. 2008b) the government opted for extensive logging, which occurred at the landscape scale. How this will affect the regional hydrology has yet to be investigated because most of the hydrological research involving logging impacts has been conducted at the plot and small watershed scale. The pine beetle have now exhausted their food source in British Columbia and the problem has now spread to large areas in western USA (Bentz et al. 2010).

In spite of the extensive logging to reduce fire risk, the number of forest fires has not declined primarily because only a small portion of infested trees could be harvested and the increased temperatures and drought conditions have affected other healthy trees that are stressed and thus more susceptible to fire. Data from the forest service in the USA and in British Columbia has shown that the annual number of fires have not increased over the past 15 years but the average size of the area affected by individual fires has increased from less than 50 ha/fire between 2002 and 2008 to 201 ha/fire in 2001, 250 ha/fire in 2014 to 190 ha/fire in 2015 in British Columbia (B.C. Wildfire Service 2015). The effects of fire on mountain streams are of course significant and highly dependent on the size of the fire and the fire intensity. The dead trees no longer evapotranspire precipitation and the hydrophobisity created from fires in conifer forests reduces the water infiltration capacity. Soil carbon that helps retain water is reduced and the changes to the soil surface conditions lead to widespread erosion particularly if intensive rain events follow drought periods. The darker soil surfaces after the occurrence of fire absorb more heat, and increase stream temperatures once the sediment reaches the stream. The protective role of the riparian zone is reduced, and the input of sediment changes the water quality, modifies the channel morphology in the downstream direction and leads to a short-term nitrogen flux in the water which usually follows fire events (Bartels et al. 2015).

5 Land Conversion and Forest Management Impacts on Water

Historic land conversions in the European Alps from forests into agriculture had many devastating impacts on the stream hydrology and stream health in the past. Over time the mountain communities have learned to restore the forests in the most sensitive sites. This protects communities and ecosystems and restores the processes that provide the key environmental services. Depopulation from mountain agricultural areas has resulted in abandonment of many fields which have now reverted back into forests by natural processes (Gehrig-Fasel et al. 2007). Also, forests are expanding into higher elevations due to climate warming and should we now consider speeding up this process by afforestation? What will be the impact on the hydrological system by both natural and human enhanced reforestation? Declining use of alpine grassland was shown to reduce water yield in the Alps (Inauen et al. 2013), while afforestation in different tropical mountains showed contradictory results (Lacombe et al. 2015). Differences in water use between degraded grassland, undisturbed broadleaved forests and pine plantation in Nepal shows that pine forests used more water thus reducing water yields (Ghimire et al. 2014). Natural upslope expansion of plants has proven to be uneven and might also be limited by geomorphic processes that enhance or restrict soil moisture conditions (Macias-Fairia and Johnson 2013). This suggests that enhancing afforestation efforts should be done with care and close attention should be placed on the water requirements of different trees and how this will affect the hydrological cycle (Zierl and Bugmann 2005).

In contrast to the land abandonment, the key relatively recent land use changes that have occurred in the Alpine regions of the world have been for summer and winter tourism. This has resulted in increases in urbanization, enhancement of the transportation infrastructure and the creation of extensive networks of trails and ski runs. Soil compaction during construction and the creation of more impervious surfaces are responsible for a larger portion of the rain to be converted into surface runoff (Fig. 14.1), resulting in more flashy stream regimes. In addition, many of the intermediate and low elevation ski areas are now engaged in snow making, which requires large quantities of water at the time in late fall when streamflow is in the declining phase. It is estimated that up to 2000 m^3 of water is needed to produce 30 cm of snow over a 1 ha area in typical mid elevation ski areas. Storing water will become a necessity and will further affect streamflow regimes.

The land use changes in the Canadian mountains watersheds have been very different from those in the Alpine region. Relatively few watersheds have been impacted by winter tourism and the main activity is commercial harvesting of trees. Over the past 50+ years some 450,000 km of logging roads have been constructed in British Columbia in order to access and harvest timber. Under Western Canadian climatic conditions it takes up to 100 years for trees to mature and much of the harvests have taken place on old stands of trees. As a result the access roads were built for forest extraction and replanting and since it takes a long time for the next generation of trees to mature, access roads are not well maintained over time. The roads not only modified the hydrological regime but are the main source of sediments in the mountain catchments. Efforts are under way to decommission some of these roads in the most sensitive environments but as a result of the pine beetle infestation more emphasis has been placed on building new access roads for harvesting the diseased trees and to reduce the fire hazards. The main challenge now is to decide what trees to replant in the harvested and pine beetle affected areas in view of climate change, length of time to mature trees and economic considerations.

6 Climate Change Adaptation Strategies to Maintain and Enhance Environmental Services in Forested Mountain Watersheds

Forests moderate the partitioning of water in the hydrological cycle and enhance most of the functions that are responsible for maintaining healthy ecosystems. It is therefore imperative to conserve and enhance headwater forests. Depending on the site conditions the precautionary principle is to address the combined effect of climate and land use change. If well managed forests are the best measures to protect environmental services in mountain watersheds then afforestation efforts should focus on the type of trees that retain precipitation, allows for maximum storage and infiltration of water, provide protection from extreme climatic events and enhance the key functions environmental services provide. Even if we protect the forest, management is required to reduce the risks imposed by climate warming which results in higher tree mortality, enhances the spreading of diseases and pests, and contributes to fires that impact the water cycle. There are a wide range of adaptation measures that need to be considered.

6.1 Advantages of Tree-Biodiversity

With all the uncertainties associated with temperature and precipitation changes and the relatively long time for trees to reach maturity, it is prudent to encourage the planting of a wide range of tree species (Thompson et al. 2009). This is often in conflict with the trend towards plantation forests of those species that have the greatest economic value. Much more emphasis should be given to the assessment of water requirements for different tree species. In areas where drier conditions are predicted trees that are water efficient should be promoted and in those areas where wetter conditions are forecasted, trees that increase water cycling should be given preferences. Having a diversity of plants will also reduce the rapid expansion of diseases and fires.

6.2 Enhancing Snow Accumulation at Appropriate Elevations

Snow accumulation is critical in mountains because it retains water longer at the surface and in soils in the spring and delays more rapid runoff when there is a shift from snow to rain. Snow accumulation is usually less efficient in forests and recent studies have found that creating small open patches in forests rather than thinning allows more snow to accumulate and be retained longer in the spring (Dickerson-Lange et al. 2015). This might be an effective short term strategy in watersheds at elevations that are experiencing the transition from snow to rain as a result of climate warming (Berghuijs et al. 2014). It is a practice that might be even more effective at higher elevations.

6.3 Maintaining Large Riparian Buffer Zones

Channelizing streams is a poor way of enhancing environmental services and, if we hope to provide enhanced habitat for aquatic organisms, wide buffer zones ranging up to 60–80 m should be promoted. This will not only allow the development of wetlands within the buffer zone but can retain sediments, remove nutrients, allow energy in the stream to dissipate and allow aquifers in the riparian zone to be recharged. These buffer zones should have a combination of grass, shrub and tree communities that can tolerate rapidly changing conditions.

6.4 Designation of Temporary Water Storage Areas

Since we are expecting more extreme flood events, building protective dykes and structures is no longer sufficient. Provisions should be made to select a wide range of appropriate topographic settings to be designated to store and retaining water during extreme storm events on a temporary basis. These can be natural areas, agricultural fields or constructed wetlands specifically designed for delaying the runoff.

6.5 Water Use Restrictions During Dry Periods

The majority of mountain communities get their drinking water from headwater streams, and in view of increased climatic variability we need better provisions in place on how to reduce water demands during extended dry periods and to make sure enough water remains in headwater streams to maintain environmental services. Water extraction permits were provided during periods when supplies were plentiful and under these new climatic conditions reallocation and conservation will be required. In Canada where water is cheap and where little consideration is given to efficient water use, the most cost effective measures are to focus on widespread water conservation initiatives. This not only applies to domestic water use but equally to irrigation in agriculture. Water use reductions of 30 or more percent are readily possible using the most efficient technologies and by changing human behaviour. Similarly, most communities now are required to develop a source water protection strategy for their domestic water sources in order to reduce the cost of water treatment. This will also help to maintain healthy water ecosystems.

6.6 Changing Land Use Management Practices

We have plenty evidence of how much water is used and what the contaminant impacts on water are from different land uses activities. Using beneficial management practices (BMP's) is now heavily promoted. BMP's in forestry focus on ways to reduce the fire risk by reducing the fuel load, providing fire breaks and modifying the tree composition and stand age. Some of these measures will also assist in snow retention. Logging activities that minimize forest floor disturbance and measures to increase soil organic matter are also heavily promoted. Building up the soil organic matter is likely the most effective way of retaining soil moisture, enhancing nutrient holding capacity and retaining CO_2. The past practice of draining wetlands and removing aboveground biomass has now also been reduced somewhat by building constructed wetland particularly in mountain communities. These buffer the runoff and reduce the contaminants contained in the urban stormwater runoff. What is equally important is to manage protected forests by selectively removing diseased trees, pay attention to fuel load changes, and to monitor mortality changes due to increased temperatures. Road decommissioning is another effective measure to minimize erosion, sediment transport and surface runoff. Restoring forests on degraded and burned land can go a long way in restoring environmental services (Chazdon 2008) since changes in different trees structures, diversity and functions increase the resilience of the forest to stress imposed by climate and human activities (Seidl et al. 2014).

6.7 Develop Better Methods to Assess the Value of Environmental Services

One of the major challenges dealing with environmental services is how to put a value on the services nature provides (de Groot et al. 2002). Most of the economic methods that have been developed to put an economic value on these services are problematic. Neither the Market Value methods (replacement or restorative value, preventative expenditures), the Shared Preference methods (non-market valuation, contingent valuation-willingness to pay, conjunct choice comparison) nor the Revealed Preference methods (hedonic pricing, benefit transfer) have proven to be totally acceptable and effective by decision makers (Engel et al. 2008). Determining the replacement or restorative cost methods are more realistic but hinge on how successful the replacement and restoration projects are to provide all the ecosystems services that occur in undisturbed environments. The additional challenges are how to determine how resilient the undisturbed environments are to emerging climatic changes and how they will adapt naturally to these changes.

7 Summary

Climate and land use changes are occurring at the same time in most watersheds and the interactions between the two accentuate the impacts on the hydrological cycle and the environmental services. In some of the more protected environments it is the climate change that has the dominant effect while in more intensively used watersheds it is the land use and human activities that need greatest attention. In mountain watersheds where a well established forest cover is present and protected, management of the forests is still essential because extreme droughts and floods can have devastating effects. The adaptation techniques that are able to enhance some of the ecological functions include all methods that enhance water storage and detention within the watershed. Soil improvement techniques that enhance organic matter retention allow more precipitation to infiltrate into the soils, recharges aquifers and helps mitigate droughts. Protecting, enhancing and constructing wetlands should be another priority since they are some of the best water storage and filtration systems nature provides. To minimize climatically induced hazards like diseases, fires and tree mortality, greater attention needs to be devoted to reducing fuel loads, creating greater forest biodiversity, and designing forests to contain trees of different ages.

In the more intensively used mountain watersheds key adaptation strategies involve changes in the way we use and allocate water, and have effective water conservation methods in place during droughts. Enhancing and protecting wide riparian buffer zones, minimizing the river channelization, and enforcing BMP's in all land use activities are other effective ways to ensure that the ecosystem functions are maintained as climatic variability increases.

References

Allen, C. D., Breshears, D. D., & McDowell, N. G. (2015). An underestimation of global vulnerability to tree mortality and forest die-off from hotter drought in the Athropocene. *Ecosphere, 68*, 129. doi:10.1890/ES15-00203.1.

Arthington, A. H., Bunn, S. E., Poff, L. N., & Naiman, R. J. (2006). The challenge of providing environmental flow rules to sustain river ecosystems. *Ecological Applications, 16*(4), 1311–1318.

Auer, I., Bohm, R., Jurkovic, A., Lipa, W., Orlik, A., Potzmann, R., Schoner, W., Ungersbock, M., Matulla, C., Briffa, K., Jones, P., Efthymiadis, D., Buretti, M., Nanni, T., Maugeri, M., Mercalli, L., Mestre, O., Moisselin, J. M., Begert, M., Muller-Westermeier, G., Kveton, V., Bochnicek, O., Stastny, P., Lapin, M., Szalai, S., Szentimrey, T., Cegnar, T., Dolinar, M., Gajic-Capka, M., Zaninovic, K., Majstorovic, Z., & Nieplova, E. (2007). HISTAPL-historical instrumented climatological surface time series of the greater Alpine region. *International Journal of Climatology, 27*, 17–46.

B.C. Wildfire Service. (2015). *Wildfire statistics*. http://bcwildfire.ca/History/average.htm

Bartels, S. F., Chen, H. Y. H., Wulder, M. A., & White, J. C. (2015). Trends in post-disturbance recovery rates of Canada's forests following wildfire and harvest. *Forest Ecology and Management, 361*, 194–207.

Bentz, B., Regniere, J., Fettig, C., Hansen, E. M., Hayes, J. L., Hicke, J. A., Kelsey, R. C., Negron, J. F., & Scybold, S. J. (2010). Climate change and bark beetles of the western United States and Canada: Direct and indirect effects. *Bio Science, 60*(8), 602–613.

Berghuijs, W. R., Woods, R. A., & Hrachowitz, M. (2014). A precipitation shift from snow towards rain leads to a decrease in streamflow. *Nature Climate Change*. doi:10.1038/NCLIMATE2246.

Chazdon, R. (2008). Beyond deforestation: Restoring forests and ecosystem services on degraded lands. *Science, 320*(5882), 1458–1460.

De Groot, R., Wilson, M. A., & Boumans, R. (2002). A typology for the classification, description and valuation of ecosystem functions, goods and services. *Ecological Economics, 41*, 393–408.

Dickerson-Lange, S. E., Lutz, J. A., Martin, K. A., Raleigh, M. S., Gersonde, R., & Lundquist, J. D. (2015). Evaluating observational methods to quantify snow duration under diverse forest canopies. *Water Resources Research, 15*, 1203–1224. doi:10.1002/2015WR017873.

Engel, S., Pagiola, S., & Wunder, S. (2008). Designing payments for environmental services in theory and practice: An overview of the issues. *Ecological Economics, 65*(4), 663–674.

Gehrig-Fasel, J., Guisan, A., & Zimmermann, N. E. (2007). Tree line shift in the Swiss Alps: Climate change or land abandonment? *Journal of Vegetation Science, 18*(4), 571–582.

Ghimire, C. P., Bruijnzeel, L. A., Lubczynski, M. W., & Bonell, M. (2014). Negative trade-off between changes in vegetation water use and infiltration recovery after reforesting degraded pasture land in the Nepalese Lesser Himalayas. *Hydrology and Earth System Sciences, 18*, 4933–4949.

Hamlet, A. F., Mote, P. W., Clark, M. P., & Lettenmaier, D. P. (2005). Effects of temperature and precipitation variability on snowpack trends in the western U.S. *Journal of Climate, 18*(21), 4545–4561.

Hamlet, A. F., Mote, P. W., Clark, M. P., & Lettenmaier, D. P. (2007). 20th century trends in runoff, evapotranspiration, and soil moisture in the western U.S. *Journal of Climate, 20*(8), 1468–1486.

Hamlet, A., McQuire-Elsner, M., Manger, G. S., Lee, S. Y., Tohver, I., & Norheim, R. (2013). An overview of the Columbia Basin: Climate change scenario project—Approach, methods and summary of key results. *Atmosphere-Ocean, 51*, 392–415.

Inauen, N., Korner, C., & Hiltbrunner, E. (2013). Hydrological consequences of declining land use and elevated CO_2 in Alpine grassland. *Journal of Ecology, 101*, 86–96.

Kurz, W. A., Dymond, C. C., Stinson, S., Rampley, G. J., Neilson, E. T., Carroll, A. L., Ebata, T., & Safranyik, L. (2008a). Mountain pine beetle and forest carbon feedback to climate change. *Nature, 452*, 987–990. doi:10.1038/nature 06777.

Kurz, W. A., Stinson, S., Rampley, G. J., Dymond, C. C., & Neilson, E. T. (2008b). Risk of natural disturbance makes future contributions of Canada's forests to the global carbon cycle highly uncertain. *Proceedings of the National Academy of Science, 1005*(5), 1551–1555.

Lacombe, G., Ribolzi, O., de Rouw, A., Pierret, A., Latsachak, K., Silvera, N., Pham Dinh, R., Orange, D., Janeau, J. L., Soulileuth, B., Robain, H., Taccoen, A., Sengphaathith, P., Mouche, E., Sengtaheuanghoung, O., Tran Duc, T., & Valentin, C. (2015). Afforestation by natural revegetation of tree planting: Examples of opposite hydrological impacts evidenced by long-term field monitoring in the humid tropics. *Hydrology and Earth System Sciences Discussions, 12*, 12615–12648.

Lindner, M., Maroschek, M., Netherer, S., Kremer, A., Barbali, A., Garcia-Gonzalo, J., Seidl, R., Delzon, S., Corona, P., Kalstom, M., Lexer, M. L., & Marchetti, M. (2010). Climate change impacts, adaptive capacity, and vulnerability of European forest ecosystems. *Forest Ecology and Management, 269*(4), 698–709.

Macias-Fairia, M., & Johnson, E. A. (2013). Warming induced upslope advance of subalpine forest is severely limited by geomorphic processes. *Proceedings of the National Academy of Science, 110*(20), 8117–8122.

Pahl-Wostl, C., Arthington, A., Bogardi, J., Bunn, S., Hoff, H., Lebel, L., Nikitina, E., Palmer, M., Poff, L. N., Richards, K., Schluter, M., Schulze, R., St-Hilaire, A., Tharme, R., Tockner, K.,

& Tsegai, D. (2013). Environmental flows and water governance: Managing sustainable water uses. *Current Opinion in Environmental Sustainability, 5*, 1–11.

Postel, S., & Richter, B. (2003). *Rivers for life: Managing water for people and nature*. Washington, DC: Island Press.

Rangwala, I., Sinsky, E., & Miller, J. R. (2013). Amplified warming projections for high altitude regions in the northern hemisphere mid-latitudes from CMIP5 models. *Environmental Research Letters, 8*(2), 024040. doi:10.1088/1748-9326/8/2/024040.

Seidl, R., Rammer, W., & Spies, T. A. (2014). Disturbance legacies increase the resilience of forest ecosystem structure, composition and functioning. *Ecological Applications, 24*, 2063–2077. doi:10.1890/14-0255.

Thompson, I., Mackey, B., McNulty, S., & Mosseler, A. (2009). *Forest resilience, biodiversity and climate change*. Secretariat of the Convention of Biological Diversity, Montreal. Tech. Series 43.

Viviroli, D., & Weingartner, R. (2004). The hydrological significance of mountains: From regional to global scale. *Hydrology and Earth System Sciences, 9*(6), 1016–1029.

Viviroli, D., Durr, H. H., Messerli, B., Meybeck, M., & Weingartner, R. (2007). Mountains of the world, water towers for humanity: Typology, mapping and global significance. *Water Resources Research, 43*(7). doi:10.1029/2006WR005653.

Zierl, B., & Bugmann, H. (2005). Global change on hydrological processes in Alpine catchments. *Water Resources Research, 41*(2). doi:10.1029/2004WR003447.

Chapter 15
Prioritizing Adaptation Needs for Ecosystem Services: A Regional Case Study in the Eastern Alps

Christin Haida, Clemens Geitner, Michiko Hama, Richard Hastik, Karl-Michael Höferl, and Katrin Schneider

1 Introduction

Ecological systems and natural resources are significant sources to generate human well-being (Millennium Ecosystem Assessment 2005; TEEB 2010; Summers et al. 2012). Climatic and socio-economic changes will have many adverse impacts on natural systems and thus threaten human well-being (Schröter et al. 2005; Metzger et al. 2006; Staudinger et al. 2012). Therefore, there is an urgent need for society to adapt to these future challenges in order to preserve ecosystem services (ES) and secure human well-being for future generations. The amount of literature concerning impact assessments of climate change on ES has been growing in recent years, with the consensus being that climate change has already an effect on the supply of ES, thus threatening human well-being and that immediate action is needed (Schröter et al. 2005; Briner et al. 2012; Elkin et al. 2013; IPCC 2014). Facing this need for action requires the integration of ES in regional and local decision making processes on climate change adaptation (Daily et al. 2009; UNEP 2012; Loft et al. 2015). Including multiple ES simultaneously into this decision making process and identifying ES of concern, which require immediate action (Metternicht et al. 2014), reduces the risk of unintended trade-offs between adaptation measures.

According to adaptation assessments as suggested by IPCC (2014), the first steps of such an integration should be an assessment of ES sensitivity to climate change,

C. Haida (✉) • M. Hama • K. Schneider
alpS – Centre for Climate Change Adaptation, Grabenweg 68, Innsbruck, Austria
e-mail: haida@alps-gmbh.com

C. Geitner • R. Hastik
Institute of Geography, University of Innsbruck, Innsbruck, Austria

K.-M. Höferl
alpS – Centre for Climate Change Adaptation, Grabenweg 68, Innsbruck, Austria

Institute of Geography, University of Innsbruck, Innsbruck, Austria

© Capital Publishing Company, New Delhi, India 2017

J. Křeček et al. (eds.), *Ecosystem Services of Headwater Catchments*,
DOI 10.1007/978-3-319-57946-7_15

followed by an identification and prioritization of adaptation needs. Adaptation needs here are understood as priority areas – i.e. priority ES – which require attention and specific adaptation action. However, the methodological foundation for prioritizing ES adaptation needs is rather limited (Daily et al. 2009; Martinez-Harms et al. 2015; Giebels et al. 2015, 2016). To date sensitivity assessments have been predominantly based on modelling and simulation approaches (Civantos et al. 2012; Briner et al. 2013; Hossain et al. 2015) or on monitoring and experiments in the field (Pauli et al. 2012; Dawes et al. 2013; López-Hoffman et al. 2013). These approaches have their drawbacks, as they mainly focus on a few non-transparently preselected ES (Haida et al. 2015), are resource intensive – regarding time, data, and personnel – and therefore are not practical for assessments of multiple ES. Involving stakeholders in participatory assessments could be another assessment option (Grêt-Regamey et al. 2013; Karrasch et al. 2014; Tuvendal and Elmqvist 2011), as they help to overcome limited resources and deliver much-needed and valuable information on the local needs and priorities. As adaptation is predominantly implemented on a local to regional scale (Tompkins and Adger 2004), it is therefore important to account for the constraints and needs of a specific region, and a sensitivity assessment should be complemented by an assessment of the regional/local importance of these ES (Martín-López et al. 2012).

Trying to further promote decision-support on adaptation needs to preserve ES, successful methods need to (i) support practitioners to prioritize adaptation needs of multiple ES, (ii) inspire and inform innovative adaptation goals and measures, (iii) be resource-efficient, pragmatic and applicable, and (iv) integrate local and scientific knowledge. Based on these requirements, a concept for ES-based adaptation assessment was developed by Haida (2016), including six steps. The first three of these steps focus on a participatory adaptation needs assessment of ES, as follows:

1. *Assessing ES relevance*: focuses on assessing the relevance of ES for a respective region and local population independent of climate change in order to identify key ES.
2. *Assessing ES sensitivity to climate change and climate change impacts*: evaluates the sensitivity of ES to climate change and anticipated impacts. Assessing sensitivity and impacts separately will help identifying ES which are affected by climatic induced stresses more effectively. Narrative impact storylines supplement this identification by describing how climate change will affect ES, based on facts, experience or rational reasoning.
3. *Prioritizing ES adaptation needs*: combines ES relevance, sensitivity and impacts in a relevance-sensitivity matrix and helps to transparently identify "hotpot ES", i.e. priority services which need future attention.

These three steps were applied and tested in a regional case study.

2 ES-Based Adaptation Needs Assessment in Practice

2.1 *Study Area*

Despite the often stated vulnerability of mountain ecosystems and watersheds to climate change (Smith et al. 2009; Andrade Pérez et al. 2010; IPCC 2014), knowledge about potential impacts on ecosystems and their services at the local and regional scale is scarce and focusses mainly on agricultural and forest ES (APCC 2014). Owing to the small-scaled diversity of topography and therefore land cover and land use within a confined space, mountain regions often provide numerous ES side by side (Grabherr 1997; Grêt-Regamey et al. 2012), causing a challenging process for decision-makers to identify and prioritize hotspot ES. The three neighbouring mountain provinces in the Alps Tyrol (Austria), Vorarlberg (Austria) and South Tyrol (Italy) provide an ideal model region to implement and test the proposed approach, due to their topographic, climatic and land cover/land use diversity (Tappeiner et al. 2008). Furthermore, projected rise in temperature and changes in precipitation pattern are expected to have severe impacts on ecosystems and the provision of ES (Auer et al. 2007; Europäische Akademie Bozen EURAC 2012; APCC 2014).

2.2 *Methods*

2.2.1 Expert Interviews

To obtain specific information regarding the influence of climate change on ES, expert interviews are a suitable method (Hagerman et al. 2010; Chowdhury et al. 2012). Making use of the expert's high level of insight and aggregated knowledge allows for assessing the sensitivity of multiple ES to climate change and the possible impacts in a resource efficient, participatory and consensus-building manner. To assess the relevance, sensitivity and impacts, interviews with 53 professionals from nine thematic fields were conducted, covering the fields of forestry, agriculture, soil science, energy, meteorology, safety, planning, tourism and environmental protection. These fields were specifically chosen so as to thematically cover the range of all ES discussed in the interviews (Table 15.1), with each field represented by six interviewees, except of soil science. The interviewees all held managerial positions, acted as decision makers and worked for governmental institutions, NGOs or in the private sector. All interviews were semi-structured and conducted on a one-on-one basis in 2011 and 2012. Henceforth, the interview partners will be referred to as experts.

At the start of the interviews 25 pre-selected ES were presented. The selection was primarily based on Millennium Ecosystem Assessment (2005) and took TEEB (2010) and Haines-Young and Potschin (2011) into consideration. In the first part of the interview, the experts were asked to rank these 25 ES according to the relevance

Table 15.1 List of ecosystem services used in the interviews with the definitions given to the experts

Ecosystem services	Definition
Provisioning services	
Fresh water	Supply and storage of fresh water
Fodder	Food for domesticated animals
Food	Food grown in wild habitats and in managed agro-ecosystems – including crops, livestock, aquaculture and wild food
Raw materials	Diversity of materials for construction, landscaping and ornaments
Medicinal resources	Resources used for biomedical products, natural medicine, pharmaceuticals, etc.
Energy	Means which can be used for energy production, e.g. hydropower, wood fuel and bio fuel from agricultural products
Regulating and maintaining services	
Water cycle	The water cycling affected by plant processes in the system
Nutrient cycle	Recycling and storage of nutrients to maintain healthy soils and productive ecosystems
Primary production	Building of biomass
Natural hazard regulation	Influence of ecosystems to moderate extreme events, e.g. storms, floods, rock falls or avalanches
Soil erosion regulation	Land cover can prevent soil erosion to maintain arable land and to prevent damage from erosion/siltation
Water flow regulation	Land cover can regulate water runoff and river discharge
Pollination	Pollination of wild plants and crops
Biodiversity	The presence or absence of selected species, functional groups of species or species composition
Habitat	The provision of suitable habitats for different species, for functional groups of species or for processes essential for the functioning of ecosystems
Biological control	Control of pests and diseases
Soil formation & fertility	Maintenance of the natural productivity of soil
Water quality	Ecosystems play a role in pollution control/detoxification and filtering of dust particles
Global climate regulation	Ecosystems play an important role in climate by either sequestering or emitting greenhouse gases
Local climate regulation	Land cover can locally affect temperature, air moisture, wind, radiation and precipitation
Air quality regulation	Maintenance of (clean) air
Cultural services	
Recreation	Natural landscapes and urban green spaces play an important role in maintaining mental and physical health
Tourism	Nature tourism provides considerable economic benefits and is a vital source of income for many countries

(continued)

Table 15.1 (continued)

Ecosystem services	Definition
Aesthetic appreciation	Attractive landscapes provide enjoyment of scenery
Spiritual values	Ecosystems are used for religious or historic purposes and can foster a local identity and sense of belonging

Based on Haida et al. (2015)

of each ES for society on a scale from 1 to 25, regardless of climate change. A detailed description and analyses of this part can be found in Haida et al. (2015). In the second part, the experts stated which of the ES are sensitive to climate change and were asked to describe in which way they are affected. To this end, the experts selected one of three pre-defined impact categories (negative, positive, neutral). All interviews were recorded and transcribed. The resulting data were analysed using the qualitative data analysis software MAXQDA 11. All statistical analyses were performed with PASW Statistics 18.

2.2.2 Data Analysis

To analyse ES relevance, Haida et al. (2015) calculated the median value for each ES based on the individual ranking of the experts. For better comparability with ES sensitivity, the original values were inverted to 1 (least relevant) and 25 (most relevant).

The analysis of the second part of the interview followed the standard procedure of deductive qualitative content analyses (Mayring 2000, 2010). For this, pre-defined key themes and sub-themes served as a coding agenda along which the interviews were manually coded, enabling a comparable and standardised frequency analysis of the coded themes. The key themes were defined by the 25 ES, and sub-themes by the three impact-categories (negative, positive, neutral). To evaluate the sensitivity of ES to climate change the key themes were assigned to the respective passage of text. Based on this, the number of experts was extracted who mentioned each of the ES, regardless of the exact content of their arguments. To test whether experts from different regions or thematic backgrounds mentioned different ES, a χ^2 test ($p > 0.05$) was used. To appraise the impact of climate change on ES in addition to the key themes assignments, the data was also coded according to the three impact categories. In a first step, the number of experts who described climate change impacts for each ES was extracted.

In case of ambiguous impact descriptions for the same ES two filters were applied, which permitted to assign every expert to one of the impact categories: (i) counting of arguments per expert and assigning the one category with the most arguments, (ii) verifying the content of the argument. This left a maximum of 53 impact trends per ES. In a second step, the data was further condensed in order to attain one distinct climate impact trend per ES. Three criteria were applied for this:

(i) removing all ES for which <10% of the experts described an impact; (ii) assigning of the trend which had been described by >50% of the experts for the respective ES, and (iii) assigning "inconsistent" to the remaining ES. For tourism, the experts described diverging arguments for winter and summer, and therefore this ES was separated into summer tourism and winter tourism. Based on the experts' arguments ES impact storylines were developed. To identify "hotspot ES", ES sensitivity was plotted against ES relevance in a sensitivity-relevance-matrix. Each ES was then attributed with one of the CC impact trends. To support decisions on how to proceed with formulating appropriate adaptation measures, the matrix was evenly divided into quadrants, which represent levels of priority intensity. The division of the quadrants was set at the mean value of the range of both ES relevance and ES sensitivity.

2.3 *Results*

2.3.1 Ecosystem Services' Sensitivity

When the experts were asked which of the 25 presented ES are sensitive to climate change, every service was mentioned at least once. Regarding three services ("natural hazard regulation", "tourism" and "energy"), there was high consensus amongst the experts that these ES are sensitive to climate change (Fig. 15.1). However, for the majority of services the experts showed no or only small consensus. A χ^2 test showed no significant difference in the perceived sensitivity for experts from different areas, except for "fresh water" in Vorarlberg and South Tyrol, with $p = 0.011$. Unlike this regional conformity, the experts showed a thematic bias (Table 15.2). In other words, the experts considered those ES as sensitive, which were associated with their respective professional background. This was the case particularly for "natural hazard regulation", "tourism", "energy" and "food". Nevertheless, only two ES showed significant differences to the remaining professions: "biological control" ($p = 0.001$) and "soil formation" ($p = 0.000$), regarding forestry and soil science, respectively.

2.3.2 ES Storylines and Climate Change Impact Trends

The assumed impacts of climate change on ES could be grouped into five classes which revealed similar trends (Fig. 15.2). Six services showed a strong tendency to be negatively affected by climate change (A). Regarding "natural hazard regulation", the experts argued that this trend is caused by a weakened protection function of forests, which in turn leads to an increase of morphodynamic processes. Rising demand caused by expanding settlements was presumed to exacerbate these impacts. This negative trend might be intensified by a weakened "biological control", as a result from invasive species and more frequent outbreaks of pests and diseases.

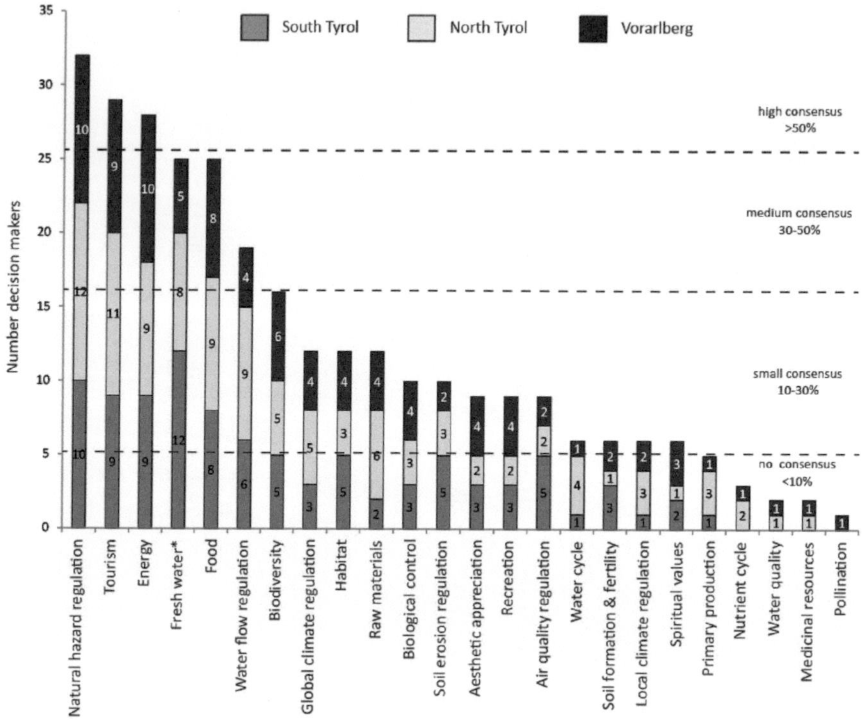

Fig. 15.1 Perceived sensitivity of ecosystem services to climate change according to the regional background of 53 experts. Shown is the number of experts who mentioned the service when asked "Which ecosystem service is sensitive to climate change?" Significance (*) tested with χ^2 test ($p > 0.05$). Levels of consensus are based on the percentage of experts

"Fresh water" supply was expected to be at risk mainly in the central Alpine-arid valleys of the study region. Retreating glaciers and thus decreasing discharge were assumed to put water supply for irrigation under pressure in the second half of the century. In addition to summer droughts, this might cause water scarcity. Predicted heavy rains after a long dry period might challenge the infiltration and retention potential of soil and vegetation and thus have a negative effect on "water flow regulation" and "soil erosion regulation". Economic benefits provided by "tourism" in winter were assumed to decline due to decreasing reliability of snow cover. "All-year tourism" and "biodiversity" showed a strong tendency that climate change will have no impact on these ES (B). In the case of biodiversity the experts argued that the overall diversity will stay the same. Loss of alpine species and habitats, caused by an altitudinal upward shift under warming temperatures, were assumed to be compensated by newly invading thermophilic species from the south.

Five ES ("food", "energy", "raw materials", "aesthetic appreciation" and "local climate regulation") showed inconsistent trends, as the impacts described for these ES were more or less evenly spread across all three categories (C). This uncertainty resulted from partially conflicting arguments, e.g. for food: expansion of cropland

Table 15.2 Perceived sensitivity of ecosystem services to climate change according to the thematic background of 53 experts

	Soil Science	Energy	Forestry	Environmental protection	Meteorology	Safety	Agriculture	Planning	Tourism
Experts (n)	5	6	6	6	6	7	6	5	6
Natural hazard regulation	40	50	66.7	50	66.7	100	50	100	16.7
Tourism	20	50	66.7	66.7	66.7	28.6	33.3	60	100
Energy	60	100	33.3	83.3	33.3	71.4	16.7	40	33.3
Fresh water	40	16.7	66.7	33.3	50	71.4	83.3	20	33.3
Food	60	33.3	50	33.3	50	42.9	100	40.0	16.7
Water flow regulation	40	16.7	66.7	66.7	50.0	57.1	16.7	0	0
Biodiversity	40	0	50	66.7	33.3	0	50	20	16.7
Global climate regulation	40	0	33.3	16.7	33.3	14.3	33.3	40	0
Habitat	20	16.7	16.7	66.7	50	0	16.7	0	16.7
Raw materials	20	16.7	50	33.3	16.7	14.3	33.3	20	0
Biological control	20	16.7	83.3*	16.7	16.7	0	16.7	0	0
Soil erosion regulation	40	16.7	50	0	16.7	0	33.3	20	0

Aesthetic appreciation	20	16.7	33.3	16.7	16.7	14.3	16.7	0	16.7
Recreation	40	0	0	33.3	33.3	14.3	16.7	0	16.7
Air quality regulation	0	16.7	33.3	33.3	33.3	14.3	0	0	16.7
Water cycle	20	16.7	16.7	0	16.7	14.3	16.7	0	0
Soil formation & fertility	80*	0	0	16.7	0	0	0	0	16.7
Local climate regulation	0	0	16.7	0	16.7	14.3	16.7	40	0
Spiritual values	0	0	16.7	0	16.7	14.3	0	40	16.7
Primary production	20	0	16.7	16.7	16.7	14.3	0	0	0
Nutrient cycle	0	0	16.7	0	16.7	0.0	16.7	0	0
Water quality	0	0	0	0	16.7	14.3	0	0	0
Medicinal resources	0	0	0	0	16.7	14.3	0	0	0
Pollination	0	0	0	0	0	14.3	0	0	0

Shown is the number of experts in percent who mentioned the service when asked "Which ecosystem service is sensitive to climate change?" Significance (*) tested with chi^2 test ($p > 0.05$)

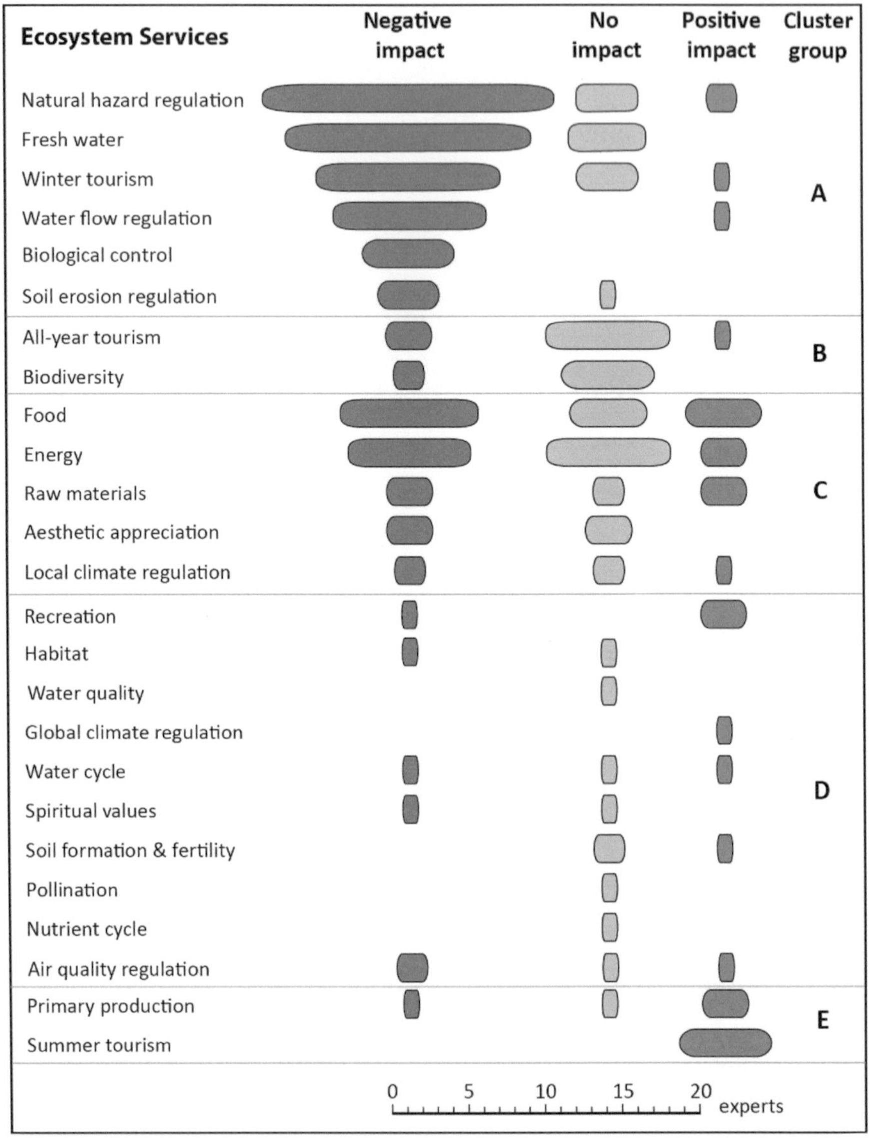

Fig. 15.2 Assumed impacts of climate change on ecosystem services according to 53 experts. Shown is the amount of experts per ES who described the respective climate change impact trend. Width of the *bars indicates* the amount of experts. Grouped into classes of similar trends

into higher altitudes versus water scarcity problems. Regarding "energy", the experts expected, on the one hand, a higher energy supply from biomass, as the tree line is predicted to shift upwards and thus increase biomass production. On the other hand, hydropower might decline due to decreasing discharge from retreating

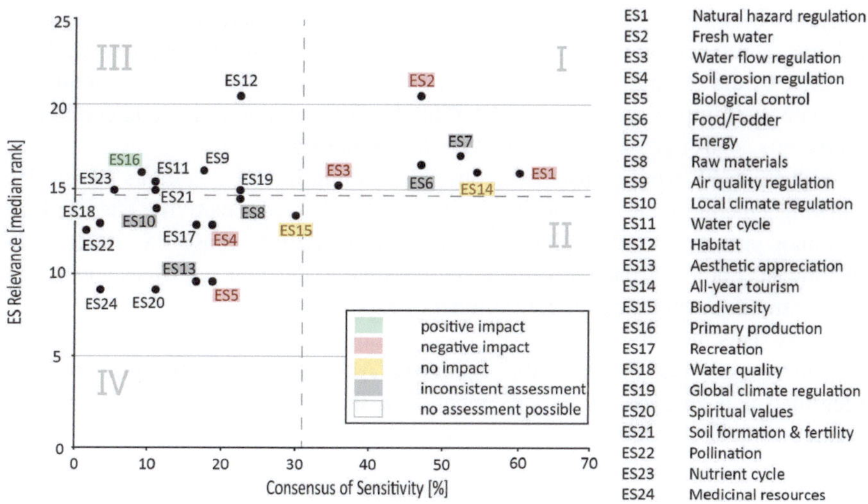

Fig. 15.3 Ecosystem services relevance-sensitivity matrix. ES relevance in median ranks from 1 (least important) to 25 (most important) based on experts' ranking regarding the perceived relevance of 25 ES from Haida et al. (2015). Consensus of sensitivity of ES in percentage of the number of experts who mentioned the services when asked "Which ecosystem service is sensitive to climate change?". Climate change impact trends are indicated in different *coloured shadings*. *Quadrants* represent levels of priority intensity from I (high priority) to IV (minor priority) and were defined by the mean value of the range of ES relevance and sensitivity values

glaciers. At the same time, demand for "energy" was supposed to rise, e.g. for cooling. This might lead to overexploitation of renewable energies, and accordingly the experts stressed the urge for more energy efficiency. "Primary production" and "summer tourism" showed a slight tendency to be positively affected (E). "Primary production" was argued to increase, caused by an altitudinal upward shift of the tree line and thus expansion of forest cover, in addition to an extension of the growing period. For the remaining ten ES, only few experts described climate change impacts (D), indicating a high level of uncertainty. This might result from missing information about climate change impacts on these ES or limited awareness of these impacts amongst decision makers.

2.3.3 Identifying "Hotspot ES"

Plotting an ES relevance-sensitivity matrix spread the 25 ES in quadrants I, III and IV (Fig. 15.3). High priority could be identified for six ES lying in quadrant I. Particularly "natural hazard regulation" and "fresh water" showed a high consensus of sensitivity, were highly relevant for the region, and had a strong tendency to be negatively affected by climate change. Regarding "energy" and "food", the experts did not describe definite trends about how climate change will affect these ES, although they were regarded as sensitive and as highly relevant. The majority of experts who mentioned "all-year tourism" stated that climate change would have positive (summer) and negative (winter) effects. However, throughout the year these impacts would neutralise. Particularly for "energy" and "food", it is essential to evaluate their impact storylines in detail to identify major concerns and conflicting arguments, so as to develop appropriate adaptation measures.

Seventeen ES were distributed along the dividing line of quadrants III and IV, showing a medium relevance with small to no consensus of sensitivity. These ES are of minor priority for adaptation assessments in relation to those ES which lie in quadrant I. For 11 of these ES it was not possible to develop storylines and define impact trends considering the low number of experts who described potential developments. This indicates a high level of uncertainty. Especially regarding "habitat", this uncertainty needs to be further examined. Although a high relevance could be identified and 26% of the experts mentioned this ES to be sensitive to climate change, only two experts were able to describe an impact trend.

2.4 *Discussion*

Climate change impacts on ES at the local to regional scale are difficult to predict (APCC 2014), and cannot always be decoupled from other global challenges, e.g. land use and socio-economic changes (de Vasconcelos et al. 2014; de Chazal and Rounsevell 2009). This difficulty is displayed in the case study results where it was not possible to describe impact scenarios for 11 ES. Moreover, for five ES the experts described several diverging scenarios, particularly, for "food" and "energy". Impacts on "food" must be differentiated since the complex topography of the study region will cause substantial changes to agricultural land-use, depending on crop type (Eitzinger et al. 2013; APCC 2014). Whilst water intensive grassland areas might decrease (Schaumberger 2012; Eitzinger et al. 2009; CH2014-Impacts, 2014), areas cultivated for thermophile crops such as wines and fruits might increase (Eitzinger et al. 2013; Europäische Akademie Bozen EURAC 2012; CH2014-Impacts 2014). Even though the knowledge base of climate change impacts on "food" production is generally high, there is still the need to combine modelled simulations with experimental research (APCC 2014) and to account for possible changes in the demand for "food" (Poppy et al. 2014). Rising demand (Burkhard et al. 2012) might also be the case for "natural hazard regulation" and "fresh water". Like the experts implied, socio-economic developments will have an influence on the changing demand for "natural hazard regulation"; for example, settlements and infrastructure expanding into zones of higher risk leads to increasing exposure and vulnerability and thus to growing demand for "natural hazard regulation" as a key component in an integrated natural hazard management.

To identify ES at risk, an approach based exclusively on interviews is prone to be biased. For tangible services, like the hotspot ES, changes are easier to observe and thus might be mentioned more frequently than less tangible ES, such as "soil formation & fertility". To interpret these results one ought to consider different motives which drive the experts' assessments. These assessments could be based on personal conviction (e.g. religion), personal experience (e.g. observed changes), logic (i.e. causal relationships), or scientifically proven information. Lack of information and high levels of uncertainty might result in maladaptation (Andrade Pérez et al. 2010), and it is therefore important to design the interviews appropriately and if possible to try to complement the interview results with generally accepted literature.

Uncertainties need to be considered with every decision which has unknown consequences (Tannert et al. 2007), and in the case of climate change adaptation these decisions might carry severe risks. This study was able to identify two types of uncertainties: ambiguities and limited or not existing knowledge. Ambiguities result out of diverging interpretative frames of interviewed experts on the same factual basis leading to a missing consensus on sensitivity, etc. This was the case for all ES which had an inconsistent assessment. For such a situation the International Risk Governance Council (2006) recommends a "participatory discourse" including civil society to discuss competing arguments, beliefs and values. The second type of uncertainty resulted from limited or not existing knowledge and information, i.e. "the unknown". ES for which it was not possible to define impact trends represent this type of uncertainty. To handle this type of uncertainty first the reason for the lack of knowledge needs to be identified. One reason is the missing exchange between stakeholders and researchers. In this case it is recommended to bridge the gap between science and practice for example by intensifying transdisciplinary research (Haida et al. 2015; Schröter et al. 2014) or engaging in innovative forms of knowledge transfer (Hauck et al. 2013; Klein et al. 2015). Another reason is general absence of knowledge and understanding of climate change processes and impacts and more research is needed (International Risk Governance Council 2006). These uncertainties need to be kept in mind when developing adaptation measure for the identified hotspot ES, for example by formulating no or low-regret measures.

In this study, the experts' statements generally conformed with scientific knowledge. Therefore the presented approach proved to be valuable to scrutinize climate change impacts on ES, and to develop adaptation strategies and measures that are specifically designed for the regional needs. To this end the approach serves as a prerequisite to develop adaptation measures which not only focus on ecological performance, resource extraction and provision of ES (Pramova et al. 2012), but also includes behavioural changes regarding consumption and demand. Stakeholders' involvement is essential for education and capacity building, to move from short-term to long-term coping strategies.

Despite the regionalization of adaptation needs and strategies (Pramova et al. 2012), the proposed process of identifying hotspot ES itself is applicable also outside mountain watersheds and is particularly valuable for regions with a low density of available data and high diversity of land cover/land use. Nevertheless, the assessment of ES sensitivity could be improved, for example, by asking the experts to rate ES sensitivity on a scale, e.g. on a Likert scale. Distinguishing the interview partners between decision makers for assessing ES relevance, and experts for ES sensitivity could be beneficial for regions in which it is difficult to recruit suitable interview partners. In case of diverging arguments focus groups or round tables could foster finding consensus and understanding (Ross et al. 2015).

3 Conclusion

This study presents and applies an approach for ES adaptation needs assessment. By linking the ES concept with climate change adaptation needs assessment this approach aids in analysing ES relevance, sensitivity and impacts for a specific region. The approach provides a novel hands-on decision support tool for practitioners, which helps with identifying and prioritizing climate change impacts issues and, based on this, with developing appropriate adaptation measures at the local to regional level. Thus it assists in responding to and preparing for climate change impacts.

The approach is particularly valuable for communities, municipalities or regional governments to address their specific needs within the range of their possibilities and resources. By including local knowledge this approach empowers community capacities and thus raises acceptance and pro-actively urges the development and implementation of adaptation measures.

As part of a more comprehensive ES-based adaptation assessment strategy (Haida 2016), the results of this adaptation needs assessment clearly indicate the urgent need to develop and implement adaptation measures regarding water related ES, such as “fresh water”, “water flow regulation” and “natural hazard regulation”.

Acknowledgements The authors are grateful to all interviewees for their input to this study. Special thanks go to Ulrike Tappeiner for her valuable input to the study design and to Andrew Greenbank for his thorough proofreading. This study was part of the project “SHIFT” funded by the Austrian COMET programme and the Central Europe Programme project “Urban_WFTP – Urban Water Footprint: a new approach for water management in urban areas”.

References

Andrade Pérez, A., Herrera Fernández, B., & Cazzolla Gatti, R. (Eds.). (2010). *Building resilience to climate change: Ecosystem-based adaptation and lessons from the field, no. 9*. Gland: IUCN.

APCC. (2014). *Österreichischer Sachstandsbericht Klimawandel 2014: Austrian Panel on Climate Change (APCC)*. Wien: Verlag der Österreichischen Akademie der Wissenschaften.

Auer, I., Böhm, R., Jurkovic, A., Lipa, W., Orlik, A., Potzmann, R., Schöner, W., Ungersböck, M., Matulla, C., Briffa, K., Jones, P., Efthymiadis, D., Brunetti, M., Nanni, T., Maugeri, M., Mercalli, L., Mestre, O., Moisselin, J.-M., Begert, M., Müller-Westermeier, G., Kveton, V., Bochnicek, O., Stastny, P., Lapin, M., Szalai, S., Szentimrey, T., Cegnar, T., Dolinar, M., Gajic-Capka, M., Zaninovic, K., Majstorovic, Z., & Nieplova, E. (2007). HISTALP – Historical instrumental climatological surface time series of the greater Alpine region. *International Journal of Climatology, 27*, 17–46. doi:10.1002/joc.1377.

Briner, S., Elkin, C., Huber, R., & Grêt-Regamey, A. (2012). Assessing the impacts of economic and climate changes on land-use in mountain regions: A spatial dynamic modeling approach. *Agriculture, Ecosystems & Environment, 149*, 50–63. doi:10.1016/j.agee.2011.12.011.

Briner, S., Elkin, C., & Huber, R. (2013). Evaluating the relative impact of climate and economic changes on forest and agricultural ecosystem services in mountain regions. *Journal of Environmental Management, 129*, 414–422. doi:10.1016/j.jenvman.2013.07.018.

Burkhard, B., Kroll, F., Nedkov, S., & Müller, F. (2012). Mapping ecosystem service supply, demand and budgets. *Ecological Indicators, 21*, 17–29.

CH2014-Impacts. (2014). *Toward quantitative scenarios of climate change impacts in Switzerland*. Bern: OCCR, FOEN, MeteoSwiss, C2SM, Agroscope, ProClim.

Chowdhury, P. D., Haque, C. E., & Driedger, S. M. (2012). Public versus expert knowledge and perception of climate change-induced heat wave risk: A modified mental model approach. *Journal of Risk Research, 15*, 149–168. doi:10.1080/13669877.2011.601319.

Civantos, E., Thuiller, W., Maiorano, L., Guisan, A., & Araujo, M. B. (2012). Potential impacts of climate change on ecosystem services in Europe: The case of pest control by vertebrates. *Bioscience, 62*, 658–666. doi:10.1525/bio.2012.62.7.8.

Daily, G. C., Polasky, S., Goldstein, J., Kareiva, P. M., Mooney, H., Pejchar, L., Ricketts, T. H., Salzman, J., & Shallenberger, R. (2009). Ecosystem services in decision making: Time to deliver. *Frontiers in Ecology and the Environment, 7*(1), 21–28. doi:10.1890/080025.

Dawes, M. A., Hagedorn, F., Handa, I. T., Streit, K., Ekblad, A., Rixen, C., Körner, C., & Hättenschwiler, S. (2013). An alpine treeline in a carbon dioxide-rich world: Synthesis of a nine-year free-air carbon dioxide enrichment study. *Oecologia, 171*, 623–637. doi:10.1007/s00442-012-2576-5.

de Chazal, J., & Rounsevell, M. D. A. (2009). Land-use and climate change within assessments of biodiversity change: A review. *Global Environmental Change, 19*, 306–315. doi:10.1016/j.gloenvcha.2008.09.007.

de Vasconcelos, A. C. F., Schlindwein, S. L., Lana, M. A., Fantini, A. C., Bonatti, M., D'Agostini, L. R., & Martins, S. R. (2014). Land use dynamics in Brazilian La Plata Basin and anthropogenic climate change. *Climatic Change, 127*, 73–81. doi:10.1007/s10584-014-1081-8.

Eitzinger, J., Kersebaum, K. C., & Formayer, H. (2009). *Landwirtschaft im Klimawandel: Auswirkungen und Anpassungsstrategien für die Land-und Forstwirtschaft in Mitteleuropa* [Agriculture in climate change – Impacts and adaptation strategies for agriculture and forestry in Central Europe]. Clenze: Agrimedia.

Eitzinger, J., Trnka, M., Semerádová, D., Thaler, S., Sovobodoá, E., Hlavinka, P., Šiska, B., Takác, J., Malatinská, L., Nováková, M., Dubrovský, M., & Žalud, Z. (2013). Regional climate change impacts on agricultural crop production in Central and Eastern Europe – Hotspots, regional differences and common trends. *The Journal of Agricultural Science, 151*, 787–812. doi:10.1017/S0021859612000767.

Elkin, C., Gutiérrez, A. G., Leuzinger, S., Manusch, C., Temperli, C., Rasche, L., & Bugmann, H. (2013). A 2 °C warmer world is not safe for ecosystem services in the European Alps. *Global Change Biology, 19*, 1827–1840. doi:10.1111/gcb.12156.

Europäische Akademie Bozen EURAC. (Ed). (2012). *Klimareport Südtirol* [Climate report South Tyrol]. Bozen/Bolzano.

Giebels, D., van Buuren, A., & Edelenbos, J. (2015). Using knowledge in a complex decision-making process – Evidence and principles from the Danish Houting project's ecosystem-based management approach. *Environmental Science & Policy, 47*, 53–67. doi:10.1016/j.envsci.2014.10.015.

Giebels, D., van Buuren, A., & Edelenbos, J. (2016). Knowledge governance for ecosystem-based management: Understanding its context-dependency. *Environmental Science & Policy, 55*, 424–435. doi:10.1016/j.envsci.2015.08.019.

Grabherr, G. (1997). The high-mountain ecosystems of the Alps. In F. E. Wielgolaski (Ed.), *Polar and Alpine tundra*. Amsterdam/Lausanne/New York: Elsevier.

Grêt-Regamey, A., Brunner, S. H., & Kienast, F. (2012). Mountain ecosystem services: Who cares? *Mountain Research and Development, 32*(S1), S23–S34. doi:10.1659/MRD-JOURNAL-D-10-00115.S1.

Grêt-Regamey, A., Brunner, S. H., Altwegg, J., Christen, M., & Bebi, P. (2013). Integrating expert knowledge into Mapping ecosystem services trade-offs for sustainable forest management. *Ecology and Society, 18*(3), art.34. doi:10.5751/ES-05800-180334.

Hagerman, S., Dowlatabadi, H., Satterfield, T., & McDaniels, T. (2010). Expert views on biodiversity conservation in an era of climate change. *Global Environmental Change, 20*, 192–207. doi:10.1016/j.gloenvcha.2009.10.005.

Haida, C. (2016). *Ecosystem services and climate change: Concept, methods, and their application to utilize ecosystem services to respond to climate change*. Dissertation, Innsbruck.

Haida, C., Rüdisser, J., & Tappeiner, U. (2015). Ecosystem services in mountain regions: Experts' perceptions and research intensity. *Regional Environmental Change, 16*, 1989–2004. doi:10.1007/s10113-015-0759-4.

Haines-Young, R., & Potschin, M. (2011, November). *Common International Classification of Ecosystem Services (CICES): Update 2011*. EEA framework contract no EEA/BSS/07/007.

Hauck, J., Görg, C., Varjopuro, R., Ratamäki, O., & Jax, K. (2013). Benefits and limitations of the ecosystem services concept in environmental policy and decision making: Some stakeholder perspectives. *Environmental Science & Policy, 25*, 13–21. doi:10.1016/j.envsci.2012.08.001.

Hossain, M. S., Hein, L., Rip, F. I., & Dearing, J. A. (2015). Integrating ecosystem services and climate change responses in coastal wetlands development plans for Bangladesh. *Mitigation and Adaptation Strategies for Global Change, 20*, 241–261. doi:10.1007/s11027-013-9489-4.

International Risk Governance Council. (2006). *Risk governance towards an integrative approach*. Geneva: International risk governance council.

IPCC. (Ed.). (2014). *Climate change 2014: Impacts, adaptation, and vulnerability*. Contribution of Working Group II to the Fifth assessment report of the Intergovernmental Panel on Climate Change. Cambridge: Cambridge University Press.

Karrasch, L., Klenke, T., & Woltjer, J. (2014). Linking the ecosystem services approach to social preferences and needs in integrated coastal land use management. A planning approach. *Land Use Policy, 38*, 522–532. doi:10.1016/j.landusepol.2013.12.010.

Klein, T. M., Celio, E., & Grêt-Regamey, A. (2015). Ecosystem services visualization and communication: A demand analysis approach for designing information and conceptualizing decision support systems. *Ecosystem Services, 13*, 173–183. doi:10.1016/j.ecoser.2015.02.006.

Loft, L., Mann, C., & Hansjürgens, B. (2015). Challenges in ecosystem services governance: Multi-levels, multi-actors, multi-rationalities. *Ecosystem Services, 16*, 150–157. doi:10.1016/j.ecoser.2015.11.002.

López-Hoffman, L., Breshears, D. D., Allen, C. D., & Miller, M. L. (2013). Key landscape ecology metrics for assessing climate change adaptation options: Rate of change and patchiness of impacts. *Ecosphere, 4*(8), art101. doi:10.1890/ES13-00118.1.

Martín-López, B., Iniesta-Arandia, I., García-Llorente, M., Palomo, I., Casado-Arzuaga, I., Amo, D. G. D., Gómez-Baggethun, E., Oteros-Rozas, E., Palacios-Agundez, I., Willaarts, B., González, J. A., Santos-Martín, F., Onaindia, M., López-Santiago, C., Montes, C., & Bawa, K. (2012). Uncovering ecosystem service bundles through social preferences. *PloS One, 7*(6), e38970. doi:10.1371/journal.pone.0038970.

Martinez-Harms, M. J., Bryan, B. A., Balvanera, P., Law, E. A., Rhodes, J. R., Possingham, H. P., & Wilson, K. A. (2015). Making decisions for managing ecosystem services. *Biological Conservation, 184*, 229–238. doi:10.1016/j.biocon.2015.01.024.

Mayring, P. (2000). Qualitative content analysis. *Forum: Qualitative Social Research, 1*(2), art.20.

Mayring, P. (2010), Qualitative Inhaltsanalyse [Qualitative content analysis]. In Mey, G., & Mruck, K. (Eds.), *Handbuch Qualitative Forschung in der Psychologie* [Handbook of qualitative research in psychology]. Wiesbaden: VS Verlag für Sozialwissenschaften/Springer Fachmedien Wiesbaden.

Metternicht, G., Sabelli, A., & Spensley, J. (2014). Climate change vulnerability, impact and adaptation assessment. *International Journal of Climate Change Strategies and Management, 6*, 442–476. doi:10.1108/IJCCSM-06-2013-0076.

Metzger, M. J., Rounsevell, M. D. A., Acosta-Michlik, L., Leemans, R., & Schröter, D. (2006). The vulnerability of ecosystem services to land use change. *Agriculture, Ecosystems & Environment, 114*, 69–85. doi:10.1016/j.agee.2005.11.025.

Millennium Ecosystem Assessment. (2005). *Ecosystems and human well-being: Current states and trends*. Washington, DC: Island Press.

Pauli, H., Gottfried, M., Dullinger, S., Abdaladze, O., Akhalkatsi, M., Alonso, J. L. B., Coldea, G., Dick, J., Erschbamer, B., Calzado, R. F., Ghosn, D., Holten, J. I., Kanka, R., Kazakis, G., Kollar, J., Larsson, P., Moiseev, P., Moiseev, D., Molau, U., Mesa, J. M., Nagy, L., Pelino, G., Puscas, M., Rossi, G., Stanisci, A., Syverhuset, A. O., Theurillat, J.-P., Tomaselli, M., Unterluggauer, P., Villar, L., Vittoz, P., & Grabherr, G. (2012). Recent plant diversity changes on Europe's mountain summits. *Science, 336*(6079), 353–355. doi:10.1126/science.1219033.

Poppy, G. M., Chiotha, S., Eigenbrod, F., Harvey, C. A., Honzak, M., Hudson, M. D., Jarvis, A., Madise, N. J., Schreckenberg, K., Shackleton, C. M., Villa, F., & Dawson, T. P. (2014). Food security in a perfect storm: Using the ecosystem services framework to increase understanding. *Philosophical Transactions of the Royal Society B: Biological Sciences, 369*(1639), art.20120288. doi:10.1098/rstb.2012.0288.

Pramova, E., Locatelli, B., Brockhaus, M., & Fohlmeister, S. (2012). Ecosystem services in the National Adaptation Programmes of action. *Climate Policy, 12*, 393–409. doi:10.1080/14693062.2011.647848.

Ross, H., Shaw, S., Rissik, D., Cliffe, N., Chapman, S., Hounsell, V., Udy, J., Trinh, N. T., & Schoeman, J. (2015). A participatory systems approach to understanding climate adaptation needs. *Climatic Change, 129*, 27–42. doi:10.1007/s10584-014-1318-6.

Schaumberger, A. (2012). *Räumliche Modelle zur Vegetations-und Ertragsdynamik im Wirtschaftsgrünland* [Spatial models for vegetation and yield dynamics for agriculturally used grassland]. Dissertation, Graz.

Schröter, D., Cramer, W., Leemans, R., Prentice, I. C., Aroujo, M., Arnell, N., Bondeau, A., Bugmann, H., Carter, T., Gracia, J., La Vega-Leiner de, A., Erhard, M., Ewert, F., Glendining, M., House, J., Kankaanpää, S., Klein, R. J., Lavorel, S., Lindner, M., Metzger, M. J., Meyer, J., Mitchell, T., Reginster, I., Rounsevell, M. D. A., Sabaté, S., Sitch, S., Smith, B., Smith, J., Smith, P., Sykes, M. T., Thonicke, K. T. W., Tuck, G., Zaehle, S., & Zierl, B. (2005). Ecosystem service supply and vulnerability to global change in Europe. *Science, 310*(5752), 1333–1337. doi:10.1126/science.1115233.

Schröter, M., van der Zanden, E. H., Van Oudenhoven, A. P., Remme, R. P., Serna-Chavez, H. M., de Groot, R. S., & Opdam, P. (2014). Ecosystem services as a contested concept: A synthesis of critique and counter-arguments. *Conservation Letters, 7*, 514–523. doi:10.1111/conl.12091.

Smith, J. B., Schneider, S. H., Oppenheimer, M., Yohe, G. W., Hare, W., Mastrandrea, M. D., Patwardhan, A., Burton, I., Corfee-Morlot, J., Magdaza, C. H., Füssel, H.-M., Pittock, A. B., Rahman, A., Suarez, A., & van Ypersele, J.-P. (2009). Assessing dangerous climate change through an update of the intergovernmental Panel on Climate Change (IPCC) "reasons for concern". *Proceedings of the National Academy of Sciences, 106*(11), 4133–4137.

Staudinger, M. D., Grimm, N. B., Staudt, A., Carter, S. L., Chapin, F. S. I., Kareiva, P., Ruckelshaus, M., & Stein, B. A. (2012). *Impacts of climate change on biodiversity, ecosystem, and ecosystem services: Technical input to the 2013 National Climate Assessment*. Cooperative report to the 2013 National Climate Assessment.

Summers, J. K., Smith, L. M., Case, J. L., & Linthurst, R. A. (2012). A review of the elements of human well-being with an emphasis on the contribution of ecosystem services. *Ambio, 41*, 327–340. doi:10.1007/s13280-012-0256-7.

Tannert, C., Elvers, H.-D., & Jandrig, B. (2007). The ethics of uncertainty. *EMBO Reports, 8*, 892–896. doi:10.1038/sj.embor.7401072.

Tappeiner, U., Borsdorf, A., & Tasser, E. (Eds.). (2008). *Alpenatlas: Mapping the alps. Society, economy, environment*. Heidelberg: Spektrum Akademischer Verlag.

TEEB. (2010). *The economics of ecosystems and biodiversity: Ecological and economic foundation*. London: The Economics of Ecosystems and Biodiversity.

Tompkins, E. L., & Adger, N. W. (2004). Does adaptive management of natural resources enhance resilience to climate change? *Ecology and Society*, *9*(2), art.10.

Tuvendal, M., & Elmqvist, T. (2011). Ecosystem services linking social and ecological systems: River brownification and the response of downstream stakeholders. *Ecology and Society*, *16*(4), art.21. doi: 10.5751/ES-04456-160421.

UNEP. (2012). *Ecosystem-based adaptation guidance: Moving from principles to practice*. Working document: April 2012.

Chapter 16
Environmental Services in Mountain Catchments Affected by the Acid Atmospheric Deposition

Josef Křeček and Ladislav Palán

1 Introduction

Considering the environmental benefits in mountain catchments, Körner and Ohsawa (2005) highlighted particularly water resources recharge (quantity and quality), recreation, timber, habitat, and beautiful scenery. Both, formulations and applications of the environmental services require the flexible and transparent decision making process that embraces a diversity of knowledge and values (Reed 2008). Mountain watersheds in central Europe are mostly forested; therefore, the forestry practices are among the most important factors of their environmental benefits. The concept of multi resource forestry (seeking the simultaneous production of several interdependent substances and services) concentrates on maintaining forests as the biophysical systems (Dudley and Stolton 2003). FAO (2006) reported several positive examples of the collaborative watershed management based on the action research, which adopted diagnostic studies, field testing and validations of improved technologies, as well as the social and environmental impact assessment.

Generally, the concept of environmental services is based on the link between flows of values to human societies as the result of the state and quantity of natural capital (MEA 2005). The assessment of ecosystem changes concentrates on the enhancement of adequate conservation and sustainable management practices. Willis (2002) categorised the main non-market costs and benefits of forestry practices upon the water supply (quantity and quality), where the principal products of mountain catchments include: (1) abstraction for potable water (for drinking and commercial uses), (2) agriculture and irrigation in downstream areas, (3) hydro-electric power generation, (4) wildlife (including recreational and commercial fisheries), and (5) other recreational uses. To identify the above benefits, in mountain catchments, unfortunately, the present estimates of hydrological characteristics

J. Křeček • L. Palán (✉)
Department of Hydrology, Czech Technical University, Prague, Czech Republic
e-mail: ladislav.palan@gmail.com

© Capital Publishing Company, New Delhi, India 2017
J. Křeček et al. (eds.), *Ecosystem Services of Headwater Catchments*,
DOI 10.1007/978-3-319-57946-7_16

(water yield, runoff timing, probability of extreme events, water quality, ecological discharge) are still affected by several important uncertainties (Foy et al. 2015).

Additionally, with rising civilisation pressures in mountain regions (acidification, declining environmental health, changes of the global climate, etc.), the identification of environmental benefits is getting even more complicated (Křeček and Hořická 2001; NIVA 2013). The aim of this paper is to discuss impacts of the acid atmospheric deposition on environmental services in forested mountain catchments of the Jizera Mountains (Czech Republic) with priority of the drinking water supply.

2 Impacts of Acidification on Ecosystem Services

The acid atmospheric deposition (initiated by emissions of sulphur and nitrogen to the air) has affected ecosystems of highly-industrialised regions in Europe already during the past 200 years, but, after the World War II, the acid atmospheric deposition has caused widespread acidification of freshwaters with damage to drinking water quality, fish populations and other aquatic organisms, particularly in headwater areas of central and eastern Europe (Křeček and Hořická 2001).

Recently, the term 'water towers' has been widely adopted, expressing the impact of mountains on the freshwater supply in downstream areas (FAO 2006). In the global scale, Viviroli et al. (2007) analysed the potential significance of mountain water resources by including the actual population in the adjacent lowlands and its water needs: 7% of the global mountain area provides essential water resources, while another 37% delivers relatively important supportive supply. Significance of mountain water resources is still increasing with expected changes in the global climate (Christensen 2005). Therefore, the acidification of headwater catchments means a serious hazard in the concept of ecosystem services supporting the drinking water supply (Křeček and Hořická 2006).

The Millennium Ecosystem Assessment (MEA 2005) recognised four categories of the ecosystem services: provisioning, regulating, supporting and cultural services. According to NIVA (2013), the provisioning services reflect namely drinking water supply, timber and fisheries; regulating services refer to carbon and bio-geochemical cycles, water quality genesis, runoff timing and flood occurrence; supporting services contribute particularly to biodiversity; and cultural services address recreation, tourism and aesthetic experiences.

The concept of ecosystem services is based on means by which all the environmental effects and their societal impacts can be evaluated together in a systematic manner. The ecosystem services can capture a wider set of costs and benefits, not traditionally valued in economic analysis (TEEB 2010). Since the 1970s, in Europe, the evaluation of economic impacts and cost-benefits by controlling the acid atmospheric load became the important issue; however, these aspirations were limited by the rather broad spectrum of uncertainties in impacts of the acid atmospheric load (Kelly-Quinn et al. 2008; NIVA 2013). Generally, acidification of freshwaters depends on two factors: the acid atmospheric deposition, and the acid-sensitivity of mountain ecosystems (acid neutralising capacity). Unfortunately, there are still large gaps in the understanding of the nature and values of the acid rain impact. Therefore, the economic-evaluation models for the acid atmospheric load do not adequately account for environmental benefits (NIVA 2013).

3 Material and Methods

3.1 The Study Area

This study was performed in the upper plain of the Jizera Mountains (North Bohemia, Czech Republic, 50°40′–50°52′N, 15°08′–15°24′E) (Fig. 16.1). The upper plain of the mountains (approx. 200 km^2) is located above the elevation of

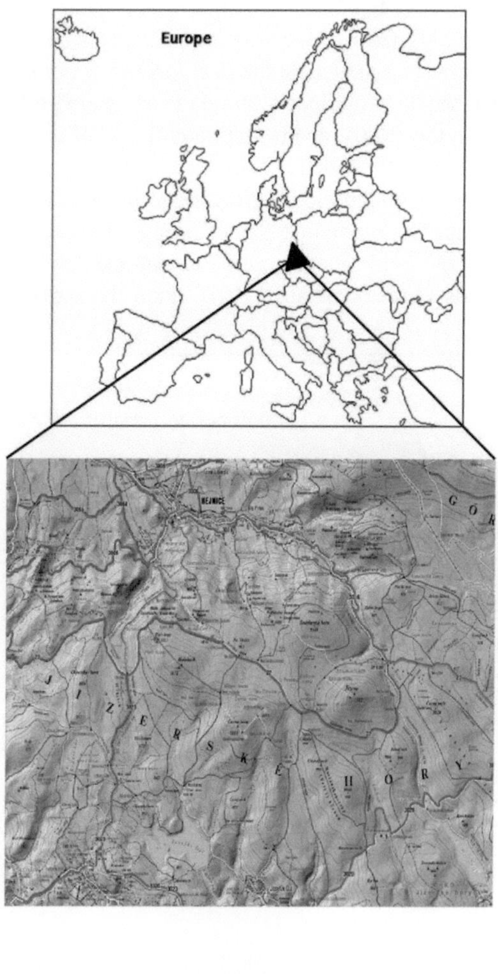

Fig. 16.1 The Jizera Mountains region (Czech Republic)

800 m (humid temperate climate, subarctic region Dfc of the Köppen climate zones), where the mean annual precipitation range from 1290 to 1400 mm and the mean annual temperature from 4 to 5 °C. In a normal year, snow cover lasts there from the beginning of November to the end of April, reaching the maximum snow-pack of 120 cm (Tolasz et al. 2007). Surface flow dominates over the runoff genesis; groundwater bodies occur only in the shallow subsurface layers and the bedrock cracks. Because of the granite bedrock and shallow poor soils, the local environments are extremely sensitive to acidification.

The upper plain of the Jizera Mountains is almost completely forested. During the time, priorities in the land-use have been changing with the civilisation level: from the early control of state boarders, big game hunting, sustainable timber harvest, downstream flood protection, to the recent preferences for the drinking water supply. The system of drinking water reservoirs has been developed there after the year of 1960. Nowadays, two reservoirs (Josefův Důl and Souš, Fig. 16.2) serve as the main providers of neighbouring cities with the drinking water. To support the water resource recharge, in 1978, *the Protected Headwater Area of the Jizera Mountains* has been declared by the decree of the Czech government (Water Act 254/2001 Coll.). But, this law still faces a serious gap in formulations of forestry services. Considering the *best forestry practices*, only the reduction of forest lands and the development of drainage areas are prohibited.

In the 1980s, headwater catchments of this region were heavily degraded by the air pollution and the acid atmospheric load (defoliation and die-back of spruce stands, and commercial forest harvest with heavy mechanisation). With the harvest of spruce plantations, the *Junco effusi-Calamagrostietum villosae* association became a dominant community on extensive clear-cut areas in the upland watersheds (Křeček et al. 2010). Since the early 1990s, with the European *Sulphur*

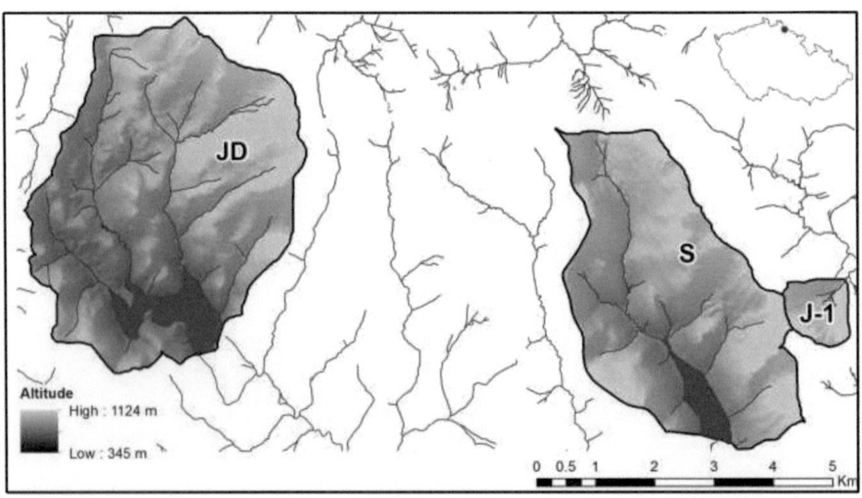

Fig. 16.2 Drinking water supply catchments Josefův Důl (JD) and Souš (S), and the experimental basin Jizerka (J-1)

Protocol (the load of sulphur decreased to approx. 40% of the year 1987) restoration activities started in catchments and lakes of the Jizera Mountains. Also, the follow-up environmentally sound forestry practices in catchments of the drinking water reservoirs started to be reconsidered (Chalupa and Křeček 1995).

3.2 Long-Term Environmental Monitoring

The long-term ecological monitoring has been carried out in catchments of the drinking water reservoirs Josefův Důl and Souš (Fig. 16.2), since 1982. This survey included observations of standard climate parameters, precipitation and runoff, acid atmospheric deposition (bulk precipitation chemistry), stream and lake water chemistry, evidence of macroinvertebrates and fish, inventory of forest stands (clear-cut area, number of trees, canopy density, vitality of trees) and herbaceous vegetation.

In the Jizerka experimental basin (Fig. 16.2), detailed observations included water flows through the forest canopy and the evidence of fog drip with the elevation. Basic characteristics of the studied catchments are in Table 16.1.

3.3 Multi-criteria Decision Analysis

The forest percentage in watersheds has rather limited information upon their hydrological regime. Several studies across the world (Daniels et al. 2004; Hawes and Smith 2005; Cambi et al. 2015; Křeček and Palán 2015; Sun et al. 2015) emphasize the importance of forestry practices in the control of water yield and other aspects of catchment hydrology (water chemistry and biodiversity, soil conservation, or stability of stream channels). The concept of multi-resource forestry requires forest maintaining as the biophysical system and the application of science-based dialogues between main stakeholders (Agnew and Goodess 2016). Triantaphyllou (2000) pointed out basic steps of the multi-criteria decision analysis (MCDA) for a collaborative watershed management: the main goal, interests and sub-interests of stakeholders, decision alternatives to reach the objectives, evaluation criteria, and outcomes associated with the alternatives.

Table 16.1 Characteristics of the catchments Jizerka (J-1), Josefův Důl (JD) and Souš (S)

Basin	Area A (km^2)	Mean elevation E (m)	Mean slope S (%)	Mean areal precipitation P (mm)	Mean areal evapotranspiration ET (mm)
J-1	1.03	927	12.00	1228	341
JD	19.64	834	11.90	1169	373
S	13.78	865	14.00	1189	362

The weighted sum model has been widely used by MCDA worldwide (Triantaphyllou 2000). Generally, this method considers m alternatives (A_1, ..., A_m) and n decision criteria (C_1, …, C_n), the best alternative $A*$ is given by Eq. (16.1):

$$A^* = \max \sum_{j=1}^{n} aij \ wj \tag{16.1}$$

where a_{ij} is the performance value (the actual value of the alternative i for $i = 1, 2, \ldots, m$) in terms of the criterion j, and w_j is the weight of importance of the criterion j ($j = 1, 2, \ldots, n$).

Peřina et al. (1977) recommended to differentiate roles of forest stands in the runoff genesis, by respecting the buffer strips around watercourses and reservoirs, stands controlling soil erosion on steep slopes, or, stands regulating water yield (particularly by controlling the evaporation loss). At present, this concept is used by FMI (2015), regulating forestry activities in protective zones of water resources on the territory of the Czech Republic.

4 Results and Discussion

4.1 Role of Forests in Watersheds of Drinking Water Reservoirs

According to the climate normal of 1961–1990, climax forests in the upper plain of the Jizera Mountains include particularly zones of (1) Norway spruce (*Picea abies*) – Common beech (*Fagus sylvatica*), where beech dominates, (2) Common beech – Norway spruce (with majority of spruce), and (3) Norway spruce (Fig. 16.3). However, during the 19th and 20th centuries, the native mixed forests were converted to the spruce plantations of a lower ecological stability.

The long-term monitoring carried out in the Jizera Mountains, 1981–2015 (Křeček and Hořická 2001, 2006; Křeček and Palán 2015) highlighted that at acid rain (pH of precipitation below 5.6), the spruce stands can significantly contribute to the acidification of headwater environments, and to the decline of water quality in reservoirs and their inlets.

In the 1980s, particularly low pH of water (4–5), high content of aluminium (1–2 mg/l), extinction of fish and drastically reduced zooplankton, phytoplankton and benthic fauna were indicated in surface waters of the studied catchments. The recent recovery of water environments in the Jizera Mountains (an increase in mean annual pH values to 5–6, a drop in aluminium concentrations to 0.2–0.5 mg/l, successful fish reintroduction) results from both, drop in the air pollution and reduced canopy by the harvest of spruce stands (partly also by the periodical liming of reservoirs and watersheds).

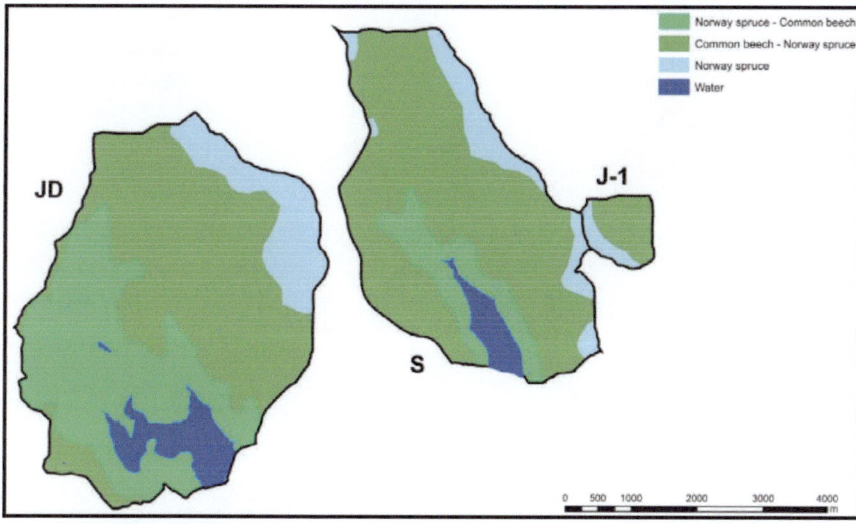

Fig. 16.3 The climax zones in watersheds of Jizerka (J-1), Josefův Důl (JD) and Souš (S)

Table 16.2 The structured forestry zones in basins of Josefův Důl (JD) and Souš (S)

Structural forestry zones	JD (ha)	JD (%)	S (ha)	S (%)
Riparian buffer strips	140	8	50	4
Soil protection	40	2	60	5
Evapotranspiration control	1090	64	770	58
Fog drip	360	21	430	33
Wetlands	70	5	0	0

By the clear-cut in the Jizerka catchment, water yield increased by 108 mm (12%) and water chemistry enhanced (pH grew from 4 to 5.5, sulphate decreased from 12 to 6 mg/l, nitrate from 5.5 to 3.5, and aluminium from 1.5 to 0.5). But, on the other hand, the commercial harvest (extensive clear-cut without buffer zones, heavy mechanisation, skidding timber by wheeled tractors) contributed to the fast (direct) runoff, soil erosion (from 0.01 to 1,34 mm/year) and sediment runoff (almost 30% of the soil volume eroded).

According to the above achievements, the structured forestry in headwater catchments should respect five watershed zones with respect of their prevailing role in the runoff genesis: (1) riparian buffer zones, (2) zones of soil protection, (3) zones of evapotranspiration control, (4) zones of significant fog drip, and (5) wetlands (Table 16.2). Their delineation in studied catchments (J-1, JD and S) is shown in Fig. 16.4, and their area is given in Table 16.2.

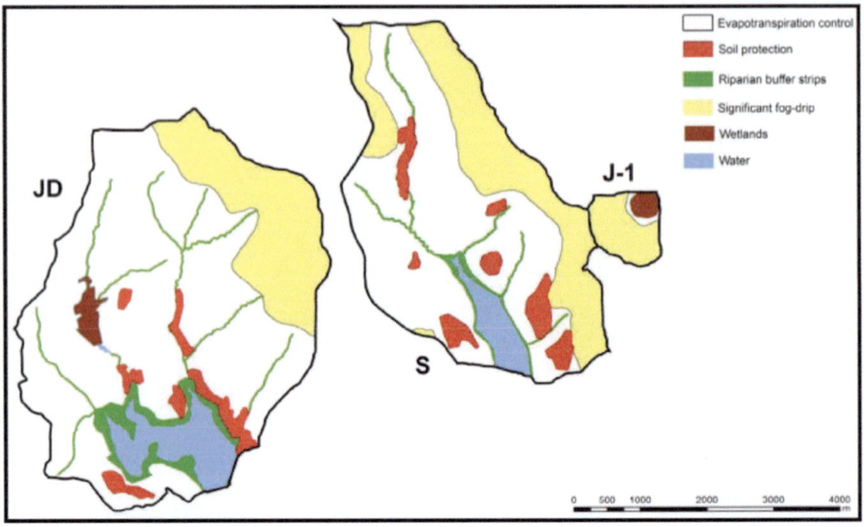

Fig. 16.4 The structured forestry in watersheds of Jizerka (J-1), Josefův Důl (JD) and Souš (S)

4.1.1 Riparian Buffer Zone

Riparian buffer strips are considered as the vital elements of watersheds with their positive role particularly in protection of water quality and biodiversity. Large number of studies documented impacts of the removal of streamside vegetation on degradation of water resources and lakes (Parkyn 2004). Hawes and Smith (2005) recommended the efficient buffer widths from 3 (bank stabilization and stream shading) to over 150 m (wildlife habitat). The most effective riparian buffers should include a mix of trees, shrubs and herbaceous plants native to the region (Jontos 2004, in Hawes and Smith 2005). In forestry systems, a buffer zone is considered as the forest area left beside the stream when the surrounding area is harvested (Parkyn 2004). In the studied watersheds, Fig. 16.4, the riparian buffer strips were allocated identically with the protection zones of the first-rank (delineated by service roads along reservoirs), and second-rank (30–100 m), claimed by the Water Act 254/2001 Coll. (FMI 2015). Activities of man there are turned down, and, silviculture restricted to selective cuttings only.

4.1.2 Zone of Soil Protection

On steep slopes (above 20% for a minimum of 10 m horizontally), the dominant role of vegetation is to control slope stability, surface runoff and soil erosion (Gray and Sotir 1996). In the Czech Republic, FMI (2015) recommended to respect slopes over 40%, where cable way techniques are required by harvesting the timber. In the catchments of Josefův Důl and Souš, the soil control segments were delineated on slopes above 30% (Fig. 16.4), according to the erosion of soil mobilized by the harvest of spruce stands. In this zone, silviculture should support humus, litter layers (consistent with FAO 2008), and deep rooting systems. Therefore, a replacement of the even-age spruce forests for stands near the native composition (mixed or beech stands) could contribute to a higher stability of slopes.

4.1.3 Zone of Evapotranspiration Control

In sites with slopes not exceeding 30% (and not located in the riparian system), forestry practices could control water yield by regulating the loss of vaporization. This zone dominates in both Josefův Důl and Souš basins (Fig. 16.4) covering 64 and 58% of their catchments (Table 16.2). In these elevations, particularly, changes in the canopy interception are expected (soil moisture deficits are not enough to limit transpiration) (Křeček et al. 2010). In similar climate, Willis (2002) reported increasing water yield (approx. 18–20% of the gross precipitation) by the shift from coniferous stands to grass. Reviewing large number of experiments, FAO (2006) reported 1.5–2% reduction of water yield per each 10% catchment recovery with mature evergreen forests. In the zones delineated in watersheds Josefův Důl and Souš (Fig. 16.4), water yield could grow up to 120 mm annually by forestry practices (forest rotation, thinning, introducing broadleaves). In this zone also, limited clear-cut could be accepted when not disturbing the soil surface.

4.1.4 Zone of Significant Fog Drip

In the catchments Josefův Důl and Souš, the fog drip was found significant in elevations above 900 m (21 and 33% of the catchment area) (Fig. 16.4). Thus, Křeček and Palán (2015) reported the rise of annual water yield by 88–106 mm (11–13%). On the other hand, in conditions of the acid rain impact (pH below 5.6), acid substances in the fog or low clouds can lead to the decline of water quality: mean observed annual load of sulphur and nitrogen by the fog drip was 1975 and 1080 kg/km^2, respectively (55 and 48% of total sulphur and nitrogen deposition registered in the bulk). Therefore, in elevations of the significant fog drip, dense coniferous stands could increase water yield, but, on the other hand, to decline the quality of water. At the current state of acidification, in the fog-drip zones (Fig. 16.4), the load of atmospheric sulphur and nitrogen could be decreased by 33% at the control of the forest canopy (or 18% at the shift of spruce stands to grass).

4.1.5 Wetlands

The peat bog *Nová Louka* of 32 ha in the JD basin belongs to the category of nature conservations according to the Nature Conservation Act (114/1992 Coll.) with restricted human activities. The area of 70 ha (5% of the catchment area) was

Table 16.3 MCDA process: performance values *a* by weights of importance *w* and criteria *C*

Criteria C	Weights w	Rating				Performance values			
		A_1	A_2	A_3	A_4	a_1	a_2	a_3	a_4
C_1. Water quality	30	5	4	2	1	150	120	60	30
C_2. Water yield	10	2	2	3	3	20	20	30	30
C_3. Flood prevention	20	4	3	2	1	80	60	40	20
C_4. Biodiversity	20	4	3	2	1	80	60	40	20
C_5. Timber yield	10	2	2	4	5	20	20	40	50
C_6. Recreation	10	3	3	2	2	30	30	20	20
Totals	100					380	310	230	150

delineated there (Fig. 16.4) including also wetland forests of its neighbourhood. Beyond the target of biodiversity (wide range of plants and animals), this plot should be considered also in a broader watershed context (catchment stabilisation, control of water quality etc.) (Křeček and Haigh 2006).

4.2 *Structured Forestry in Water Supply Catchments*

Core participants of the stakeholder dialogue on forestry practices to be preferred in water supply catchments included sectors of water management, forestry and nature conservation. Next, municipalities, recreation business and NGOs were involved (Chalupa and Křeček 1995). The main goal was to find optimum forestry practices supporting the water resource recharge in reservoirs Josefův Důl and Souš (according to the Water Act 254/2001 Coll.). Consequently, preferences in forestry (Forest Act 289/1995, Coll.), nature conservation (Nature Conservation Act 114/1992 Coll.), local business and environment were taken into account. In the MCDA procedure, decision criteria *C* (main stakeholder interests) included: C_1 – water quality, C_2 – water yield, C_3 – flood prevention, C_4 – biodiversity, C_5 – timber production and C_6 – recreation potential. The discussed alternatives *A* of watershed management involved: A_1 – five forestry zones (Table 16.2), A_2 – two zones (riparian buffer strips and fog drip zones), A_3 – commercial forestry with only riparian buffer strips, and A_4) commercial forestry with environmental sound harvest technology (required in "protective forests" of headwater regions). According to the weight of importance *w* (ordinal 5-point scale) for each criterion (formulated on the stakeholder dialogue), the calculated performance values *a* are in Table 16.3. The concept of five structured forestry zones (A_1) obtained the highest performance value a_1, followed by the alternative two-zone approach (A_2 – respecting only riparian buffers and fog drip sequences) with the score a_2.

The risk of uncertainties by the MCDA method, both internal (lack of complete knowledge by decision makers judgements) and external (unsatisfactory knowledge of watershed processes and forestry action consequences), depend on probability and magnitude of adverse effects (Triantaphyllou 2000). In this study, the applied

MCDA (Table 16.3) involve uncertainties reflecting the probability of benefits considered by the structured forestry practices in the delineated zones (Fig. 16.4). While environmental benefits of riparian buffer zones are well documented, the consequences between forestry practices and fog-drip or evaporation are complicated with possible compensations by the understory vegetation or levels of the air pollution. The risk of uncertainties in effects of the structured forestry practices was evaluated by the expert team: from ±5% (riparian buffers and wetlands), ±10% (soil protection), ±20% (evapotranspiration control) and ±30% (fog drip regulation). The priorities should be modified with a progress in air pollution control, and changes in the climate. Thus, the main goal could shift from water quality to water yield, or, minimum seasonal runoff. Moreover, the future climate could also affect the distribution of climax forest zones in studied watersheds (Fig. 16.3). Therefore, the weighted uncertainty risk by accepting the alternative with five zones (A_1) in watersheds Josefův Důl (JD) and Souš (S) is 20 and 22%, respectively. Additionally, there are still principle doubts: the criteria (C) and estimated weights of importance (w) reflect on the recent level of acidification. The priorities could be modified with a progress in air pollution control, and also with expected changes in the climate, probably, from water quality to water yield, or to a minimum seasonal runoff. Moreover, the future climate could also affect the distribution of climax forest zones in studied watersheds (Christensen 2005; Sun et al. 2015).

The acidified environment in the reservoirs Josefův Důl and Souš already brought out some economic consequences. Evident are, particularly, the rising operational costs by the treatment of drinking water, or, the reconstruction of the water treatment plant in Bedřichov (using raw water from the Josefův Důl reservoir) in amount of approx. 120 million CZK (Křeček and Palán 2015). However, the relatively high risk of the uncertainties mentioned above still complicates an adoption of the payment for ecosystem services in their watersheds.

5 Conclusions

Results of this study confirmed the importance of mountain forestry in water resources control and drinking water supply. The concept of 'protective forests' of the *Protected Headwater Areas* in the Czech Republic (controlling the forest-land percentage, drainage and clear-cut areas) is not adequate to the present and future changes in mountain environment.

In the water supply catchments, a system of structured forestry practices should respect five main priorities: protection of surface waters (buffer strips along reservoirs and streams), conservation of biodiversity (wetland-spots), soil protection (slopes over 30%), fog drip maintenance (elevations above 900 m), and evapotranspiration control (rest of the catchment). The delineation of these partial zones within a watershed depends on morphological configurations.

In conditions of significant 'acid rain' impacts, the environmental services should modify the acid atmospheric load of the catchment by controlling the canopy area and its roughness. But, the priorities could be corrected with a progress in the air pollution control, or, changes in the climate.

An adoption of the payment for environmental services is still complicated with a relatively high risk of uncertainties, increasing with the changing climate and non-point pollution. Consequences between the structured forestry practices and hydrological outputs also face problems by fully understood roles of the herbaceous vegetation or soil conditions.

Acknowledgements This research was supported by the Earthwatch Institute (Oxford, UK, Project on Mountain Waters of Bohemia), and by the Czech Technical University in Prague (Project SGS16/140/OHK1/2T/11).

References

Agnew, M., & Goodess, C. (2016). *Stakeholder guidance note. CIRCE RL11: Integrating case studies*. Norwich: Climatic Research Unit, University of East Anglia.

Cambi, M., Certini, G., Fabiano, F., Foderi, C., Laschi, A., & Picchio, R. (2015). Impact of wheeled and tracked tractors on soil physical properties in a mixed conifer stand. *iForest, 9*, 89–94.

Chalupa, M., & Křeček, J. (1995). *National debate on management of mountain watersheds in the Jizera Mountains (Czech Republic)*. Prague: Forest-Water Foundation.

Christensen, J. H. (2005). *Prediction of regional scenarios and uncertainties for defining European climate change risks and effects (PRUDENCE)*. Final Report, Danish Meteorological Institute, Copenhagen.

Daniels, B., McAvoy, D., Kuhns, M., & Gropp, R. (2004). *Managing forests for water quality: Forest roads* (Utah Forests Facts, NR/FF/010). Utah State University Forestry Extension.

Dudley, N., & Stolton, S. (2003). *Running pure: The importance of forest protected areas to drinking water, The arguments for protection series, research report of the World Bank and WWF Alliance for Forest Conservation and Sustainable Use*. Gland: WWF International: IUCN The World Conservation Union, cop.

FAO. (2006). *The new generation of watershed management programmes and projects, FAO Forestry Paper 150*. Rome: Food and Agriculture Organization of the United Nations.

FAO. (2008). *Forests and water, FAO Forestry Paper 155*. Rome: Food and Agriculture Organization of the United Nations.

FMI. (2015). *Protection of water resources* (In Czech). Operating Methodology in Forestry, Forest Management Institute, Brandýs n.L. (Czech Republic).

Foy, C., Arabi, M., Yen, H., Asce, A. M., Gironás, J., & Bailey, R. T. (2015). Multisite assessment of hydrologic processes in snow-dominated mountainous river basins in Colorado using a watershed model. *Journal of Hydrologic Engineering*, 20, 04015017.

Gray, D. H., & Sotir, R. B. (1996). *Biotechnical and soil bioengineering slope stabilization: A practical guide for erosion control*. New York: Wiley.

Hawes, E., & Smith, M. (2005). *Riparian buffer zones: Functions and recommended widths*. New Haven: Eightmile River Wild and Scenic Study Committee, Yale School of Forestry and Environmental Studies.

Kelly-Quinn, M., Cruikshanks, R., Johnson, J., Matson, R., Baars, J. R., & Bruen, M. (2008). *Forestry and surface water acidification*. Research Report, University College Dublin.

Körner, C., & Ohsawa, M. (2005). Mountain systems. In R. Hassan, R. Scholes, & N. Ash (Eds.), *Ecosystems and human well-being: Current state and trends*. London: Island Press.

Křeček, J., & Haigh, M. J. (2006). *Environmental role of wetlands in headwaters, NATO science Series, 63*. Dordrecht: Springer.

Křeček, J., & Hořická, Z. (2001). Degradation and recovery of mountain watersheds: The Jizera Mountains, Czech Republic. *Unasylva, 52*, 43–49.

Křeček, J., & Hořická, Z. (2006). Forests, air pollution and water quality: Influencing health in the headwaters of Central Europe's "Black Triangle". *Unasylva, 57*, 46–49.

Křeček, J., & Palán, L. (2015). Effects of fog precipitation on water resources and drinking water treatment in the Jizera Mountains, Czech Republic. *Civil Engineering Journal, 10*, 1–7.

Křeček, J., Nováková, J., & Hořická, Z. (2010). Ellenberg's indicator in water resources control: The Jizera Mountains, Czech Republic. *Ecological Engineering, 36*, 1112–1117.

MEA. (2005). *Ecosystems and human well-being: Synthesis, Millenium Ecosystem Assessment*. Washington, D.C: Island Press.

NIVA. (2013). *Effects of long range transported air pollution (LRTAP) on freshwater ecosystem services, ICP-Waters report 115/2013*. Oslo: Norwegian Institute for Water Research.

Parkyn, S. (2004). *Review of riparian buffer zone effectiveness, Ministry of Agriculture and Forestry, Technical Paper No: 2004/05*. Wellington: MAF Information Bureau.

Peřina, V., Krečmer, V., Šach, F., Dressler, M., Réman, Z., Křeček, J., Skýpala, J., Beneš, J., & Jařabáč, M. (1977). Advanced research project to demonstrate the role of forests in water resource recharge of drinking water reservoirs (In Czech). Report of the Forest Research Institute, Opočno Research Station (Czech Republic).

Reed, M. S. (2008). Stakeholder participation for environmental management: A literature review. *Biological Conservation, 141*, 2417–2431.

Sun, F., Yihe, L., Bojie, F., & Jian, H. (2015). Hydrological services by mountain ecosystems in Qilian Mountain of China: A review. *Chinese Geographical Science, 26*, 1–14.

TEEB. (2010). *The economics of ecosystems and biodiversity*. London: Ecological and economic foundations. United Nations Environment Programme. Earthscan Ltd..

Tolasz, R., et al. (2007). *Climate atlas of Czechia*. Prague: Czech Hydrometeorological Institute.

Triantaphyllou, E. (2000). *Multi-criteria decision making methods: A comparative study, Applied Optimization Series, 44*. Dordrecht: Springer.

Viviroli, D., Dürr, H. H., Messerli, B., Meybeck, M., & Weingartner, R. (2007). Mountains of the world, water towers for humanity: Typology, mapping, and global significance. *Water Resources Research, 43*, W07447.

Willis, K. G. (2002). *Benefits and costs of forests to water supply and water quality. Social and Environmental Benefits of Forestry*. Report to Forestry Commission, Phase 2, Centre for Research in Environmental Appraisal and Management, University of Newcastle.

Chapter 17
Impacts of Reclaimed Opencast Coal-Land on Headwater Ecosystem Services

Martin Haigh

1 Introduction

Opencast coal-mining is a controversial and always damaging land-use choice, to the degree that it is now suspended in some countries (e.g. National Assembly for Wales 2015; ECO 2016; Davenport 2016). Mine operations cause many offsite problems that include pollution to air through dust, water contamination through leachate and sediment release, to transport systems through traffic (because of the volume of material moved) and to visual quality, because of the unsightly appearance of the mine workings (Younger 2004). Of course, the use of coal is also famously polluting and it is one of the key sources of greenhouse gas emissions (Metz et al. 2007).

Land reclamation processes can also be problematic. Pollution is often caused by the release of chemically charged sediments into neighbouring water-courses, not least the runoff of nutrients applied large-scale during land reclamation plantings, and by long-term visual pollution caused by any shortfall in the imagination, skill and resources provided to land reclamation engineers (Kuter 2013; Haigh 2000). However, once the land has been 'reclaimed', not all of the impacts of former opencast coal-mine sites are wholly negative. This chapter explores the post-land reclamation impacts of opencast (surface) coal-mining on the ecosystem services provided by headwater catchments in the Southeast Wales.

M. Haigh (✉)
Department of Social Sciences, Oxford Brookes University, Oxford, UK
e-mail: mhaigh@brookes.ac.uk

© Capital Publishing Company, New Delhi, India 2017

J. Křeček et al. (eds.), *Ecosystem Services of Headwater Catchments*,
DOI 10.1007/978-3-319-57946-7_17

2 Land Reclamation After Surface Coal Mining—A Typology

Inevitably, opencast (i.e. surface) coal mining causes the total destruction of the land that existed before the mine, including the bulk of its ecological properties and any cultural and or historical artefacts that were not consciously removed, like the famous church-on-wheels in Most, N. Bohemia (Glassheim 2007). Similarly, despite any attempt to preserve pre-existing top-soils, land reclamation creates an entirely new landscape: new landforms, new watercourses, new vegetation and new soil constructed from new materials, many imported for the purpose (Haigh 2000).

This new landscape is the focus of this chapter; not what happens during mining or reclamation but what remains thereafter. This recognises that this new landscape is the greatest impact and effect of opencast coal-mining. The logic is simple. Even a small economic rent (or benefit) multiplied through a very long (infinite) timescale accumulates to outweigh one that is of short duration. Mining and reclamation processes rarely extend across more than a few years. Reclaimed land is supposed to sustain or, ideally, self-sustain and improve its qualities across many centuries (Haigh 2000). Soil generation, the creation of a self-sustaining soil system, has been called *"the primary driver for a durable future. Without it there can be no viable economy"* (Todd et al. 2010). However, this key issue rarely receives the attention it deserves (Table 17.1).

Heyd (2007) admits three alternatives for land reclamation agencies: to leave the land alone, to restore it to a condition resembling its original state, or to transform them into 'artworks' and goes on to wonder whether such 'artistic reclamation' can really redeem these blots on the landscape (cf. Kuter 2013). In practice, land reclamation schemes may be classified into one of five categories, examples of which are visible in most of the coalfields of the UK as elsewhere. These categories are: (1) Sustainable Reclamation, (2) Self-sustainable Reclamation, (3) Cosmetic Reclamation, (4) Failed Reclamation, and (5) No Reclamation (Table 17.1). Generally, the news media and environmental literature portray all of the impacts of coal-mining, not least the provision of coal for combustion and the consequent release of greenhouse gas etc., as negative in environmental terms and, by extension, this includes the post-mining landscape and reclaimed lands, especially those where reclamation is imperfect. In reality, of course, while very many of those impacts are negative, some seriously negative, there can also be positive consequences from the presence of former opencast (surface-mined) coal-lands in headwater catchments, some of which are suggested in Table 17.1. The Ecosystem Service approach has many problems, some of which are critiqued below, but it does provide a crude means of evaluating the impacts of particular landscape types. This chapter aims to scope the effect of each type of reclamation on ecosystem services using examples from the headwater, 'Heads of the Valleys', region of Southeast Wales, mainly the Boroughs of Torfaen and Blaenau Gwent, as its principal exemplars.

Table 17.1 Classification of observed land reclamation outcomes

1.	*Sustainable Land Reclamation* projects are designed either to permit a new economic land-use that will create sufficient investment for repair and maintenance or upon the assumption that the local community will take on future costs of the repair and maintenance, especially of runoff control structures and channel-ways (e.g. Ebbw Vale Garden Festival site or Cwm Llamarch, Blaenau Gwent).
2.	*Self-sustainable Land Reclamation* projects are designed with the intent to return the geosystem to the control of natural processes. This 'wilding' or 'wild-becoming' approach recognises that only nature is capable of creating self-sustaining geosystems (e.g. Cradle for Nature Project, Torfaen, see: Plamping et al. 2016).
3.	*Cosmetic Land Reclamation* projects are designed to meet legal specifications, especially to be in peak condition on the date the project is given its final inspection and handed over to other owners. The fate of these sites depends largely upon chance and they may end up in categories 1, 2 or 4.
4.	*Failed Land Reclamation* projects emerge from any of the three categories above where either repair and maintenance has been neglected or the original design was inadequate (e.g. Pwll Du, Torfaen). Sometimes, the stressed geoecological circumstances of failed or failing sites providte habitats for unusual ecological communities (e.g. Waun Hoscyn, Torfaen).
5.	*No Reclamation*. There are many orphan projects where the land was left very much as it was when the mine operation ceased (e.g. Henryetta, Oklahoma). Not always is this 'dishonest' neglect (East Pit, Margam, see: Monbiot 2015). For example: sometimes land, left much as it was when mining ceased, provides informal or by-design zones for recreational use by mountain bike and motorcycle enthusiasts (e.g. Pwll Du, Torfaen and, by design, at Buffalo Rock, Illinois, a site later reshaped for art (Friederici 1992). These sites may remain unvegetated and/or toxic wastelands but where, by chance, the surface spoils include materials that are not inimical to plant colonisation, they can be reclaimed by nature and become important reserves for wildlife.

3 Ecosystem Services: A Problematic Concept

It may be said that 'Ecosystem Services' are the services that the environment provides to the life-support system that is Nature, things that are normally taken for granted such as clean air, water, fertile soil, fisheries, land, etc. This concept is, of course, anthropocentric (utilitarian) rather than ecocentric; it asks 'what has our environment ever done for us and why should we pay'?

However, the concept of ecosystem services emerged in the 1970s as a way of drawing attention to the wider economic benefits of ecological conservation (Westman 1977). Of course, there are fundamental problems in linking any ecological system with the short-term thinking that is inherent to socio-economic and political analysis—even if the motivation is the defence of that ecosystem's qualities, rather than simply justifying its exploitation or destruction, which is often the case. Inevitably, the Ecosystem Service concept is a two-edged sword. Its problems were exemplified by a fore-runner of this notion, 'Contingent Valuation', which asked the nonsense question—'how much is it worth to you to keep this aspect of your environment—how much the blue sky, how much the soil microorganism', etc. (cf. Hausman 1993)? However, inevitably such ideas are popular with economists and planners seeking ways of assessing the economic value of a particular aspect of the land, for whatever reason (Gómez-Baggethun et al. 2010).

The Ecosystem Services approach was much popularised by the Millennium Ecosystem Assessment (Millennium Ecosystem Assessment 2003). This began by trying to focus on human well-being and the ways that this is affected by human actions, arguing that: *"The ways in which ecosystems are affected by human activities will have consequences for the supply of ecosystem services—including food, fresh water, fuelwood, and fibre—and for the prevalence of diseases, the frequency and magnitude of floods and droughts, and local as well as global climate. Ecosystems also provide spiritual, recreational, educational, and other nonmaterial benefits to people. Changes in availability of all these ecosystem services can profoundly affect aspects of human well-being—ranging from the rate of economic growth and health and livelihood security to the prevalence and persistence of poverty"* (Millennium Ecosystem Assessment 2003). However, while accepting that ecosystems have more values than can be expressed in conventional economic terms, later the report capitulates to the easy economic option arguing that: *"The purpose of economic valuation is to make the disparate services provided by ecosystems comparable to each other, using a common metric"* based on "use-values" on "non-use" values in order to *"assess the relative impact of alternative actions so as to help guide decision-making*" (Millennium Ecosystem Assessment 2003). The report's final paragraph does concede the self-evident fact that *"Of course, the importance of ecosystems goes beyond their role for human wellbeing"*. Indeed, it is more than this, as Baird Callicott (1994) points out: the *"difference between natural and moral limitations on human behaviour entails a fundamental consequence"*, possibly enforceable by law in the latter case but, potentially, more dramatically enforced by Nature in the former. However, the core message of the Ecosystem Services literature is about the commodification of the natural world; it remains an attempt to squeeze consideration of its role into a narrow, short-term, 'me'-centred, economic valuation. As such, the whole ecosystem services enterprise resolves to a crude *"campaign to define the world as an immense collection of service commodities"* (Robertson 2012), a dispiriting and, ultimately, self-destructive worldview.

This said, the approach may have a value in land reclamation, which depends hugely on the disbursement of appropriate funds to ameliorate the damage already done by human actions. So, while this chapter takes only the first steps toward assigning cash-value to services that the reclaimed coal-land environment provides, it recognises that much poor quality land reclamation is caused by underfunded reclamation (Ellison and Griffiths 2014). Its focus is the application of a kind of medical-checklist aiming to describe and diagnose the several and widely disparate ways that different ecosystem services and the welfare of the whole habitat are affected by the presence of reclaimed opencast coal-lands, especially in southeast Wales.

4 Ecosystem Services in Headwater Catchments

The ecosystem services concept links the processes and species of natural ecosystems with human welfare by detailing the benefits or disbenefits that human populations derive from their functioning (Costanza et al. 1997). This species of 'Ecological Economics' evaluates benefits that are tangible but also intangible in economic terms, such as aesthetic, cultural or spiritual value. The Millennium Ecosystem Assessment (2005) emphasised biodiversity but classified Ecosystem Services into four classes: Supporting (nutrient cycling, soil formation, primary production, etc.), Regulating (climate, flooding, water quality, disease, etc.), Provisioning (food, freshwater, wood and fiber, fuel, etc.) and, finally, Cultural (aesthetic, spiritual, educational, recreational). In their view, Provisioning had the greatest impact and the greatest potential for 'socio-economic mediation', Regulating came next with still great impacts but less potential for 'socio-economic mediation', while Cultural had the weakest impacts and potential. They proposed no direct links between Supporting activities and human well-being, which is assessed in terms of Security, Material well-being and 'Good Social Relations'. The scheme provided the foundations for the later—the European Common International Classification of Ecosystem Services (EEA 2016; CICES 2016)—and for TEEB (2010) 'The Economics of Ecosystems and Biodiversity', a rival system built on the same four foundations.

TEEB (2010) uses an exploratory tiered approach that runs from 'Recognizing' value, the emphasis here, which may or may not include monetary value but certainly includes biodiversity values, through 'Demonstrating' value, for example through cost benefit analysis, to 'Capturing' value, which involves creating mechanisms that incorporate the values of particular ecosystems into decision making through "incentives and price signals". TEEB argues that sound ecosystem and biodiversity management can be enhanced by including its estimates of natural capital with governmental and business accounting schemes, arguing, for example, that investing U$ 45 billion into protected areas would produce U$5 trillion of ecosystem service benefits. Inevitably, the methods for generating such figures are highly suspect. Of course, the last two phases are most problematic and TEEB's plea for 'mercy' towards Nature will no doubt, normally, be dwarfed by competing financial incentives such as profit. Mining and forests both feature among TEEB exemplars (TEEB 2010). Haines-Young and Potschin (2013) estimates that there is a 75% overlap between TEEB and CICES (2016). However, this also applies to the criticisms of a system, memorably and aptly dubbed: "terrible economics, ecosystems and banking" by Splash (2011), who argues that "short-termism, anthropocentrism, systems control and narrow species preferences seem to dominate" (Splash 2011). TEEB's greatest weakness is that it allows the financial and banking sector to 'capture values' and that it allows human preferences for nature and species to affect their valuation. Human preference is very different to the actual functional value of a species or other ecosystem attribute to a natural geoecological system. Nevertheless, this chapter uses the TEEB classification of ecosystem services for the evaluations of the role of ecosystem services provided by Reclaimed Coal-land and of the ways they differ from those of the pre-mining environment (Table 17.2).

5 Discussion

The key feature of reclaimed opencast coal-land is that it is completely different from the land that existed before mining. As Larondelle and Haase (2012) argue: *"it is not practical or possible to restore nature or ecosystems that existed in the pre-mining rural landscape; instead, one can identify future land use solutions that provide places to live, produce food, and supply recreational areas or that can mitigate climate change and temperature rise…"*. In other words, it is necessary to make the best of what can be done in the new situation and work with the anthropogenic landscape and the so-called 'novel ecosystems' it may develop or that may be developed upon it (cf. Collier 2014).

The negative impacts of opencast coal mining on environment tend to emerge from their large land-take, their transformation of the geosystem (including their modification of topography as well as surface and ground water flow), their poor quality soils and the possible creation of contaminated land, hence: their release of pollutants and contaminants, especially in waters, their disturbance of both flora and fauna and complete obliteration of previous cultural features and consequent relocation of residential areas and transport routes (Kuter 2013). The success of the reclamation in economic, aesthetic and biological, including biodiversity, terms is dependent upon the goals, resources, skills and capabilities of the project team in charge of restoration, upon the specific and local challenges and properties of the mine spoils (especially those left near the ground surface), and upon the aftercare and management afforded to the site (especially in its first decades of existence).

Most critically, it is necessary to discover and set in place effective and sustainable post-mining land uses that benefit both future generations in general and future landowners in particular. The character of reclamation can be affected by many factors beyond the resource of the reclamation contractor. Ideally, opencast mining involves a compact between the mine operators and the local community, who, ideally, much influence the land reclamation project's goals (Haigh 1999). These goals should be measured against the actual capacity of the land to sustain reclamation to a particular land use. It is not unusual for mining contractors in Wales to offer to give the local community exactly what it wants—in exchange for permission to mine. However, in many cases, even when the reclamation is properly funded and delivered effectively, the results may not be sustainable because of maintenance costs, as was the case, in SE Wales of several features associated with the reclamation work performed for the Ebbw Vale Garden Festival (1992), such as its "Stairway to the Stars" cascade (Parker 2008). The UK's Mineral Planning Guidance (MPG7 1989) countenanced three different after-uses for opencast coal-land—agriculture—especially grassland (DEFRA 2005), forestry and amenity, meaning for recreation. The later UK Government's Mineral Technical Advice Note 2: Coal

Table 17.2 Ecosystem Services according to TEEB (2010)

General criteria	For reclaimed coal-land and SE Wales
Provisioning Services are ecosystem services that describe the material outputs from ecosystems that include food, water and other resources	
Food: Ecosystems provide the conditions for growing food—in wild habitats and in managed agro-ecosystems	Reclaimed land is, generally, of very low productivity but it can support some grazing and forestry. Productivity on land restored as lower grade agricultural land can be only slightly lower than that of comparable undisturbed lands where modern restoration techniques have been applied, although soil compaction and the need for special training of farmers remain problems now as in Tomlinson's day (Tomlinson 1980). In West Virginia, where mine waters are severely affected by AMD (Acid Mine Drainage, *see* Haigh 2003), attempts have been made to rear trout in the waters at AMD treatment plants (Miller 2008). In places like S. Wales, where AMD is not a problem, aquaculture remains a possibility.
Raw materials: Ecosystems provide a great diversity of materials for construction and fuel	Reclaimed lands contribute little by way of raw materials, their coal is exhausted. However, these lands are all considered suitable for biofuel development, perhaps using willow or grasses (cf. Litterick and Booth 2009). Already, several sites have been scoped (Royal Society 2008).
	In line with UK National Policy, attempts were made to use reclaimed opencast coal-land for forestry (e.g. Rhigos, near Merthyr Tydfil, known locally as Bonsai Woods, and Maesgwyn Cap, near Glynneath) but tree growth has been very slow and output of little commercial value. Maesgwyn, which closed in the late 1980s,t is also famous for having been worked by the world's largest walking dragline, the Ransomes and Rapier W1800. For many years, Maesgwyn was used by the Forestry Commission (UK) as a test-bed for the disposal and utilisation of sewage sludge as a fertiliser, despite the risks of land contamination.
	Between 2009 and 2011, this and neighbouring sites were redeveloped as wind farms.
Fresh water: Ecosystems provide surface and groundwater	Reclaimed land provides water, mainly as surface runoff, which may or may not be contaminated. In SE Wales, compacted layers in the immediate subsurface of former opencast coal land convert incident rainfall to runoff very quickly. Of course, opencast spoils are locally contaminated (and widely 'border-line contaminated') with metals, especially where the surface cover of applied topsoil and vegetation is absent, perhaps because of erosion. However, high rainfall and rapid through-flow ensures dilution and that the offsite impacts are local.
	In SE Wales, the main impact of reclaimed opencast coal-lands on river water quality is likely to come through diffuse pollution from the zone of aeration and water table fluctuation zone in the body of the buried spoils. This zone may be mainly detached from the surface hydrological system and deprived of rainwater infiltration by the impermeability of the surface layers on these sites. However, the lateral percolation of groundwater from offsite through the crushed rocks of the spoil is very likely to pick up metals from chemically active inclusions, often former deep-mine spoils, within the opencast mine fill. Even so, the effect on river water quality is not great. It may involve slightly higher flood peaks, elevated levels of iron, and slightly elevated levels of lead, perhaps arsenic, copper and zinc, as well as ammonia N, although to be fair, opencast sites are only one among several possible sources for such metals, which include natural ground waters and other types of industrial waste (Haigh and Kilmartin 2015).

(continued)

Table 17.2 (continued)

General criteria	For reclaimed coal-land and SE Wales
Medicinal resources: Plants used as traditional medicines and by the pharmaceutical industry	Reclaimed lands offer little in these respects although, in the case-study area, they provide illegal recreational drugs in the form of 'Magic Mushrooms' (e.g. *Psilocybe semilanceata* (Fr.), P. Kumm. (1871)).
Regulating Services are the services that ecosystems provide by acting as regulators, e.g. regulating the quality of air and soil or by providing flood and disease control	
Local climate and air quality regulation: Trees provide shade and remove pollutants from the atmosphere. Forests influence rainfall. Carbon sequestration and storage: As trees and plants grow, they remove carbon dioxide from the atmosphere and lock it away in their tissues	Reclaimed opencast coal-land in Wales has negative impacts on both water quality and human health. Dust can be a major problem during both opencast working and reclamation as at the Ffos-y-fran Land Reclamation Scheme, Merthyr Tydfil (Dedikova 2014). Of course, coal from opencast sites in Wales is used to run coal-fired power stations that contribute to Climate Change, which is one reason why there is currently a moratorium on new opencast mining and protests mounted at those in operation like Ffos-y-fran (Morris 2016). However, tree planting on former reclaimed opencast land is an effective way of sequestering carbon as trees and new soils, albeit, in much smaller amounts than were released through coal burning (Plamping et al. 2016). Planting trees can also help mitigate metal contamination of the mine-spoils (Desai 2013).
Moderation of extreme events: Ecosystems and living organisms create buffers against natural hazards such as floods, storms and landslides	Reclaimed coal-lands tend to be much less effective than the lands they replace in the moderation of runoff. In Appalachia, Ferrari et al. found that flood magnitudes increase linearly with mined area. In Wales, where the spoil surface is largely impermeable, there are short lag-times between rainfall and runoff. Runoff occurs as surface and near-surface flows through a shallow active topsoil layer. The result is a steeply rising, flashy, hydrograph and increased flood peak. Similarly, when rainfall ceases, there is little soil water storage, so the hydrograph falls quickly and flows diminish to nothing soon after rainfall ceases, which is why most channels on former opencast lands are ephemeral streams (Haigh and Kilmartin 2015).
Waste-water treatment: Micro-organisms in soil and in wetlands decompose human and animal waste, as well as many pollutants	Elsewhere in the world, retention ponds are used in attempts to moderate and treat runoff from former opencast mines, often employing reed beds for the purpose. However, there is little such activity in SE Wales, although lakes feature on several of the sites reclaimed for recreation, such as Garn Lakes and Bryn Bach Parc. As for the soil, its microbial ecosystem tends to be severely depleted and, estimates suggest, it may take more than 100 years for the soil ecosystem to develop sufficiently to recycle carbon effectively. This means that, especially under forested sites, there tends to develop a surface soil horizon of undecomposed organic matter. This is also why these Anthrosols are also frequently classed as special kinds of immature Entisols (e.g. Hartman et al. 2004; Sencindiver 1977).

(continued)

Table 17.2 (continued)

General criteria	For reclaimed coal-land and SE Wales
Erosion prevention and maintenance of soil fertility: Soil erosion is a key factor in the process of land degradation and desertification	The value of these sites for erosion control depends largely on the success of the revegetation (Haigh 2000, Haigh 1978). In SE Wales, sediment production rates can be high from exposed spoil surfaces, which are often on steeper slopes (as at Pwll Du, north of Blaenavon), from active gully channels during the early years of restoration (as at Waun Hoscyn, south of Blaenavon), and from sites affected by acidity where the vegetation cover is being destroyed, such as at Cwm Llamarch, Blaenau Gwent (Haigh 1978, 1980).
Pollination: Some 87 out of the 115 leading global food crops depend upon insect pollination including important cash crops	Once again, the contribution of reclaimed coal-land depends on either good design or the fortuitous success of Nature. In SE Wales, the quality of reclaimed land ranges from completely barren land through prairie-style open grassland to wooded parkland. Blaenau Gwent (2015) has sown pollinator patches on roadside verges and similar work could be done on reclaimed sites. However, many support flowering plants such as Ling, Gorse and *Gallium saxatile*.
Biological control: Ecosystems are important for regulating pests and vector-borne diseases	Reclaimed sites are young ecosystems associated with disturbed ground with low natural fertility, which often have been amended with organic material, seed-mixtures and plantings of trees, etc. Sites undergoing reclamation treatment and degraded reclaimed lands provide ecological niches that, potentially, favour invasive species. Fortunately, studies at the Varteg Opencast suggest this problem is not severe. The chief invader is Ragwort (*Senecio* spp.), which is toxic to livestock (National Assembly for Wales 2006).
Habitat or Supporting Services underpin almost all other services. Ecosystems provide living spaces for plants or animals; they also maintain a diversity of different plants and animals	Commonly, not inevitably, reclaimed land offers less diverse living spaces with lower bio-productivity than the lands they replace. For example, objecting on behalf of the Gwent Wildlife Trust to the proposed opencast coal mine at Varteg, Torfaen, Jones (2014) emphasises the pre-mining area's coherent habitat mosaic that includes acid unimproved grassland, heathland, marshy grassland and acid flushes as well as post-industrial land supporting a diverse range of native and archaeophyte non-woody plant species, at least 20 indicator species. The reclaimed post-mining landscape on Varteg is relatively bland in topographic terms; however, many of these habitats are to be found on opencast coal land reclaimed from the 1960s in the same area. The proposed mining contractors response to detailed ecological surveys on the site is that disruption would be temporary and of minor negative long-term significance, moderate in the case of the acid flushes, but positive in the case of the ponds and ditch habitats that would follow site remediation to plan (Harmer Ltd 2014). Of course, sadly, not all site remediation proposals actually go to plan and there have been recent 'rip and run' scandals in Wales (Monbiot 2015).

(continued)

Table 17.2 (continued)

General criteria	For reclaimed coal-land and SE Wales
Habitats for species: Habitats provide everything that an individual plant or animal needs to survive. Migratory species need habitats along their migrating routes	Commonly, not inevitably, reclaimed land has a much lower biodiversity than the land it replaces. The reclaimed post-mining landscape is a relatively bland mix of (often artificially introduced) species. However, where a reclamation is failing, the stressed habitats that result on otherwise abandoned land can offer niches to unusual species and, where a patchwork of habitats has developed or been developed, places that meet the needs of migratory birds (Blaenau Gwent County Borough 2015). Sometimes such habitats are created by design. The Welsh Government has launched the Wales Nature Recovery Plan (2014–2020) that aims to reverse the national decline in biodiversity, using reserves on post-industrial sites, including Parc Slip near Bridgend. In Blaenau Gwent County Borough (2015) such sites support species including: Lapwing, Peregrine Falcon, Brown Hare, Blue Tailed Damselfly, A Lady's Mantle and
	Peltigera Lichen species. Breeding birds identified on the former opencast land at Varteg include Skylark, Meadow Pipit, Wheatear and Winchat as well as Red Grouse, Dunnock, Wren and Reed Bunting (Gwent ornithological Society 2005). Lichens thrive on degraded former opencast lands and land improvement is cited as one of the greater threats to such species. A total of 96 lichens, including eight rated nationally scarce, were recorded on the site of the proposed Varteg Opencast site, which includes extensive areas of reclaimed opencast land (Hope 2014). In addition, some of the less successfully reclaimed sites like Pwll Du, north of Blaenavon, support extensive areas covered by unusual cryptogamic soil crusts, including some dominated by filamentous algae.
Maintenance of genetic diversity: Genetic diversity distinguishes different breeds or races, providing a gene pool for developing crops and livestock	Reclaimed opencast sites are not significant reservoirs of genetic diversity, partly because their land surfaces are of recent construction. Their vegetation usually has two components—seeded or planted vegetation from commercial sources or adventive colonisers from neighbouring habitats. They may, of course, provide very unusual habitats, perhaps due to the seepage of acid waters or heavy loadings of metals, which may provide refuges for unusual species.
Cultural Services include the non-material benefits people obtain from contact with ecosystems. These include aesthetic, spiritual and psychological benefits	
Recreation and mental and physical health: Natural landscapes and urban green space have a role in maintaining mental and physical health	Health is a major negative issue during both the mining and reclamation phases of opencast mining. Problems include dust, air quality, noise, vibration, and visual pollution as well as loss of amenity (Chadderton et al. 2011). Many of these vanish when reclamation is completed.
	In SE Wales, much reclaimed land is unenclosed, often common; these sites are often used by recreational walkers. Elsewhere, Llanilid West Revised site, restored 1997–1999, was sold for use as the Valleywood Film Studio. Despite many problems, construction began in 2007 and filming from 2009.

(continued)

Table 17.2 (continued)

General criteria	For reclaimed coal-land and SE Wales
Tourism: Nature tourism provides considerable economic benefits and is a vital source of income for many countries	Reclaimed opencast sites, apart from those developed as parks, such as Bryn Bach Parc, Tredegar, or Garn Lakes, Torfaen, which is also used for an annual County Fayre (Roberts 2013), or those developed for commercial entertainment like golf courses or race-tracks (e.g. Bryn Bach Parc, Tredgar, cf. Most, Czechia), have little touristic value although they often provide open space for recreational walking, biking, off-road driving or dog-walking.
Aesthetic appreciation and inspiration for culture, art and design: Language, knowledge and appreciation of the natural environment have been intimately related throughout human history	Every vestige of the landscape before mining is usually destroyed in the mining and reclamation process. The aesthetic qualities of the subsequent reclaimed coal-lands are usually negative, although sometimes less negative than that of the industrial landscape they replace. However, occasionally, the artistry of landscape design has created landscapes that are attractive and valued by local people. This is especially true of sites restored for recreational purposes such as Bryn Bach Park, Tredegar (Golf, water sports, showground, camping and picnic areas, airstrip, etc., *see* Probert 1986), and the Kays and Kears site, Torfaen, (mined between 1995–1997), which was reclaimed as public open space, a wooded country park called Garn Lakes.
	Sometimes sites restored to Nature also develop acceptable aesthetic qualities. For example the Park Slip Nature Park reclamation (1990–1992), a joint initiative with The Wildlife Trust of South and West Wales (which now supports uncommon invertebrates and breeding lapwings) has been recognised by an award from Campaign for the Protection of Rural Wales (2002).
	More often, these reclaimed lands span the aesthetic range banal-bland to eyesore-ugly (Heyd 2007). This is why, like other local authorities, Torfaen County Borough's Local Development Plan (1996) included all of the former opencast coal-land as 'Environmental Improvement Areas' albeit within the context of larger 'Special Landscape Areas' designated for conservation.
Spiritual experience and sense of place: Nature is a common element of all major religions; natural landscapes also form local identity and sense of belonging	During repeated debates about new mining proposals, often framed as schemes that concern 'Environmental Improvement' or 'Reclamation', supporters regularly comment that coal-mining is part of the regions cultural heritage, its raison-d'etre and so fundamental to its identity. Certainly, large tracts of former opencast coal-land of this area lie within the Blaenavon Industrial Landscape World Heritage Site, Torfaen, which was recognised as a 'cultural' World Heritage Site of 'Outstanding Universal Value' by UNESCO in December 2000, because of its relics from iron, steel and coal working during the early years of the Industrial Revolution. The larger Blaenavon Landscape of Outstanding Historic Interest is also indexed by the Register of Landscapes of Historic Interest in Wales (Countryside Council for Wales and International Council on Monuments and Sites 1998) as *"one of the best preserved, relict industrial landscapes in Wales, containing a vast concentration and diversity of historical and archaeological features. The whole area is covered by early, coal opencasts and it survives as probably the only sizeable, abandoned, multiple period, opencast mineral working in South Wales"* (Torfaen County Borough. 2013). Certainly, opencast coal mining began at Pwll Du, north of Blaenavon, as early as 1942 (Appleyard and Curry 1946) and has continued, sporadically, ever since although at a much smaller scale since the last large British Coal site in the area, Blaenant, Blaenau Gwent, was reclaimed after 1978.

(MTAN2 2009) advises that mining may be allowed where it benefits the community by *"The removal of hazards arising from previous underground working; the restoration of land to public amenity and open space; the creation of areas for nature conservation; the remediation of damaged land; demonstrable employment or economic benefits, or the preparation of land suitable for future built development"*. In the USA, Skousen and Zipper (2014) list after-uses as: (1) farmland, (2) hay land and pasture, (3) biofuel crops (e.g. Skousen et al. 2013) despite their contribution to carbon emissions, (4) forestry, (5) wildlife habitat, and (6) building site development, despite the problems of land contamination. Table 17.3 explores the issues linked to such after-uses.

Many problems arise because of the immaturity of the land with its associated natural systems and their consequent accelerated dynamics that include chemical and physical processes linked to the weathering of the newly fragmented and exposed mine-spoils, which, in this case, contain water-unstable components. In SE Wales, these problems include auto-compaction through accelerated weathering, which effectively reduces the active layer to the depth of any applied topsoil plus a few centimetres of the underlying mine-spoils, where it forms a layer of spoil with densities ranging upwards to that of concrete (Haigh and Sansom 1999). The restoration of a functioning soil system has often been called the primary ecological issue in the restoration of formerly mined lands (Haigh 2000). Unfortunately, this is a low visibility and very time-consuming business requiring timespans that long outlast legal requirements for aftercare. These rarely exceed a single decade where a century might be more appropriate. Comment on the lack of long-term post-restoration monitoring and research, which adds another handicap to land reclamation theory and practice.

In SE Wales, most post-mining land is used as rough grazing or pasture. Often, this land is held in traditional ownership as Common. One problem of this is that land management tends to be neglected or reduced to the lowest common denominator. Commoners have rights to graze on the land, which they defend with vigour, but the notion of accompanying responsibilities is taken less seriously. This is one reason why the most degraded land of the former opencast sites in SE Wales are on Common land, although another is that these lands are, inherently, of poor quality due to their topographic position and climate. Where land has been enclosed and managed more intensively, land degradation is less of an issue, except where acid seeps to the surface or accidental damage occurs due to excess runoff or the trafficking of vulnerable soils or drains. After-use land managers are rarely trained in handling the vulnerabilities of reclaimed land with its thin subsoils and un-associated, often impermeable, buried mine-spoil subsoils.

In the USA, Maczkowiack et al. (2012) recognise that land management is a key factor in the sustainability of pasture, especially stock-rate decisions and the commitment of the land user to respecting the relatively vulnerable condition of the artificial soils on reclaimed sites. Several areas of Blaenant Opencast Reclamation, Blaenau Gwent, as also the neighbouring Walters Mine Reclamation, have suffered damage, over the years, due to overgrazing and trafficking in wet conditions, which allowed wheel ruts to cut through the new soils and exposed the mine-spoils beneath

(Haigh 2010). However, few coal-mine operators think of providing training to after-users of the land, although making such future land-users aware of the special properties of reclaimed opencast land, such as there being just a thin soil layer over, possibly toxic, mine-spoils, would seem to be essential.

The immaturity of the new land and its geoecological system is one reason why, in Appalachia, the preference for after-uses have shifted from forest, originally, through grassland, and then back to forest again with a new emphasis on ecosystem functioning (Burger 2015). In SE Wales, Haigh et al. (2015) propose the use of the ancient and traditional means of long rotation forest fallowing to restore the degraded reclaimed lands of Cwm Afon Lwyd, Torfaen. Their methodology, they describe as building a 'Cradle for Nature', which means using temporary forest plantings as bioaccumulators that will help new, 'natural', self-sustaining, geo-ecological systems establish and develop upon the new land in the shelter of their trees.

Reflecting on Tables 17.1, 17.2, and 17.3, it is possible to summarise the Ecosystem Service contribution made by reclaimed opencast land in SE Wales, with especial reference to the land it replaces. This is summarised in Table 17.4. This shows that, when designed for the purpose and completed effectively, reclaimed opencast coal-lands can have positive benefits for biodiversity and nature conservation, for recreation and amenity, and for industrial and other building development, and that, even when not properly undertaken, reclaimed land can have benefits for recreation, amenity and possibly nature conservation. However, reclaimed opencast sites also make many negative contributions to ecosystem services. These include hydrological issues associated with accelerated runoff, increased flood hazard, reduced water quality and increased contamination, although such impacts are ameliorated by dilution in larger catchments (Haigh and Kilmartin 2015). They also include negative impacts on biodiversity, to nature conservation through habitat destruction, and to agricultural production by reason of these land's relatively low bioproductivity and immature, sometimes contaminated, soils. Inevitably, reclaimed lands replace all of the historical and cultural artefacts in the pre-mining landscape and, usually, not universally, provide instead a bland, artificial-looking replacement of low aesthetic value.

An additional negative is the unreliability of the land reclamation process. It is hard for communities to guarantee that, once the coal and its revenue is gone, the mining contractors will produce work that lives up to their promises. In the UK's Parliament, a local MP described how judges ruled that the Celtic Energy coal-mining company did nothing wrong by leaving a deadly water-filled pit at an old opencast site and advised that Wales' abandoned mines *"will cost hundreds of millions of pounds to fix"* (McCarthy 2015). Nor are such problems only 'things of the past', the risks remain. In 2008, there were eight operational opencast mining sites in the South Wales Coalfield; none in S.E. Wales, which produced about 1.6 million tonnes of coal. (Coal demand in Wales in 2007 was in excess of 6 Mt; some 30% for coke production and blast furnace use, 70% for electricity generation, *see* MTAN2 2009). In 2014, Ellison and Griffiths identified ten active opencast coal mines, plus four in 'aftercare' and three where applications to mine were submitted. This rather

Table 17.3 Critique of some possible alternative for post-mining land-uses

	Post-mining land-use	Issues in S.E. Wales and associated reclamation types
1	Agriculture (arable, pasture or garden)	Productivity tends to be lower than pre-mined land and there are risks of the development of toxicity through the release of metals from the mine-spoil subsoils but the land use is usually sustainable, given appropriate management (Chapman 2001). Subsurface compaction may be a serious problem (Haigh and Sansom 1999).
2	Forestry (commercial or aesthetic/recreational establishment of semi-natural woodland)	Productivity tends to be lower than pre-mined land but there is evidence that the growth of trees helps reduce levels of metal toxicity in the soils (Desai 2013). This land-use is usually sustainable, given appropriate management, and potentially self-sustainable if allowed to revert to Nature. The approach builds soil, sequesters carbon, and may be a good way of preparing former coal-land for other uses as above (Burger 2015; Plamping et al. 2016).
3	Hydroreclamation (water supply, flood water retention, recreational— sailing, swimming, fishing)	Despite the potential for acidity and leaching of toxins, if the water body is large enough, such effects may be reduced by dilution. Water filled voids may be useful, in cosmetic terms, for recreation, also for water retention; such sites may develop self-sustainable biosystems in due course and aquaculture is a possibility (Miller 2008).
4	Recreation (museum or exhibition, sports facility such as racetrack or golf-course, dirt-bike reserve, open green space, hunting reserve)	This is a typical means for the rapid disposal of coal-mine disturbed land, often the approach is cosmetic but, potentially, the land use generates sufficient income for maintenance and hence may be classified as sustainable. For example, the Ffos Las site, Trimsaran, restored between 1997 and 2000 was developed as a new 2.4 km National Hunt horse racing track, Garn lakes as a Country Park, and Bryn Bach Parc, Tredegar for a variety of recreational uses.
5	Construction of residential, commercial (e.g. shopping centre), industrial, educational (e.g. school), other buildings and gardens	This can be problematic because of the presence of toxins, especially metals, in the mine-spoils, so may be judged potentially unsuitable for domestic gardens and school grounds. Where contamination is not a problem, residential development is possible. For example, Oak Regeneration proposed the construction of a 2500 home 'Garden City' on the former Margam Opencast near Port Talbot in SW Wales (Whiteley and Turner 2013). Elsewhere, development for industrial or commercial enterprises provides a sustainable low cost solution.

(continued)

Table 17.3 (continued)

Post-mining land-use		Issues in S.E. Wales and associated reclamation types
6	Wildlife conservation	In many cases, if the mine-spoils are suitable, former coal-land will regenerate when left to Nature, although this may be a very slow process and may leave toxic areas un-vegetated for decades or longer. Naturally, the low-cost do-nothing options are popular with many in the industry and also with ecologist supporters interested in creating wildlife refuges. Frequently, the approach benefits from some initial management and ground-work to overcome the low fertility and other stresses of such sites. Such sites may develop unusual biological communities and so contribute to local biodiversity and also provide refuges for key species (Hitchings 2016). The approach is, of course, self-sustainable. Unfortunately, it more often appears as a consequence of 'rip-and-run' land abandonment or failed/poor quality cosmetic reclamation than of careful design.
7	Landfill by pit backfilling	If properly sealed by a durable membrane and properly monitored and maintained, then the large voids left by opencast mining are attractive sites for waste disposal. The risk, especially in headwater sites, is the escape of leachate and the contamination of both surface and ground-waters. This is why in *"the waste management hierarchy, landfilling is the least preferable option and should be limited to the necessary minimum. Where waste needs to be landfilled, it must be sent to landfills which comply with the requirements of Directive 1999/31/EC"* (European Commission 2016) although this does not apply to inert waste used for redevelopment or restoration work. In the headwaters of SE Wales, at Pwll Du, the proposed establishment of a site for the disposal of ash in an opencast void on the hillcrest above the Usk Valley caused protests that included the blocking of access roads—some of which remain blocked more than a decade after the plan was shelved.

Adapted from: Narrei and Osanloo (2011) and Mborah et al. (2016)

myopic report explains the problem of reclamation failure on inadequate financial bonding for reclamation (cf. Miller 2005). They cite the example of Scottish Coal Ltd., which had gone into liquidation leaving 20 Scottish opencast sites awaiting reclamation, and they blame the inadequate guidance provided to local authorities. Of the 10 operational sites, only four were deemed to pose no risk that reclamation would be underfunded. However, five offered some degree of risk because of potential reclamation funding gaps. One further site, Dynant Fawr, in SW Wales, was deemed effectively abandoned due to the liquidation of the mining company, Carmarthen Mining Ltd., which had left the Local Authority with insufficient financial bonding to cover the costs of reclaiming this 33 ha site (Ellison and Griffiths 2014). In April, 2015, the Welsh National Assembly passed a motion instigating a "*moratorium on opencast mining across Wales, in order to ascertain whether planning law and current guidance provides sufficient protection for communities affected by opencast mining"* (National Assembly for Wales 2015).

Of course, the problems caused by the underfunding and skimping of land reclamation do provide some kind of justification for the Ecological Services methodology. Reclamation costs money and so it is worthwhile knowing how much money has to be spent to achieve the goals desired by future land users (Miller 2005).

Of course, the concept of ecosystem services is a crude attempt to commodify Nature (Robertson 2012). As in medicine and health care, economists try to reduce the holistic view of welfare and well-being to a single figure despite having to compare, in their medical check-up's list, very different things (Millennium Ecosystem Services Assessment 2003). Here too, an arbitrary economic value reduces to absurdity of the more complex message expressed in terms of blood pressure, body mass, heart-rate, blood chemistry, immune system functioning, infections, bone density, psychology, etc. that together contribute to the holistic concept of wellness. However, such numbers are required by those who offer health insurance. Similarly, this chapter has attempted to perform the larger, medical-health check-up style analysis for headwater ecosystems but stopped at the TEEB (2010) Ecosystem Service diagnostic checklist. This may be weighted towards biodiversity issues rather than ecosystem functioning but TEEB's first phase does attempt to evaluate the geoecological system as a whole before proceeding to issues connected to human disposal or decision making. Table 17.2 uses this checklist and its box-filling motif to evaluate many of the issues connected to the assessment of the ecosystem service provided by reclaimed coal-lands. Completing it underlines the self-evident facts that the quality and type of the ecosystem services rendered by reclaimed opencast sites depends, firstly, upon the competence and goals of the land reclamation process, and, secondly, upon the various limitations and challenges presented by the newly created land with its immature soil and ecological systems. It also reinforces the conclusion that economic Ecosystem Service analysis may be appropriate in land reclamation contexts where so much depends on a sufficient amount of money being available to the land reclamation planners and where there is a long history of land reclamation being underfunded, skimped, or dodged (Monbiot 2015).

6 Conclusion

Five different styles of reclamation are recognised in SE Wales (Table 17.1), which reflect the different ambitions and capabilities of the land reclamation agencies mitigated by the inherent problems of site and mine-spoil materials. Even in the present day, more than half of the opencast coal-mines awaiting reclamation are threatened by shortfalls in funding for reclamation and there is no doubt, that apart from a few showcase sites, much previous reclamation was skimped or poorly conceived. Despite the qualms of those who see the Ecosystems Services methodology as a crude attempt to commodify Nature and to place an anthropocentric veil over

Table 17.4 Ecosystem Services provision of reclaimed opencast coal-lands (Wales)

Ecosystem service	Positive impacts	Negative impacts
Water supply and runoff	None	Accelerated runoff, higher flood peaks, possible contamination by chemicals and sediments causing reductions in water quality, although the impacts tend to be local to the reclaimed opencast site (Haigh and Kilmartin 2015).
Biodivtersity and nature conservation	Creation of complex and unusual habitat mosaics by neglect of reclamation or by design.	Typically bland landscapes of sown/planted species, sometimes offering niches for invasive species.
Agriculture (pasture)	None unless the reclamation involves the removal of previously damaged industrial land.	Land can be reclaimed for agriculture and improved grassland but inherent reduced productivity requires additional fertilisation and great care in trafficking and tillage.
Food and fuel crops	None—but the low productivity land may be suitable for biofuel production.	Replacement of more productive with less productive land for both food and fuel production along with the associated threat of contamination.
Industry and economy	Design may create flat-land suitable for development as industrial or commercial units and, if soil contamination is not an issue, for recreation, housing and other urban uses. Even un-reclaimed, neglected, sites may sometimes be used for golf courses, bike circuits and similar off-road activities. Thus Torfaen County Borough's Local Development Plan describes the area containing the former opencast coal lands as "*remote and relatively isolated but [with] a dramatic and unique environment rich in ecological diversity …" but also as one earmarked for "small-scale regeneration opportunities that can enhance service provision and environmental improvements*".	There is a much larger area of reclaimed land than can be developed for industrial, urban and official recreational uses. There remain the problems inherent in the potential toxicity of the buried mine-spoils.

(continued)

Table 17.4 (continued)

Ecosystem service	Positive impacts	Negative impacts
Recreation and amenity	Creates open sites that by design or neglect provide open spaces, not always green, for recreation and leisure activities.	Technically none, although neglected lands (and some designed lands) can encourage fly tipping and vandalism.
Cultural and spiritual	None, except, almost uniquely, around Blaenavon, Torfaen, where the range of reclaimed opencast sites of different ages is considered a valued cultural artefact.	The original land with all of its cultural and spiritual attributes has been obliterated and lost to future generations.
Aesthetic qualities	When properly designed, reclaimed opencast sites in Wales can have higher aesthetic qualities than the lands (especially the derelict industrial and coal-mine disturbed lands) that they replace.	Due to poor quality reclamation design, most of the reclaimed coal-lands in SE Wales are bland in appearance while their drains and watercourses usually look artificial and, occasionally, rank as hazardous, e.g. 'Outlet C', Blaenant, Blaenau Gwent (Haigh 2010).

attempts to exploit and despoil natural landscapes, it may be applicable in the specific case of land reclamation. Here, adequate funding is needed and, more often than not, not available to create high quality post-mining landscapes and sustainable land uses. Land reclamation and after-care has to be taken into account and failure to do this effectively has serious consequences that are all too visible in the South Wales landscape.

This study has applied the first phase of the TEEB (2010) Ecosystem Services methodology to the evaluation of the problem of considering the contributions to ecosystem services of reclaimed lands in SE Wales. It has found that this box-filling approach does encourage a comprehensive overview of ecosystem services, albeit one weighted to biodiversity issues (Table 17.2). The method also provided a useful springboard for the consideration of land uses for reclaimed land, which is a key concern in reclamation planning (Tables 17.3 and 17.4).

This study finds that, sometimes by design and occasionally by accident, reclaimed opencast coal-land in SE Wales can have positive benefits for biodiversity, nature conservation and for the provision of amenity and recreation. When appropriately designed and managed, they can provide sites suitable for industrial and other building development and even for agricultural uses, mainly pasture, potentially biofuel, despite the poor fertility status of the soils.

Reclaimed opencast sites also have many negative impacts. The historical and cultural artefacts of the pre-mining environment are exchanged for a new, often bland, artificial-looking, replacement. Most sites contribute to accelerated runoff, increased flood hazard, and to decreased water quality and contamination, at least locally (Haigh and Kilmartin 2015). Equally, reclaimed opencast sites are sometimes, not universally, associated with reduced biodiversity but, locally, can contribute positively to nature conservation. Their key problems often emerge from the low bioproductivity of their immature, sometimes contaminated soils.

References

Appleyard, K. C., & Curry, G. (1946). Open-cast coal production in wartime. *Journal of the Institution of Civil Engineers, 26*, 331–376.

Blaenau Gwent County Borough. (2015). *Blaenau Gwent Local Biodiversity Action Plan 2015*. Ebbw Vale, Blaenau Gwent County Borough Council. Retrieved May 29, 2016, from http://www.blaenau-gwent.gov.uk/documents/Documents_Environment/LBAP_2015.pdf

Burger, J. (2015). Mined land reclamation in the Appalachian coalfields: A case for an ecosystem reclamation approach. In M. Tibbet (Ed.), *Mining in ecologically sensitive landscapes*. Leiden/Clayton South: CRC Press/Balkema/CSIRO.

Callicott, J. B. (1994). *Earth's insights: A survey of ecological ethics from the Mediterranean Basin to the Australian outback*. Berkeley: University of California.

Chadderton, C., Elliott, E., & Williams, G. (2011). *A guide to assessing the health and wellbeing impacts of opencast mining*. Cardiff: National Health Service (NHS) Wales, HIA Support Unit. Retrieved May 26, 2016, from Google cache of: www.wales.nhs.uk/sites3/Documents/522/OpencastguidanceFinal.pdf

Chapman, R. (2001). Competition and succession in re-created botanically diverse grassland communities. In P. G. Tow & A. Lazenby (Eds.), *Competition and succession in pastures*. Wallingford: CABI Publishing.

CICES. (2016). *Towards a common classification of ecosystem services*. Brussels, European Environment Agency. Retrieved May 25, 2016, from http://cices.eu/cices-structure/

Collier, M. J. (2014). Novel ecosystems and the emergence of cultural ecosystem services. *Ecosystem Services, 9*, 166–169.

Costanza, R., d'Arge, R., de Groot, R., Farberk, S., Grasso, M., Hannon, B., Limburg, K., Naeem, S., O'Neill, R. V., Paruelo, J., & Raskin, R. G. (1997). The value of the world's ecosystem services and natural capital. *Nature, 387*, 253–260.

Countryside Council for Wales and International Council on Monuments and Sites. (1998). *Register of landscapes of outstanding historic interest in Wales: Cofrestro Dirweddauo Ddiddordeb Hanesyddol Eithriadol Yng Nghymrs*. Cardiff: Cadw Welsh Historic Monuments.

Davenport, C. (2016). Politics: In climate move, Obama halts new coal mining leases on public lands. *New York Times (New York Edition)* January 15, 2016, p. A17. Retrieved February 27, 2016, from http://www.nytimes.com/2016/01/15/us/politics/in-climate-move-obama-to-halt-new-coal-mining-leases-on-public-lands.html?_r=0

Dedikova, N. (2014). *Opencast coal mining in Wales: Development of the industry in terms of social and environmental impacts*. Cardiff: Friends of the Earth Cymru.

DEFRA. (2005). *DEFRA guidance for successful reclamation of mineral and waste sites*. Bury St Edmunds, Department for Environment, Food and Rural Affairs, National Land Management Team. Retrieved May 24, 2016 from http://webarchive.nationalarchives.gov.uk/20090306103114/http:/www.defra.gov.uk/farm/environment/land-use/reclamation/guidance-full.pdf

Desai, M. (2013). *Phytoremediation: A tool for restoring land degraded due to open cast coal mining*. Ph.D. thesis. Oxford Brookes University, Oxford, England.

ECO. (2016). *No new coal mines: China announces three year ban*. Wellington: Environment and Conservation Organisations of New Zealand. Press Release: Sunday, 3 January 2016.

EEA. (2016). *Common International Classification of Ecosystem Services (CICES Version 4.3, 2011)*. Nottingham/Copenhagen: University of Nottingham/European Environment Agency (EEA). Retrieved January 15, 2016, from http://cices.eu

Ellison, R., & Griffiths, G. (2014). *Research into the failure to restore opencast coal sites in South Wales*. Cardiff: Welsh Government. Retrieved May 31, 2016, from http://gov.wales/docs/desh/research/140424failure-to-restore-opencast-coal-sites-in-south-wales-en.pdf

European Commission. (2016). *Environment: Waste*. Bruselles: European Commission. Retrieved February 21, 2016, from http://ec.europa.eu/environment/waste/landfill_index.htm

Friederici, P. (1992, May 21). Art under fire: Tourist attraction closed due to flying bullets. *Chicago Reader*. Retrieved May 14, 2016, from http://www.chicagoreader.com/chicago/art-under-fire-tourist-attraction-closed-due-to-flying-bullets/Content?oid=879752

Glassheim, E. (2007). Most, the town that moved: Coal, communists and the 'gypsy question' in post-war Czechoslovakia. *Environment and History, 13*, 447–476.

Gómez-Baggethun, E., De Groot, R., Lomas, P. L., & Montes, C. (2010). The history of ecosystem services in economic theory and practice: From early notions to markets and payment schemes. *Ecological Economics, 69*(6), 1209–1218.

Gwent Ornithological Society. (2005/2014). Ornithological survey of proposed development area east of Varteg. In: Glamorgan Power (Ed.), *Varteg Planning Application*. Bargoed: Glamorgan Power, Item 26. Retrieved May 30, 2016, from http://www.planapps.torfaen.gov.uk/Varteg/

Haigh, M. J. (1978). *Evolution of slopes on artificial landforms, Geography Research Paper 183*. Blaenavon/Chicago: University of Chicago.

Haigh, M. J. (1980). Slope retreat and gullying on revegetated surface-mine dumps, Waun Hoscyn, Gwent. *Earth Surface Processes, 5*(1), 79–82.

Haigh, M. J. (1999). Land reclamation and the community. In G. G. Pivnyak & R. K. Singhal (Eds.), *Mine planning and equipment selection '99 and mine environmental and economical issues '99*. Dnipropetrovsk: National Mining University of Ukraine.

Haigh, M. J. (Ed.). (2000). *Reclaimed land: Erosion control, soils and ecology*. London/Rotterdam: Taylor and Francis/CRC Press/Balkema.

Haigh, M. J. (2003). Acid mine drainage and its control. In A. L. Ramanathan & R. Ramesh (Eds.), *Recent trends in hydrogeochemistry*. New Delhi: Capital Publishing Company.

Haigh, M. (2010). Exploring sustainability in the context of land reclamation: An exercise for environmental management trainees. *International Journal of Environment and Sustainable Development, 9*(1–3), 255–268.

Haigh, M., & Kilmartin, M. P. (2015). Reclaimed opencast coal lands in southeast Wales: Impacts on water quality. In: Subramanian, V. S. (Ed.), *Surface and sub-surface water in Asia – Issues and perspectives*. Amsterdam: IOC Press. doi:10.3233/978=1=61499-540-1-16. Retrieved May 28, 2016, from https://books.google.co.uk/books?hl=enandlr=andid=zmZrCgAAQBAJandoi=fndandpg=PR1anddq=Subramanian,+V.S.+(ed).+Surface+and+Sub-surface+Water+in+Asia+-+Issues+and+Perspectives.+Amsterdam,+IOC+Press,andots=ts2CWlqbFaandsig=dF1XloZ9iPI6mnqxDohvf1VcHKs#v=onepageandqandf=false

Haigh, M. J., & Sansom, B. (1999). Soil compaction, runoff and erosion on reclaimed coal-lands (UK). *International Journal of Surface Mining, Reclamation and Environment, 13*, 135–146.

Haigh, M., Reed, H., Flege, A., D'Aucourt, M., Plamping, K., Cullis, M., Woodruffe, P., Sawyer, S., Panhuis, W., Wilding, G., Farrugia, F., & Powell, S. (2015). Effect of planting method on the growth of *Alnus glutinosa* and *Quercus petraea* in compacted opencast coal-mine spoils, South Wales. *Land Degradation and Development, 26*, 227–236.

Haines-Young, R., & Potschin, M. (2013). *Common International Classification of Ecosystem Services (CICES): Consultation on Version 4*. August–December 2012. EEA Framework Contract No EEA/IEA/09/003. Download at www.cices.eu and a full spread sheet showing the classification.

Harmers Ltd (2014). *Varteg planning application (14/P/00637): Glamorgan Power Company Limited*. Environmental Statement: Proposed Land Reclamation/Coal Recovery Scheme a

Varteg Hill, Torfaen November 2014. Retrieved May 26, 2016, from http://www.planapps.torfaen.gov.uk/varteg/

Hartman, B. A., Ammons, J. T., & Hartgrove, N. T. (2004). A proposal for the classification of anthropogenic soils. In: ASMR (Ed.), *National meeting of the American Society of Mining and Reclamation and 25th West Virginia surface mine drainage task force, proceedings*. Lexington: American Society of Mining and Reclamation.

Hausman, J.A. (Ed.). (1993). *Contingent valuation: A critical assessment*. Amsterdam, North Holland (Contributions to Economic Analysis 220)/Elsevier Science.

Heyd, T. (2007). Reflections on reclamation through art. *Ethics, Place and Environment, 10*, 339–345.

Hitchings, E. (2016, Spring). Greening the brownfields. *Nature's Home, the RSPB Magazine*, 2016, 58–64.

Hope, J. C. E. (2014). Lichen survey for planning application at Varteg, Torfaen, S. Wales, Watsonian vice-county 35 – Monmouthshire; NGR: c. SO 26 06. In Glamorgan Power (Ed.), *Varteg planning application*. Bargoed: Glamorgan Power, Item 24. Retrieved May 30, 2016, from http://www.planapps.torfaen.gov.uk/Varteg/

Jones, S. (2014). *14/P/00637 opencast mine at Varteg*. Monmouth: Gwent Wildlife Trust. Retrieved May 26, 2016, from: www.planapps.torfaen.gov.uk/.../Gwent%20Wildlife%20Trust%20comments%2006-14/P/00637

Kuter, N. (2013). Reclamation of degraded landscapes due to opencast mining. In Özyavuz, M. (Ed.), *Advances in landscape architecture*. Rijecka: InTech. Retrieved February 27, 2016, from http://www.intechopen.com/books/advances-in-landscape-architecture/reclamation-of-degraded-landscapes-due-to-opencast-mining

Larondelle, N., & Haase, D. (2012). Valuing post-mining landscapes using an ecosystem services approach: An example from Germany. *Ecological Indicators, 18*, 567–574.

Litterick, A.M. and Booth, E.J. (2009). *The potential for biofuel crop production on a former coal washing site in Kinglassie, Fife*. Banbury: WRAP. Retrieved May 29, 2016, from http://www.wrap.org.uk/sites/files/wrap/approved%20%20-%20kinglassie2008%20report%20-%20formatted.pdf

Maczkowiack, R. I., Smith, C. S., Slaughter, G. J., Mulligan, D. R., & Cameron, D. C. (2012). Grazing as a postmining land-use: A conceptual model of the risk factors. *Agricultural Systems*, *109*, 76–89. http://dx.doi.org/10.1016/j.agsy.2012.03.002

Mborah, C., Bansah, K. C., & Boateng, M. K. (2016). Evaluating alternate post-mining land-uses: A review. *Environment and Pollution, 5*, 14–22.

McCarthy, J. (2015). Wales' abandoned mines: 'It will cost hundreds of millions of pounds to fix our landscape', warns MP. *Walesonline*, 13, 48, 7 February, 2015. Retrieved May 28, 2016, from http://www.walesonline.co.uk/news/wales-news/wales-abandoned-mines-it-cost-8598165

Metz, B., Davidson, O. R., Bosch, P. R., Dave, R., & Meyer, L. A. (Eds.). (2007). *Contribution of working group III to the fourth assessment report of the intergovernmental panel on climate change*. Cambridge: Cambridge University Press. Retrieved February 27, 2016, from https://www.ipcc.ch/publications_and_data/ar4/wg3/en/contents.html

Millennium Ecosystem Assessment. (2003). *Ecosystems and human well-being: A framework for assessment*. Washington, DC: Island Press. Retrieved February 26, 2016, from http://www.millenniumassessment.org/en/index.html

Millennium Ecosystem Assessment. (2005). *Ecosystems and Human Well-being: Synthesis*. Washington, DC: Island Press. Retrieved May 26, 2016 from http://www.millenniumassessment.org/documents/document.356.aspx.pdf

Miller, G. (2005). *Financial assurance for mine closure and reclamation*. London, International Council on Mining and Metals. Retrieved May 31, 2016, from https://www.icmm.com/document/282

Miller, D. (2008). Using aquaculture as a post-mining land-use in West Virginia. *Mine Water and the Environment*, *27*: 122–126. http://dx.doi.org/10.1007/s10230-008-0038-4

Monbiot, G. (2015). Big Coal's big scam: Scar the land for profit, then let others pay to clean up. *The Guardian On-line*, 28 April 2015. Retrieved February 27, 2016, from http://www.theguardian.com/commentisfree/2015/apr/28/big-coal-keep-it-in-the-ground-energy-opencast-mines

Morris, R. (2016). Climate protesters invade UK's largest opencast coalmine. *Theguardian.com* 3 May 2016. Retrieved May 27, 2016, from http://www.theguardian.com/environment/2016/may/03/climate-protesters-invade-uks-largest-opencast-coal-mine

MPG7. (1989). *The reclamation of mineral workings*. Cardiff: Welsh Office, Department of the Environment, Minerals Planning Guidance 7. Retrieved May 26, 2016, from http://gov.wales/docs/desh/publications/890101mpg7en.pdf

MTAN2. (2009). *Minerals technical advice note (MTAN) Wales 2: Coal*. Cardiff: Welsh Government. Retrieved May 26, 2016, from http://gov.wales/docs/desh/policy/090120coalmtanen.pdf

Narrei, S., & Osanloo, M. (2011). Post-mining land-use methods optimum ranking, using multi attribute decision techniques with regard to sustainable resources management. *OIDA International Journal of Sustainable Development, 2*(11), 66–76.

National Assembly for Wales. (2006). *Subordinate legislation: Plant health Wales: The code of practice to prevent and control the spread of ragwort*. Cardiff: NAW. Retrieved May 26, 2016, from http://www.assembly.wales/Laid%20Documents/LD5671%20-%20The%20Code%20of%20Practice%20to%20Prevent%20and%20Control%20the%20Spread%20of%20Ragwort%202006-23052006-20224/bus-GUIDE-N000000000000000000000000044593-English.pdf

National Assembly for Wales. (2015). *Plenary, votes and proceedings, meeting date: Wednesday, 22 April 2015, Item 4*. Cardiff, NAW. Retrieved February 27, 2016, from http://www.senedd.assembly.wales/documents/g3134/Decisions%20Wednesday%2022-Apr-2015%2013.30%20Plenary.pdf?T=2

Parker, M. (2008). *The garden Festival now. Ebbw Vales, garden Festival Wales*. Retrieved May 30, 2016 from http://www.gardenfestivalwales.co.uk/index.php?p=now

Plamping, K., Haigh, M., Reed, H., Woodruffe, P., Fitzpatrick, S., Farrugia, F., D'Aucourta, M., Flege, A., Sawyer, S., Panhuis, W., Wilding, G., Cullis, M. and Powell, S. (2016). Effects of initial planting method on the performance of mixed plantings of alder and oak on compacted opencast coal-spoils, Wales: 10-year results. *International Journal of Mining, Reclamation and Environment*, Published online: 10 Feb 2016, 1–15. doi:10.1080/17480930.2016.1139528.

Probert, G. (1986). New life for wasteland. *Landscape Design, 159*, 54–55.

Roberts, K. (2013). *A local biodiversity action plan for Torfaen 2013–2023: Towards the sustainable management of our natural resources*. Cwmbran: Torfaen Borough Council.

Robertson, M. (2012). Measurement and alienation: Making a world of ecosystem services. *Institute of British Geographers, Transactions, 37*, 386–401.

Royal Society. (2008). Sustainable biofuels: Prospects and challenges. London, Royal Society Policy Document 01/08. Retrieved May 29, 2016, from https://royalsociety.org/~/media/Royal_Society_Content/policy/publications/2008/7980.pdf

Sencindiver, J. C. (1977). Classification and genesis of minesoils. Ph.D. dissertation, Morgantown, WV, West Virginia University, (abstract 77-22746).

Skousen, J., & Zipper, C. E. (2014). Post-mining policies and practices in the Eastern USA coal region. *International Journal of Coal Science and Technology*, *1*, 135–151. Retrieved May 28, 2016, from http://www.asmr.us/Publications/Journal/Vol%202%20Issue%201/Skousen-WV.pdf

Skousen, J., Keene, T., Marra, M., & Gutta, B. (2013). Reclamation of mined land with switchgrass, Miscanthus, and Arundo for biofuel production. *Journal of the American Society of Mining and Reclamation, 2*, 177–191.

Splash, C. (2011). Editorial: Terrible economics, ecosystems and banking. *Environmental Values, 20,* 141–145. Retrieved May 25, 2016, from http://docserver.ingentaconnect.com/deliver/connect/whp/09632719/v20n2/s2.pdf

TEEB. (2010). *The economics of ecosystems and biodiversity: Mainstreaming the economics of nature: A synthesis of the approach, conclusions and recommendations of TEEB*. Nairobi:

United Nations Environment Programme. Retrieved May 23, 2016, from http://www.unep.org/pdf/LinkClick.pdf

Todd, J., Doshi, S., & McInnes, A. (2010). Beyond coal: A resilient new economy for Appalachia. *Solutions, 1*(4), 45–52. Retrieved May 28, 2016, from http://www.thesolutionsjournal.com/node/706/cite

Tomlinson, P. (1980). The agricultural impact of opencast coal mining in England and Wales. *Minerals and the Environment, 2*(2), 78–100.

Torfaen County Borough Council. (1996). *Local plan (1996–2006): Deposit version*. Ponypool: Torfaen County Borough Council.

Torfaen County Borough Council. (2013). *Local development plan (to 2021) adopted plan – Written statement, December, 2013*. Cwmbran: Torfaen County Borough Council.

Westman, W. (1977). How much are nature's services worth? *Science, 197*, 960–964.

Whiteley, A., & Turner, R.. (2013, March 26). Garden city with 2,500 homes planned at former opencast mine. *Wales Online*. Retrieved May 27, 2016, from http://www.walesonline.co.uk/news/wales-news/garden-city-2500-homes-planned-2020906

Younger, P. L. (2004). Environmental impacts of coal mining and associated wastes: A geochemical perspective. *Geological Society, London, Special Publications, 236*, 169–209.

Chapter 18
The Effect of Land Degradation on Ecosystem Services

Ádám Kertész

1 Introduction

Land degradation leads to very severe damages on the Earth. Global climate change and land use change are always present in the media and in scientific publications, though much less attention is paid to the degradation of the Earth surface. Land degradation is an interrelated process; other processes like global climate change, land use change and even population growth are strongly linked to it.

The word "degradation" means a change to a lower quality. The following definition refers directly to ecosystem services: "Land degradation is the reduction in the capacity of the land to provide ecosystem goods and services and assure its functions over a period of time for the beneficiaries of these" (Woodfine 2013). See further definitions below.

The objective of the paper is to introduce the problem of land degradation, the concept of ecosystem services and the relationship between them as well as to present a case study on the effects of soil erosion on ecosystem services.

2 Land Degradation

According to the UNEP (1992) definition "land degradation leads to a significant reduction of the productive capacity of land". According to Barrow (1991) land degradation can be defined as "the loss of utility or the reduction, loss or change of features or organisms which cannot be replaced". The land is degraded when "it suffers a loss of intrinsic qualities or a decline in its capabilities" (Blaikie and

Á. Kertész (✉)
Research Centre for Astronomy and Earth Sciences, Hungarian Academy of Sciences, Budapest, Hungary
e-mail: kertesza@helka.iif.hu

© Capital Publishing Company, New Delhi, India 2017

J. Křeček et al. (eds.), *Ecosystem Services of Headwater Catchments*,
DOI 10.1007/978-3-319-57946-7_18

Brookfield, 1987). The UNEP (1992) definition emphasizes the reduction of the potential of natural resources as a result of processes acting in the landscape. Johnson and Lewis (1995) underline the role of human interventions in land degradation and they put the reduction of biological productions and/or utility of an area into focus. It means the reduction or loss of biological productivity and a negative effect on the functioning of the land and related ecosystems (Hudson and Alcántara-Ayala, 2006). Functioning of the land involves the interaction of environmental factors and also connections between landscape components like hillslopes and floodplains.

The concept of land degradation originates from soil degradation and it is quite frequently used as a synonym of soil degradation. It is evident that if the soil is degraded, it has a huge impact on the land and on the landscape because the degradation of the soil prevents or reduces the possibilities of plant growth. Land and soil are not identical notions but they are used, more or less, almost with the same meaning. Landscape degradation means much more than just the degradation of the uppermost layer of the Earth crust; it means the decline of the whole landscape. The term "desertification" is applied for land degradation processes in arid, semi-arid and dry sub-humid areas.

The degradation of the land and the landscape has been taking place all over the history of the Earth. Before the appearance of remarkable human activities, i.e. roughly up to the beginning of the Holocene era land degradation was the result of the interaction of natural factors. In the twentieth and twenty first centuries the importance of anthropogenic influence on degradation processes became more and more significant. Climate and land use change as main global processes have already been mentioned above. The most damaging changes are deforestation and the removal of natural vegetation. In addition to land use change unsustainable agricultural land use, poor soil and water management practices, floods, drought, landslides also contribute to land degradation of anthropogenic origin.

Land degradation affects 33% of the Earth's land surface, hitting more than 2.6 billion people in more than 100 countries where 40% of agricultural area of the world is degraded (Kapalanga 2008).

Land degradation processes include soil erosion by water and wind, chemical degradation (acidification, salinization/alkalization processes, leaching) as well as physical degradation of soils (soil compaction, crusting structural damage, degradation due to the extreme soil moisture regime) and biological degradation. The main driving force is in most cases the water regime.

3 Ecosystem Services

In the last decades the number of publications on ecosystem services has been increasing exponentially (Báldi 2011). Ecosystem services are "the benefits people obtain from ecosystems" (Millennium Ecosystem Assessment 2006). As it is well known, the four categories of ecosystem services are: supporting, provisioning, regulating and cultural.

The supporting services provide the basic conditions for life (e.g. fertile soils, nutrient cycling, primary production). 30–50% of primary production is consumed by humans, mainly by agriculture and forestry and the rate is growing (Foley et al. 2011). Regulating services provide the functioning of ecosystems (e.g. clean air, flood and disease control, pollination). Supporting services belong to the third group (e.g. food, fuel, wood) and the cultural services include recreational and cultural benefits.

Recognizing the importance and critical state of ecosystem services an international platform was created: Intergovernmental Science-Policy Platform on Biodiversity and Ecosystem Services (IPBES) which is an intergovernmental body assessing the state of biodiversity and of the ecosystem services. IPBES is placed under the auspices of four United Nations entities: UNEP, UNESCO, FAO and UNDP and administered by UNEP. Its secretariat is hosted by the German government and located on the UN campus, in Bonn, Germany. The present assessment of IBES was preceded by the Millennium Ecosystem Assessment (Millennium Ecosystem Assessment 2006). Thematic assessment of land degradation and restoration is deliverable 3(b)(i) of the IPBES working programme (Montana 2016.)

All ecosystem services are endangered by land degradation. We will just take the example of food. One of the most important supporting services i.e. food and food security for a tremendously growing population is also endangered by land degradation. The increasing demand for food will lead to the intensification of agriculture. A more intensive agriculture may cause more serious land degradation problems as well. Future land degradation problems will be, of course, different in various parts of the world. According to Crosson (1997) an accelerating rate of degradation (0.4%) can be assumed for the coming 30 years, accompanied by a decline of altogether 17% of agricultural activity.

Land degradation has a negative effect on productivity. According to Pimentel et al. (1993) soil erosion causes 15–30% less production. Nutrient depletion because of erosion results in 29% decline of crop production and 19% loss in total production. Data based on empirical evidence show lower values. The negative effect is also manifested in yield reduction (e.g. yield reduction in Africa due to past soil erosion may range from 2 to 40%, with a mean loss of 8.2% for the continent, Eswaran et al. 1999). The relationship between erosion and productivity can be expressed by a negatively exponential curve, i.e. production diminishes very rapidly in the early stages of erosion. Different soil types show, of course, different patterns (Tengberg and Stocking 1997).

4 Case Study

Soil erosion is one of the most destructive land degradation processes also concerning ecosystem services (food provision). It is therefore necessary to identify areas with various rates of sensitivity to soil erosion on one hand, and to determine ecosystem services affected by soil erosion, on the other. For the investigation of the relationship between soil erosion and ecosystem services a study area was selected in SW Hungary. The study area is the natural macro-region of the Transdanubian Hills. With an area of 12,000 km^2 this landscape unit is mainly a hilly country, one third of it is lowland and 5% are mountains of medium height (up to 700 m). The overwhelming part of it is covered by loose sediments prone to soil erosion. The soils developed on these loose sediments and the agricultural use leaves the surface uncovered after harvest so that huge areas are sensitive to soil erosion. A considerable part of the forest was cut a few centuries ago to gain new territories for agricultural use. The percentage of forests in 2010 was only 25 and 60% is used for agriculture.

The case study consists of two parts. A soil erosion sensitivity analysis was carried out first in order to identify the areal differences in sensitivity. The relationship between soil erosion and ecosystem services was then treated in the second part of the case study.

4.1 Sensitivity to Soil Erosion

Soil erosion sensitivity is determined by applying a sensitivity index. Sensitivity classes are determined for each factor. The value of a sensitivity index will be obtained by the geometric mean of the weighted factors. The factors involved are as follows: soil types, soil parent material, slope gradient and land use. Sensitivity maps were prepared for each factor and these maps were also evaluated.

The soil erosion sensitivity map (Fig. 18.1) was prepared with the application of sensitivity maps of the above mentioned factors. The evaluation of the soil erosion sensitivity map delivered the following results. The most sensitive areas are valley side slopes (category 5). In the western part, arable fields where the soil parent

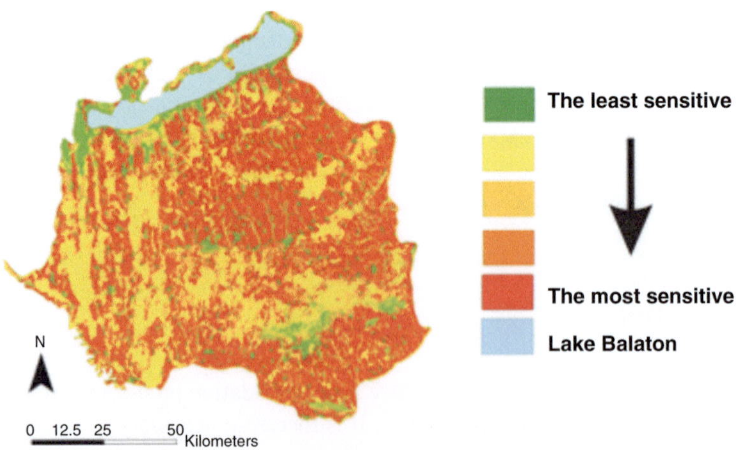

Fig. 18.1 Soil erosion sensitivity map of the Transdanubian Hills

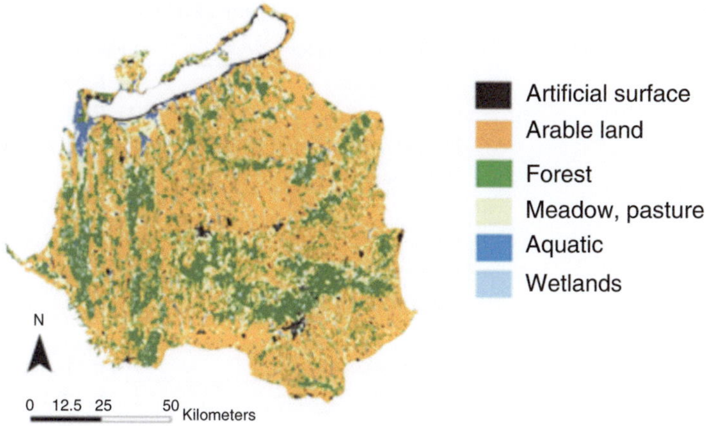

Fig. 18.2 Land cover map of the Transdanubian Hills

material is sand are also very sensitive. An overwhelming part of the area, 44%, belongs to category 4. Here loess is the soil parent material and in spite of gently slopes, the land use type (arable land and vineyards) gives the explanation for relatively high sensitivity. Medium sensitivity (category 3) is on meadow while pasture and forested areas have low sensitivity (category 2). Wetlands along Lake Balaton and some forested parts in the mountains (South-Eastern part) are the least sensitive, also because the soil parent material is mostly the regolith of a hard rock. The soil erosion sensitivity map helps to identify the areas with different sensitivity from the aspect of ecosystem services.

4.2 *Soil Erosion and Ecosystem Services*

As already discussed, land degradation processes reduce the quality of land in multiple ways. Improper land use leads to long-term losses of ecosystem functions and productivity. The land cover map of the study area is presented in Fig. 18.2 and Table 18.1 shows the ecosystem services influenced negatively by soil erosion indicating the different land cover types which provide them. With the help of Table 18.1 ecosystem services for each spot on the map can be identified.

5 Conclusions

As one third of the Earth surface is affected by land degradation, the rate and quality of ecosystem services are reduced here. The case study presented in the paper revealed the relationship between soil erosion and ecosystem services.

Table 18.1 Ecosystem services influenced by soil erosion, provided by various land cover types

Ecosystem service categories	Ecosystem services	Land cover types
Supporting	Soil formation	Ar, F, M
	Nutrient cycling	Ar, F, M, Aq, W
	Primary production (biomass)	Ar, F, M, Aq, W
Regulating	Water quality	F, M, Aq, W
	Erosion control	Ar, F, M, W
Provisioning	Production of food	Ar, F, M, Aq, W
	Fiber	Ar, F, M
	Fresh water	Aq, W
	Fuel (energy sources)	Ar, F
	Genetic resources	Ar, F, M, Aq, W
	Ornamental resources	F, M, Aq, W
Cultural	Recreation and ecotourism	Ar, F, M, Aq, W, S
	Aesthetic values	Ar, F, M, Aq, W, S

Ar Arable, *F* Forest, *M* Meadow, *Aq* Aquatic, *W* Wetland, *S* Settlement

The application of a sensitivity index proved to be suitable to characterize sensitivity to soil erosion. The sensitivity map was validated and the areas belonging to various sensitivity classes are in accordance with the present situation in the study area.

Ecosystem services endangered by soil erosion were identified for various land cover types. The results are shown on two maps which can be compared. This comparison identifies those areas which are sensitive to erosion and dispose off a potential damage to ecosystem services.

Acknowledgement This study was supported by the National Research, Development and Innovation Office (NKFIH), Project No. 108 755 and this support is gratefully acknowledged here.

References

Báldi, A. (2011). Your money, or your life? (in Hungarian). *Magyar Tudomány, 172*, 774–779.

Barrow, C. J. (1991). *Land degradation: Development and breakdown of terrestrial environments*. Cambridge: Cambridge University Press.

Blaikie, P., & Brookfield, H. (1987). *Land degradation and society*. London: Methuen & Co..

Crosson, P. R. (1997). Will erosion threaten agricultural productivity? *Environment, 39*(4–9), 29–31.

Eswaran, H., Lal, R., & Reich, P.F. (1999). Land degradation: An overview. In Bridges, E. M., Hannam, I. D., Oldeman, L. R., Pening de Vries, F. W. T., Scherr, S. J., & Sompatpanit, S. (Eds.), *Responses to land degradation. Proceeding of 2nd international conference on land degradation and desertification, Khon Kaen, Thailand*. New Delhi: Oxford Press.

Foley, J. A., DeFries, R., & Regory, P. A. (2011). Global consequences of land use. *Science, 309*, 570–574.

Hudson, P. F., & Alcántara–Ayala, I. (2006). Ancient and modern perspectives on land degradation. *Catena, 65*, 102–106.

Johnson, D. L., & Lewis, L. A. (1995). *Land degradation*. Oxford: Creation and Destruction.
Kapalanga, T. S. (2008). *A review of land degradation assessment methods*. Reykjavík: Land Restoration Training Program. http://www.unulrt.is/static/fellows/document/taimi.pdf
Millennium Ecosystem Assessment. (2006). *Ecosystems and human well-being, Millennium Ecosystem Assessment*. Washington, DC: Island Press.
Montana, J. (2016, May 10). *How IPBES works: The functions, structures and processes of the intergovernmental platform on biodiversity and ecosystem services* (C-EENRG Working Papers, no. 2), University of Cambridge.
Pimentel, D., Allen, J., & Beers, A. (1993). Soil erosion and agricultural productivity. In D. Pimentel (Ed.), *World soil erosion and conservation*. Cambridge: Cambridge University Press.
Tengberg, A., & Stocking, M. (1997, December 8–12). *Erosion-induced loss in soil productivity and its impacts on agricultural production and food security*. Paper presented to the FAO/AGRITEX expert consultation on integrated soil Management for sustainable agriculture and food security in Southern and Eastern Africa, Harare, Zimbabwe.
UNEP. (1992). *World Atlas of desertification*. London: UNEP – Nairobi and Edward Arnold.
Woodfine, A. (Ed.). (2013). *Land degradation in drylands: Methodology and results*. Rome: FAO. http://www.fao.org/3/a-i3241e.pdf

Chapter 19
Extreme Climate Events and Erosion Control in Headwater Catchments of Serbia

Stanimir Kostadinov, Olivera Košanin, Ana Petrović, and Slavoljub Dragićević

1 Introduction

Floods are the most frequent natural catastrophic events worldwide (Berz et al. 2001; Barredo 2007; De Moel et al. 2009; Bissolli et al. 2011). In Serbia, the risk of torrential floods is the most common natural hazard, and a permanent threat of ecosystems, local and national economy, and social life (Kostadinov 1996; Dragićević et al. 2011). The average annual economic loss due to natural hazards over the world has been estimated at 40 billion EUR (MRG 2003), and, particularly, flash floods caused several serious loss of life and economic damage.

A torrent is a consequence of intensive erosion processes within the catchment. Generally, torrents are characterized as steep stream channels (above 1%), relatively small catchment, steep and irregular sections, extreme runoff variations, high sediment transport and sedimentation (Kostadinov 2007). The torrential character of a stream regime is manifested by a large range in the annual discharge, and by a short duration of flood waves; ratio between the mean annual discharge and peak flows during flood situations is the order of 1:1000.

In Serbia, about 75% of the territory is endangered by water or wind erosion, and more than 12,000 torrents are registered there. The area of torrential watersheds exceeds 1000 km^2 (Gavrilović 1975). The increasing frequency of torrential floods, registered in the last decades, can be explained by the global climate change (Petrović et al. 2014). The aim of this study was to analyse processes of water

S. Kostadinov (✉) • O. Košanin
Faculty of Forestry, University of Belgrade, Belgrade, Serbia
e-mail: Stanimir.kostadinov@sfb.bg.ac.rs

A. Petrović
Geographical Institute "Jovan Cvijić", Serbian Academy of Science and Arts, Belgrade, Serbia

S. Dragićević
Faculty of Geography, University of Belgrade, Belgrade, Serbia

© Capital Publishing Company, New Delhi, India 2017

J. Křeček et al. (eds.), *Ecosystem Services of Headwater Catchments*,
DOI 10.1007/978-3-319-57946-7_19

erosion in headwater basins of Serbia, to bring the inventory of torrential floods in the period 1915–2014, and to discuss possibilities of erosion control in mountain catchments.

2 Material and Methods

The inventory used the existing erosion map, outcomes of former torrential flood research, analysis of the current state of erosion and torrential floods, and possibilities of their prevention. Considering the archival information, for the period of 1915–1970, the inventory of torrential floods reported by Gavrilović (1975), and, for the period of 1971–2014, the archive documentation of the newspaper "Politics" were applied. Several flood events have been reconstructed using the marks in stream channels, or, the data on precipitation and catchment morphology. The inventory included the date, watershed name, municipalities affected, damage, casualties, and additional information available. In detail, the torrential floods were investigated in the Fruska Gora Mts. and Vršačka Brda Hills, south of the Danube and Sava rivers. The soil characteristics of investigated sites are provided by Popović (1995), and the land-use evidence by Dragićević et al. (2011). The method of Gavrilović' (1972) was applied to estimate potential loss of the soil by erosion processes, and the approach of Gavrilović (1975) to estimate flood frequencies of not gauged torrents in Serbia. Changes of the climate have been manifested by changes in temperature and precipitation (WMO 2001), and the future climate scenarios are based on regional models of the scale 50 × 50 km (IPCC 2007).

3 Results and Discussion

3.1 Erosion and Torrential Floods

Considering the genesis of runoff and erosive processes in a catchment, the most important characteristics are relief, soil properties, land use, and climate conditions. In Serbia, the land use evidence is shown in Table 19.1.

Almost all the territory of Serbia is affected by soil erosion; the distribution of erosion risk categories (from low to excessive erosion) is shown in Table 19.2. The method of Gavrilović' (1972) was applied to estimate potential loss of the soil by erosion processes. The mean annual loss of soil in mountain catchments of Serbia is 37 million m^3 (i.e. 422 $m^3 \times km^{-2}$), while the annual sediment runoff is 9 million m^3 (i.e. 106 $m^3 \times km^{-2}$).

In 1951–2008, trends in annual and summer air temperatures on the territory of Serbia are presented in Fig. 19.1 (Popović et al. 2009). The results of regional climate models show that in 2071–2100, the mean annual temperature in Serbia will

Table 19.1 Land use in Serbia (without the Autonomus province Kosovo)

Land use	Area	
	ha	%
Forests	2,252,400.0	29.1
Other wooded land	382,400.0	4.9
Barren land	92,000.0	1.2
Agricultural land	3,594,800.0	46.4
Meadows and pastures	1,029,600.0	13.3
Built-up area	312,000.0	4.0
Inland water	85,200.0	1.1
Total	7,748,400.0	100.0

Table 19.2 Aerial extend of the erosion risk in Serbia

Category	Erosion scale	Area	
		km^2	%
I	Excessive	2888.0	3.27
II	Intensive	9138.0	10.34
III	Medium	19,386.0	21.94
IV	Weak	43,914.0	49.78
V	Very weak	13,035.0	14.75
Total		88,361.0	100

increase by 2.6 °C in average; namely with warmer summers (3.5 °C), in the rest of the year, the increase in temperature might vary from 2.2 to 2.5 °C. Contrary to the air temperature, precipitation show a decreasing trend from 3 to 10 mm per month, but the intensity of rainstorms has increased. Concerning soil erosion, particularly summer rainstorms cause a serious risk of flash floods; however, the estimates of such an increase in 2071–2100 (in flood frequency and magnitude) still remain uncertain (IPCC 2007). Generally, Pruski and Nearing (2002) reported that in the future, with changing climate and land-use, increasing erosion of soil by 1.7% could be expected with each 1% rising amount of precipitation.

The performed analysis of average annual torrential rains in 1928–1968 included 312 gauging stations (100 stations on the territory of Vojvodina, 188 in Central Serbia and 24 in Kosovo). The highest average torrential rain in Serbia for that period was 45 mm day^{-1}. In the last 30 years, several torrential floods dramatically exceeded events of 1928–1968 as a result of much serious rainstorm (220 mm day^{-1} in 1988, 218 mm day^{-1} in 2014, or 120 mm day^{-1} in 1982). In the 1980s, the torrential floods inventory recorded the most destructive events, particularly, in the Ljeshterska Valley, the Sejanichka River, and the Vlasina River. Considering the seasonal occurrence of torrential floods, the largest number was registered in June (233 events, i.e. 27.5%) and May (178 events, 21%), followed by July (88, 10.4%), March (71, 8.4%) and February (67, 7.9%). These results correspond to earlier analysis of peak-flows in torrents of Serbia presented by Petrović et al. (2014) (Table 19.3).

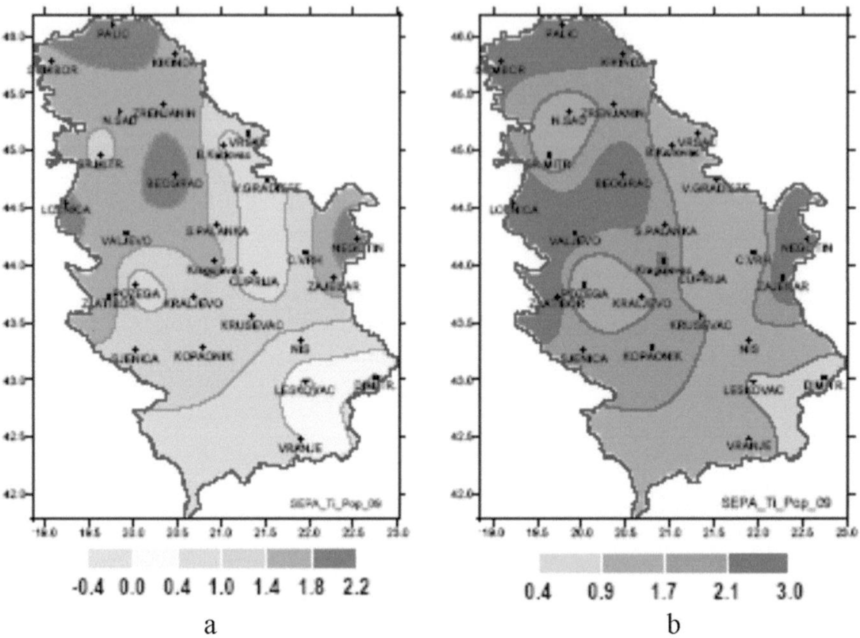

Fig. 19.1 Trends in annual (**a**) and summer (**b**) temperatures of the air in Serbia (Popović et al. 2009)

Table 19.3 Registered torrential floods in Serbia

No.	Period	Number of floods recorded	Annual mean	Casualties
1.	1915–1930	33	2.06	>22
2.	1931–1960	175	5.83	>51
3.	1961–1990	384	12.80	36
4.	1991–2014	297	12.38	27
Total: 100 years		889	9.89	136

Comparing the first (2) and the last periods (12), the average number of registered torrential floods has increased six times. More frequent flooding could be explained namely by the changing climate. Most probably, the observed rising trend in torrential floods will continue with even more extreme rainstorms. The torrential floods of May 2014 showed the effect of more intensive rainstorms (nature's ability), but also, consequences of the lack in torrent control (prevention) and communal preparedness (Kostadinov et al. 2013).

3.2 Torrent Control

The torrential floods (flash floods) occur quickly responding to intensive rainstorms in a small basin. Therefore, the most effective control of torrential floods is prevention which consists of erosion control works within a catchment, and constructions in streams channels (an integrated torrent control). The integral system of torrent control should include all important components of protection: strategy, methods, measures and constructions. There are two principal possibilities to reach the general aim of torrent control: first, to reduce the frequency, size and intensity of torrential events; second, to reduce the number of human beings, property and socioeconomic activities endangered. Also, very important is to increase the resistance of property, materials and constructions in endangered zones. In practice, two groups of measures are considered "event-concerned" and "object-concerned". The first group includes the classic forest-engineering measures: Biological oriented to re-vegetation, (mainly forestation), management measures to increase the canopy surface or stabilize the soil/vegetation complex, and finally engineering measures or constructions (Hattinger 1986).

The second group concerns on land use planning (including information, incentives and legal measures on the basis of expertises and danger zones plans), and, then measures to increase the resistance of the endangered constructions and cultivated land. In summary, the integral torrent control is an interdisciplinary approach which combines various measures and methods of ecology, economy, sociology and technical science to an integrated "Eco-socio-engineering system" with the aim to get a maximum security of humans within the endangered zones.

Focusing on ecosystem services in mountain catchments, particularly, biological engineering (afforestation of bare lands, grassing, amelioration of degraded forests, meadows and pastures, establishment of orchards on the terraces) are considered in the integral torrent control. In addition, administrative measures (adopting regulations and restrictions) and educational (or advertising) must be included (Kostadinov 2003, 2007). In Serbia, the most important administrative measures are: "Erosion Area Designation Plans" and "Operational Plans for Flood Control on II Category Water Streams" (torrents), which according to the Water Law of the Republic of Serbia from 2010, developed by municipalities (local governments). Forest cover has been widely assumed as the most effective measure in bioengineering (Kostadinov 1996). An example of the bare soil afforestation, and reducing intensity of soil erosion, on steep slopes in the Vlasina River watershed is given in Fig. 19.2.

In the past, with the political and economic transition, companies responsible for erosion and torrent control were privatized, and shortly afterwards liquidated. Recent problems related to flash floods and torrential erosion showed well the need of an effective integral torrent control. Namely, the recent catastrophic torrential floods (May and September 2014) had warned Serbia to be extremely prone to erosion and torrential flood risks. In March 2016, fifteen municipalities in Serbia declared "the state of natural disaster", and later the Serbian government declared "a country status of emergency". In order to meet the task of integral torrent control

Fig. 19.2 Afforestation of bare land in the Vlasina River watershed (Photo: V. Milojević)

in Serbia, it is necessary to prepare the following plans: National strategy for erosion and torrent control, erosion map, torrent cadastre, and cadastre of erosion control works.

4 Conclusions

Serbia is extremely prone to erosion and torrential flood risks. In the last 20 years, the intensity and frequency of torrential floods have increased significantly. The prevention and control of torrents is a consistent challenge. In identified torrential catchments, the integrated watershed management (both land erosion and torrent erosion control) has to include technical interventions in stream channels and soil bioengineering at critical slopes. Considering an effective administrative framework to control torrential hazards, the following steps in mapping and planning are necessary: (1) National strategy for erosion and torrent control, (2) Erosion map of Serbia, (3) Torrent cadastre, and (4) Cadastre of erosion control works.

The main goal is to achieve maximum security of people living in the mountain regions or nearby lowland cities. But, additionally, the torrent control should respect the environmental security, drinking water supply, rural development, sustaining biodiversity, etc. With the expected climate change impacts, particularly, rising extreme precipitation and more intensive torrential processes are expected. Implementation of this fact might significantly increase the efficiency of the torrent control systems.

Acknowledgement This study was supported by the project "Climate change impacts on environment: Influence monitoring, adaptation and mitigation" financed by the Serbian Ministry of Education and Science (43007), 2011–2016.

References

Barredo, J. I. (2007). Major flood disasters in Europe: 1950–2005. *Natural Hazards, 42*, 125–148.

Berz, G., Kron, W., Loster, T., Rauch, E., Schimtschek, J., Schmieder, J., Siebert, A., Smolka, A., & Wirtz, A. (2001). World map of natural hazards – A global view of the distribution and intensity of significant exposures. *Natural Hazards, 23*, 443–465.

Bissolli, P., Friedrich, K., Rapp, J., & Ziese, M. (2011). Flooding in eastern central Europe in May 2010: Reasons, evolution and climatological assessment. *Weather, 66*, 147–153.

De Moel, H., van Alphen, J., & Aerts, J. C. H. (2009). Flood maps in Europe – Methods, availability and use. *Natural Hazards Earth System, Science, 9*, 289–301.

Dragićević, S., Filipović, D., Kostadinov, S., Ristić, R., Novković, I., Živković, N., Andjelković, G., Abolmasov, B., Šećerov, V., & Djurdjić, S. (2011). Natural hazard assessment for land-use planning in Serbia. *International Journal of Environmental Research, 5*, 371–380.

Gavrilović, S. (1972). Engineering of torrents flows and erosion. *Journal "Izgradnja"* (Spec. ed.). Belgrade (in Serbian).

Gavrilović, S. (1975). *Torrents in Serbia, Belgrade. Fund for water and Institute for water managemnt of eroded regions*. Faculty of Forestry, Belgrade University (in Serbian).

Hattinger, H. (1986). Proposals for an integrated protection against torrents. In *Contributions to research on torrent erosion and avalanches*. Wien: IUFRO, Subject Group, S1.04–00 Mittelungen der Forstlichen Bundesversuchsanstalt.

IPCC. (2007). *Climate change 2007. Fourth assessment report of the Intergovernmental Panel on Climate Change*. Cambridge: Cambridge University Press.

Kostadinov, S. (1996). Soil erosion and sediment transport depending on land use in watersheds. In J. Křeček et al. (Eds.), *Hydrological problems and environmental management in highlands and headwaters*. New Delhi: IBH Publishing.

Kostadinov, S. (2003). Erosion and torrent control in mountainous regions of Serbia. In *Proceedings of International year of mountainous conference*. Belgrade: Faculty of Forestry, Belgrade University.

Kostadinov, S. (2007). *Erosion and torrent control in Serbia: Hundred years of experience*. Key note paper, International conference "Erosion and torrent control as a factor in sustainable river basin management", September 25–28, 2007, Belgrade, Abstract book.

Kostadinov, S., Košanin, O., Borisavljević, A., & Milčanović, V. (2013). Water erosion and torrential floods – significant factor in land degradation. In Proceedings of the 1st International congress on soil science, September 23–26, 2013, Belgrade.

MRG. (2003). *Annual review: Natural catastrophes 2002*. Munich: Munich Re Group.

Petrović, A., Kostadinov, S., & Dragićević, S. (2014). The inventory and characterization of torrential flood phenomenon in Serbia. *Polish Journal of Environmental Studies, 23*, 3–11.

Popović, V. (1995). *Pedological end soil erosion characteristics of space of Republic of Serbia*. In Spatial Plan of the Republic of Serbia. Faculty of Geography, Funded by Ministry of Environment and Spatial Planning, Belgrade (in Serbian).

Popović, T., Đurđević, V., Živković, M., Jović, B., & Jovanović, M. (2009). *Climate change and expected effects*. Belgrade: RHSS.

Pruski, F. F., & Nearing, M. A. (2002). Runoff and soil-loss responses to changes in precipitation: A computer simulation study. *Journal of Soil and Water Conservation, 57*, 7–16.

WMO. (2001). *Statement on the status of the global climate in 2001*. Geneva: World Meteorological Organization.

Part IV
New Challenges for Environmental Education and Active Citizenship

Chapter 20
Ecosystem Services, European Union Policies, and Stakeholders' Participation

Pier Carlo Zingari and Giovanna Del Gobbo

1 Introduction

There is a growing science and policy consensus on the need of reconsidering the inextricable links between natural and cultural capital. These links are made of the unfolding combinations of local knowledge and scientific evidences, of skills and practices, of the multiple values assigned by individuals and groups to ecosystems, of the supply and demand for ecosystem services and, finally, of the corporate and national systems of environmental accounting (Posey 1999; Convention on Biological Diversity 2010, 2012; Commission of the European Union 2014; Council of the European Union 2014).

Headwater catchments are, at the same time, drivers of key ecological processes, natural places providing ecosystem services, and sites of cultural and spiritual values. They are relevant examples of the inextricable links between natural and cultural capital showing that the provision of ecosystem services largely depends on balancing between the resilience of nature and the influences of individuals, communities and groups (Hill et al. 2014).

The long-term and short-term human influence on ecosystem conditions in catchments is positive in many cases and negative in others. Challenges for environmental education and active citizenship lie in every catchment—as each one of us is living in a catchment—and within an inclusive, positive and participatory dialogue between individuals, communities and groups, representing the diversity of "holders" involved and acting as both providers and beneficiaries of ecosystem services.

P.C. Zingari (✉)
Italian Botanical Society, Florence, Italy
e-mail: pzingari@gmail.com

G. Del Gobbo
University of Florence, Florence, Italy

© Capital Publishing Company, New Delhi, India 2017

J. Křeček et al. (eds.), *Ecosystem Services of Headwater Catchments*,
DOI 10.1007/978-3-319-57946-7_20

2 Ecosystems and Ecosystem Services: Reconnecting Natural and Cultural Capital

In 1992, the Convention on Biological Diversity (CBD), in an effort to sustainably connect environment and development issues, defined the ecosystem as "*a dynamic complex of plant, animal and micro-organism communities and their non-living environment interacting as a functional unit*" (article 2). In 2000, the Fifth Conference of Parties—COP CBD—endorsed the society- and culture-oriented ecosystem approach, along with its operational guidance, as "*a strategy for the integrated management of land, water and living resources ... It recognizes that humans, with their cultural diversity, are an integral component of many ecosystems*". The Principles of the ecosystem approach reiterate the complementary of nature and society (CBD-Decision V/6):

- economic, cultural and societal needs and choices expressed as clearly as possible;
- local people and communities rights and interests fully considered;
- cultural and biological diversity as central interconnected components;
- management of ecosystems for their intrinsic values and for the tangible or intangible benefits for humans, in a fair and equitable way;
- decentralisation of management involving all stakeholders and balance local interests with wider public interests, greater responsibilities, ownership, accountability, participation, and use of local knowledge besides scientific disciplines;
- effects of management on adjacent and other ecosystems;
- conservation of ecosystem structure and functioning, in order to maintain ecosystem services;
- recognition of change as inevitable; and
- use of all forms of relevant information, including scientific and local knowledge, innovations, and practices.

In this articulated framework, re-connecting the natural and cultural capital, recognising and investing on human capital, is an international commitment, an ongoing process and a way forward. Headwater catchments, with the vital services they provide and the multiple values they represent, are at the centre of this principle, process and perspective. But how can we define natural, cultural and human capital in this framework? Although originally the concept of capital belongs to economics, it is today commonly used under different angles. Some definitions are provided here.

The term "natural capital" has been proposed by David Pearce as a way to underline the role of nature in supporting the economy and human well-being (Pearce et al. 1989). Robert Costanza (2008) used the term in 1992, and in 2008 provided the following definition: "*the extension of the economic notion of capital—manufactured means of production—to environmental goods and services*". The natural capital includes all natural resources needed by human societies for their survival and well-being. Maes et al. (2013), along with other authors, highlights the biotic and abiotic resources as parts of one single natural capital. Both the stock and the services are embodied in the concept of natural capital: for example the living ecosystems of any headwater are stocks providing flows of goods and services to people and society. Since the flow of services from natural stocks requires that they function as whole systems, their structure and diversity are important components of natural capital. The stocks and the services can be depletable or non-depletable, renewable or non-renewable. Ecosystems and their services are renewable and depletable. They always depend on their resilience and stability (Holling 1973).

The term "cultural capital" has been originally introduced by Pierre Bourdieu (1979, 1986) within sociology, referring to the forms of knowledge, capacities, skills, education and advantages that a person holds, giving him or her a higher status in society. Carl Folke and Fikret Berkes (1992), from an environmental and economic perspective, use the term "cultural capital" as the adaptive capacity of human populations to deal with and modify the natural environment. They also note the fundamental dynamic aspect of human-environment interactions, which may be viewed as a coevolutionary interrelationship in which the two sides change one another continuously by mutual feedback. David Throsby (1999) finally formalises the concept of cultural capital within the broad discourse of economics, through a preliminary consideration of culture and cultural values. Cultural capital is an asset that contributes to cultural value. More precisely, cultural capital is the stock of cultural values embodied in an asset. This stock gives rise to a flow of goods and services over time, i.e. to commodities that themselves have both cultural and economic value. The asset may exist in tangible or intangible forms.

It is a definition that also recalls the concept of human capital (Kiker 1966). Human capital is associated with the ability to produce, accumulate and exchange knowledge in order to generate innovation and to sustain the competitiveness of a system (Bramanti and Odifreddi 2006). According to this position, the main output of human capital is to be found in the creation of ideas, theories, activities that result in the ability to interpret and transform the environment and create culture. In this sense, the human capital is the engine of development insofar, valuing the resources and endogenous potential—natural and cultural—it is able to perform an active role for sustainable processes. Human capital is generated thanks to the ability to learn and it is sustained by education.

When working on catchments the links between natural and cultural capital, between people and their territories, are indeed inextricable as they are in all other land resources and in all ecosystem services frameworks (FAO 2013). Local and traditional knowledge, adaptive technical skills and agro-silvo-pastoral practices have been working hand in hand with natural resources for ages shaping and maintaining in most cases productive and resilient ecosystems. Direct and indirect pressures have been also taking place undermining the components of ecosystems, from water, soils, climate and vegetation. Although the long-term human influence on ecosystem conditions in catchments is positive in most cases, it has been and still is negative in some cases and situations, such as economic crises and interests, conflicts, demographic pressures and technical mismanagement. The overall provision of ecosystem services largely depends on balancing the resilience of nature with the

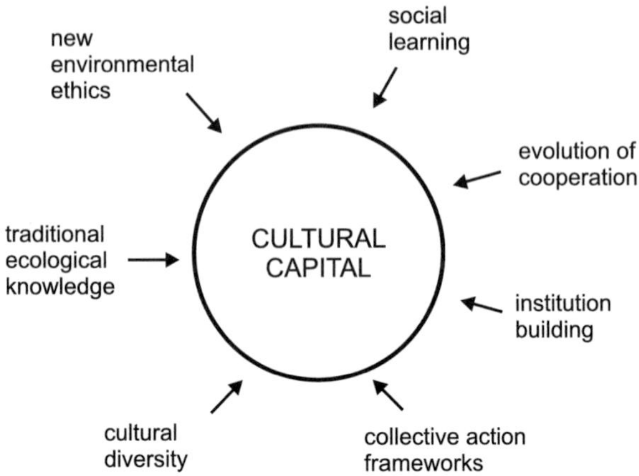

Fig. 20.1 Ways in which cultural capital may be conserved and enhanced (Folke and Berkes 1992)

influence of human activities. The positive links and balances between natural and cultural capital have been largely neglected by sectoral policies and institutions, by vertical scientific (Fig. 20.1) disciplines and by specialised territorial planning, including catchments land use (Nelson 2009).

3 From International Initiatives to National Policies: The Case of Europe

CBD is the international legally-binding reference Convention for ecosystems, their services and interactions with the social and cultural spheres. Nevertheless other conventions include commitments on nature as related to culture. The Convention for the Safeguarding of the Intangible Cultural Heritage (2003) defines the intangible cultural heritage, as "*transmitted from generation to generation, is constantly recreated by communities and groups in response to their environment, their interaction with nature and their history, and provides them with a sense of identity and continuity, thus promoting respect for cultural diversity and human creativity*". The United Nations Framework Convention on Climate Change (1992) explicitly refers to human activities, individuals and groups, as drivers of change. Founded on evidences, data and scientific research, the critical step for ecosystem assessment and services has been made by the comprehensive Millennium Ecosystem Assessment (2005). The assessment moves from the perspective of past, current and future changes in ecosystems and their services, their causes and effects, and options for conservation. Basically, ecosystem services are seen as the linkages between ecosystems and human well-being and simply defined as "*benefits people obtain from ecosystems*" (Millennium Ecosystem Assessment 2005). Once again natural and cultural connections are rather explicit in the ecosystem services categories:

- provisioning: food, water, timber and fibres;
- regulating: climate, floods, disease, wastes and water quality;
- supporting: soil, photosynthesis and nutrient cycling; and
- cultural: recreation, aesthetics and spiritual benefits.

The water-related dimension of ecosystem services has been recently reviewed in a comprehensive global scientific perspective integrating knowledge from different disciplines, such as hydrological modelling, environmental economics and consolidating current thinking, involving disciplines such as psychology, sociology, anthropology, law and human rights (Martin-Ortega et al. 2015).

The European Union and its current 28 Member States adopted the overarching EU Biodiversity Strategy to 2020 (2011) responding to the double mandate of the CBD-COP 10, held in Nagoya, Japan:

(a) the global Strategic Plan for Biodiversity 2011–2020; and
(b) the Nagoya Protocol on Access to Genetic Resources and Benefits Sharing.

The strategy aims at orientating each European country towards six main targets and twenty specific actions. Within Target 2. Maintain and restore ecosystems and their services, the Action 5 calls for improving the knowledge of ecosystems and their services in the EU and focuses on the following objective: "*Member States, with the assistance of the Commission, will map and assess the state of ecosystems and their services in their national territory by 2014, assess the economic value of such services, and promote the integration of these values into accounting and reporting systems at EU and national level by 2020*". To this end, Europe has been carrying on a research and policy process to support the development of a coherent analytical framework to be applied by the European Union and its Member States in order to ensure consistent approaches and harmonised data. This process is called Mapping and Assessment of Ecosystems and their Services, usually referred to as the MAES process (Maes et al. 2013). So far a number of documents and synthesis reports have been issued identifying: the analytical framework, the assessment indicators, the spatial extent of ecosystems and the supply and use of ecosystem services at the European scale between 2000 and 2010, the science contribution to environmental policy, and a synthesis on concepts, data, implementation and challenges.

In line with the Millennium Ecosystem Assessment, the objective of the MAES is to provide a critical evaluation of the best available information for guiding decisions on complex public issues. It is structured around a conceptual framework that links human societies and their well-being with the environment. More specifically the typology of ecosystems to be assessed and mapped is proposed making use of the Common International Classification of Ecosystem Services (CICES), developed for environmental accounting purposes. Currently the MAES is a growing participatory science-policy effort involving 28 countries, including governments, academia, NGOs, private and public bodies at different levels, various Commission services, the European Joint Research Centre, the European Environmental Agency and dedicated Topic Centres, the European Statistical Agency. Today the main results of the MAES process are monitored and evaluated through a centralised website (MAES 2016). The MAES process focuses on water, watersheds, green (and blue) infrastructure with particular attention to three main axes directly derived from the Water Framework Directive of 2000 and the Floods Directive of 2007:

(a) River basin management plans (2012a);
(b) Water scarcity and drought policy (2012b);
(c) Blueprint to safeguard water resources (2012c). Essential scientific and technical tools for water and catchments conservation and management are being delivered. Some among there are:

    - pan-European ecosystem map for terrestrial, freshwater and marine ecosystems (European Environmental Agency 2016a);
    - distribution data for species and habitats—Habitats Directive. 10 × 10 km ETRS 89 grid cells. (Commission of the European Union 2016a);
    - conservation status at the national biogeographical level and the EU biogeographical level (European Environmental Agency 2016b);
    - facts and figures on the state and trends of the different biodiversity and ecosystem components (European Environmental Agency 2016c);
    - ecosystem function and capacity of ecosystem service provision;
    - potential impacts of major pressures on forest, wetland, and heathland ecosystems;
    - biophysical assessment and economic valuation of water ecosystem services (Commission of the European Union 2016b);
    - ecosystem services concepts used in the River Basin Management Plans of the Water Framework Directive in five case studies in Europe: Italy, Scotland, Romania, Belgium and Portugal (OpenNESS 2016a);
    - pilot study on the valuation of green infrastructure for water pollution control (Italy) using ecosystem services concepts and local stakeholders involvement;
    - methodological framework for natural capital accounting (European Environmental Agency 2016d);
    - research projects involving countries in Europe and beyond, such as Operas on Ecosystem Science for Policy & Practice (Operas 2016), OpenNESS aiming at operationalisation of natural capital and ecosystem services (OpenNESS 2016b), and ESMERALDA Enhancing ecosystem services (Fig. 20.2) mapping for policy and decision-making (ESMERALDA 2016).

As far as MAES is concerned with water-related policies, and especially with catchment planning and management, some key general and specific considerations can be drawn. The quality and quantity of ecosystem services are always related to the state of conservation of the living stocks from which they are originated, i.e. the ecosystems. This consideration is particularly relevant for water, which is, at the same time, a strategic resource, a public good and an ecosystem service, often coming from private land, as it is the case in most of the European countries. At the catchment level the provision of ecosystem services is connected to the demand by

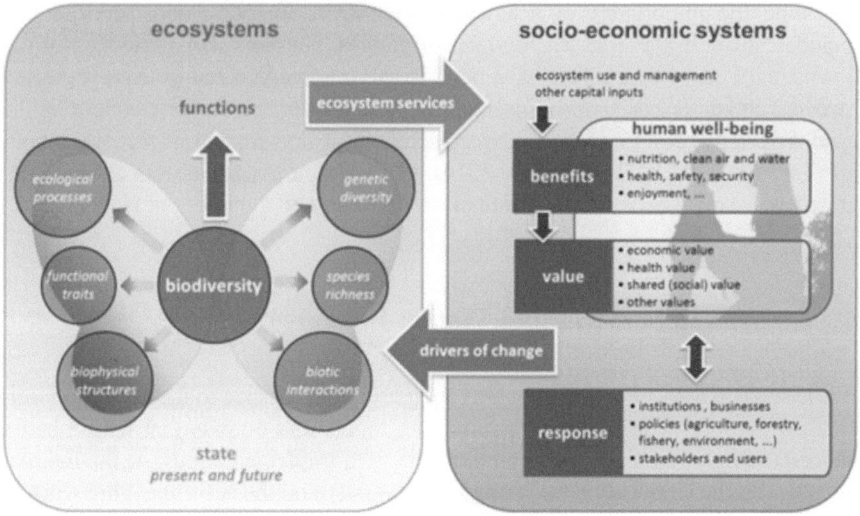

Fig. 20.2 Conceptual framework for the EU wide ecosystem assessment (From MAES 2013)

local communities, as well as by non-local groups and individuals. The full exercise of identification, analysis, data collection, mapping, assessment, monitoring and evaluation is a two-way, top-down and bottom-up, and a multi-scale-related process. The top-down corresponds to the overall framework provided by the European and national policies, strategies, methods and data. The bottom-up or local scale (e.g. sub-catchments) takes the lead either in terms of stakeholder's engagement and in terms of field data collection. The "holders" in the context of catchments are a large variety. They vary from the communities and groups, to the individual stakeholders as interest-holders, but also share-holders (e.g. private and public landowners), local and traditional knowledge- and skill-holders, provision-holders, value- and memory-holders (e.g. people living and working in the catchment), as well as final decision-holders and cost-holders. Their engagement is directly related to their perceptions, interests, values, intrinsic motivation and good communication. Their commitment is linked to their knowledge and skills as well as the awareness of their role. Participation *per se* is insufficient if there is no selection and real engagement of the key "holders". Moreover a stakeholders-based approach is not effective unless it is able to generate social learning processes and to go towards a collaborative governance (Pahl-Wostl et al. 2007).

This is particularly relevant to the payment for ecosystem services (PES). In Europe and in the United States PES schemes for farmers and, consequently, for each individual catchment, have been applied through the agri-environmental policies (Baylis et al. 2008). Currently MAES is developed also as a further tool to tackle PES along with European guidelines for designing them at the national level, identifying beneficiaries and adequate regulatory targets following the Water Framework Directive implementation and its outcomes. MAES contributes to upscaling the importance of headwater catchments for ecosystem services that includes not only the area where water originates, but the wider protection of the downstream ecosystems. This have been largely underestimated or even neglected by research and policies, and consequently by management, as, for example, in the case of the EU Water Framework Directive that limits its scope to surface water bodies (Takashi et al. 2002; Hill et al. 2014). MAES, as a horizontal tool for the integrated assessment ecosystems and their services, can fill a number of current gaps.

4 Ecosystem Services and Water: Lessons Learned

While science and policy are attaching increasing importance to the ecosystem services, a number of initiatives are providing inputs and outputs that help a better understanding of methods and their concrete field implementations. At the international level, the Convention on Biological Diversity and the accompanying work of the International Platform on Biodiversity and Ecosystem Services (IPBES) are catalysing interests and exercises to translate the concepts into practices. While the concepts are now rather clear, at least at the expert level, the practices should be further explored, communicated and transmitted. In this sense it has been suggested to listen to the voices of experience: "*simply generating and communicating scientific knowledge is not sufficient. The focus on biodiversity knowledge that has been scientifically validated might be attractive to 'elite actors', from natural scientists to national governments, but it omits many other important stakeholders and knowledge-holders, including indigenous people, businesses, farmers, community partnerships and fishers. What counts as legitimate knowledge, and how it is generated, influences its practical effectiveness*" (Turnhout et al. 2012).

Consequently one of the main general lessons learned on ecosystem services is the need to re-conciliate the natural and cultural capital by bringing together local experience and scientific knowledge. In policy terms this means that strategies and plans to restore and maintain good ecological conditions must go through a constructive dialogue between local stakeholders and different-level institutions. In many countries, the success stories of PES are the result of bridging experience and science in the field while combining private and public interests (Porras et al. 2013). In the light of decades of PES successes and failures, most of the lessons learned are basically related to the criteria of the ecosystem approach, such as:

- clearly expressed economic, cultural and societal needs and choices;
- full consideration of local people and communities knowledge, skills, practices, rights and interests;
- cultural and biological diversity interconnection;
- intrinsic values and tangible or intangible benefits fairly considered in management; and
- good conservation status of ecosystems in their dynamics and change.

The example of using the ecosystem approach in the implementation of the EU Water Framework Directive "*demonstrates its potential for a systemic thinking as it can provide a consistent framework for identifying shared aims and evaluating alternative water management scenarios in decision making. Allowing for a broad consideration of the benefits, costs and trade-offs that occur in each case, this approach can further improve the economic case for certain measures, and can also help restore the shift in focus from strict legislative compliance towards a more holistic implementation that can deliver the wider aims and intentions of the Directive*" (Vlachopoulou et al. 2014).

5 Ecosystem Services: Changes and Challenges

Dramatic processes and changes are taking place globally and locally in a growing populated and urbanised world: conflicts, migrations, water crisis, floods, desertification, erosion, pollution, biodiversity and heritage loss, ecosystems degradation, among others. These negative processes and changes are causing extensive damages and generating unbearable economic and social costs. The ecosystem approach and assessment allow to accounting for the economic dimension of benefits from nature and nature-based solutions. Headwater catchments are of paramount importance as natural and cultural places providing vital services and values to face these processes and changes. Consequently, a number of challenges today can be identified within science, international commitments and policies. The main overall scientific challenges for ecosystems services is the understanding of natural and cultural connections: "*concepts and methods traditionally developed independently within the respective disciplines of ecology and social science are not sufficient to address the interrelated nature of ecosystems services. Within the ecosystems services community, the need for better integration of social and ecological science has mostly been framed in terms of cooperation between ecologists and economists; however, for cultural ecosystems services in particular, the cooperation must be extended to broader domains of environmental and social sciences*" (Daniel et al. 2012). In the same way international commitments related to biological and cultural diversity, such as CBD and UNESCO Conventions, have established challenging joint programmes on the integration of environment and people, that are waiting for concrete implementation.

One of the main challenge for environmental policies, as other policies, is the shift from the negative paradigm of "polluter pays" and "footprint" to the positive one of "provider gets" and "handprint" (Mauerhofer et al. 2013). The way this policy challenge can be met is to bring together the diversity of actors and "holders"—public, private, NGOs, local communities, groups and individuals—in an exercise of collaborative governance based on data, knowledge, motivation, participation, consultation and consensus (Emerson et al. 2012). Finally the big societal challenge is the educational approach on the supply and demand of ecosystem services, especially water, for a more active citizenship and participation.

6 Educational Approaches and Stakeholders Participation

Headwater catchments are of vital importance in a changing world with a growing demand for water and a limited supply. If we consider that less than 0.5% of the total freshwater of the planet is actually available, the reconsideration of the importance of headwater catchments is an obligation for all, as we all live in a catchment and, consequently we all depend on its water. Consequently, how can we educate citizens and "holders" to better act for water? What are the most effective educational, communication, learning and training approaches to raise the necessary awareness and secure a full and committed participation of people and groups?

Innovative trends have emerged in the understanding and management of social-ecological systems (Paavola and Hubacek 2013), and this new emphasis on sustainability requires increased social engagement, along with a participatory learning process (Pahl-Wostl et al. 2007). Often, attempts to build participatory governance systems are not able to deliver the expected results: they do not produce a collective programme, due to a lack of community awareness, commitment and effective participation, as well as the absence of mechanisms to support community involvement in the processes of natural resources management (Fernández 2015). Building a participatory and consultative framework means to invest in learning and awareness of communities: it means not only to underline need of education for sustainability, but properly, the need of sustainable learning. Sustainability means that all the societal expectations, ideals and needs are balanced with the goods and services provided by ecosystems, under "appropriate" management, in a dynamic and constantly adjusting process.

To implement a participatory management of water resources does not mean building a project network, aimed exclusively at distributing functions and tasks, but it means giving life to a strategic partnership. It is a strategy necessarily based on "social learning" for sustainable development. Social learning as an approach to the understanding and management of environmental issues has become a prominent interpretative framework in the assessment and management of natural resources. In resource assessment and management, the notion of social learning coincides with the thrust for public participation and the growing importance given to sustainability. New methodological approaches such as participatory integrated assessment and sustainability science involves the consideration that public participation is necessary in any attempt to build a robust knowledge capable of dealing with the challenges, complexities, and uncertainties of sustainable development.

This position includes the need of a distributed and shared common learning at local community level, considering sustainability as a result of a co-constructed learning process through the involvement of different institutions, groups, networks and individuals. Any local community, in the framework of social learning, learns by an interactive and negotiated process among the various actors, towards functional integrated solutions of problems and the development of a widespread culture of sustainability. It is understandable that the same management actions can take the value of action to support learning of all those involved in different ways. This is indeed a sustainable approach to the problem of management, aiming to highlight the educational action in its horizontal and global democratic and ecological sense, and in the light of an "educational ecology", which perceives the value of learning in its different aspects: formal, non-formal and informal.

The UN Decade of Education for Sustainable Development (ESD) 2005–2014 has been providing a number of valuable inputs, specifically directed to water with the aim "*to help people to develop the attitudes, skills, perspectives and knowledge to make informed decisions and act upon them for the benefit of themselves and others, now and in the future* "(United Nations 2003). Education is the strategic entry point for water, implying that:

- learning encourages behavioural changes and provides the skills required for participation in water governance;
- schools and other educational environments promote water sustainability; and
- educational structures, policy and management provide guidance, oversight, coordination, monitoring and evaluation to ensure an effective, sustainable and institutionalized educational response to water governance challenges.

Turning into a practical good example, the Scotland's Centre of Expertise for Waters (CREW), launched in 2011, is significantly working to connect research, policy, education and participation on water-related ecosystem services. CREW identified how and where the refinement of the ecosystem services approach improves the delivery of the EU Water Framework Directive, developing a set of recommendations, which put stakeholders participation at the core of the exercise (Spray and Blackstock 2013).

The example of using the ecosystem approach in the implementation of the EU Water Framework Directive "*demonstrates its potential for a systemic thinking as it can provide a consistent framework for identifying shared aims and evaluating alternative water management scenarios in decision making. Allowing for a broad consideration of the benefits, costs and trade-offs that occur in each case, this approach can further improve the economic case for certain measures, and can also help restore the shift in focus from strict legislative compliance towards a more holistic implementation that can deliver the wider aims and intentions of the Directive*" (Vlachopoulou et al. 2014).

7 Conclusion

Ecosystem service is a rapidly evolving paradigm under implementation worldwide with a strong interdisciplinary and intersectoral significance, and a multilevel pertinence. Ecosystem services are not goals *per se*, but means to achieve well-being (Costanza et al. 2014). They are identified by, and made for people's values. Possibly nowhere else than in the headwater catchments they are a challenge for local people and for broader downstream communities, activities and settlements.

This paradigm helps to reconnect nature and culture with benefits for both, and hopefully reducing pressures on the environment, valuing and accounting for the positive externalities. Since the early 2000, the European Union started a process from the nature and water legislation to the biodiversity strategy towards the integrated management of natural resources with and for people, with their cultural diversity, following the principles of the ecosystem approach.

Besides, a new approach to heritage is proposed with its many dimensions: cultural, physical, digital, environmental, human and social. Its value—both intrinsic and economic—is a function of these different dimensions and of the flow of associated services (Commission of the European Union 2014).

New challenging perspectives for headwater catchments are those provided by the social and ecological integration, as proposed by the ecosystem approach. Radical changes in resource management require social learning as sustainable learning that increases the capacity of those who manage, in an integrative and organic way of the whole social-ecological system of which they are part. Improve the co-adaptive systemic capacity of agents to anticipate and deal with the unintended, undesired, and irreversible negative effects of development is a strategy, vision and perspective. The institutional frameworks should help, facilitate and assist connections between sectors, disciplines and groups. Europe is making this effort developing its MAES with a link to the CBD process. Governments should follow at the national level. But much is in the hands of local "holders" who needs further systemic thinking supported by education, learning, training, sharing and participating.

References

Baylis, K., et al. (2008). Agri-environmental policies in the EU and United States: A comparison. *Ecological Economics, 65*, 753–764.

Bourdieu, P. (1979). Les trois états du capital culturel. *Actes de la Recherche en Sciences Sociales, 30*, 3–6.

Bourdieu, P. (1986). The forms of capital. In J. F. Richardson (Ed.), *Handbook of theory of research for the sociology of education*. Westport: Greenword Press.

Bramanti, A., & Odifreddi, D. (2006). *Capitale umano e successo formativo. Strumenti, strategie, politiche*. Milano: Franco Angeli Editore.

Commission of the European Union. (2011). *Communication on "Our life insurance, our natural capital: An EU biodiversity strategy to 2020"*. COM (2011) 244.

Commission of the European Union. (2012a). *Report on River Basin management plans*. COM (2012) 670.

Commission of the European Union. (2012b). *Report on the review of the European water scarcity and droughts policy*. COM (2012) 672.

Commission of the European Union. (2012c). *Communication on a blueprint to safeguard Europe's water resources*. COM (2012) 673.

Commission of the European Union. (2014). *Communication on "Towards an integrated approach to cultural heritage for Europe"*. COM (2014) 477.

Commission of the European Union. (2016a). *Distribution data for species and habitats Habitats Directive 10 × 10 km ETRS 89 grid cells*. http://ec.europa.eu/environment/nature/knowledge/rep_habitats/index_en.htm. Accessed 14 Mar 2016.

Commission of the European Union. (2016b). *Cook-book for water ecosystem service assessment and valuation*. Luxembourg: Joint Research Centre. 135 pp. http://publications.jrc.ec.europa.eu/repository/bitstream/JRC94681/lbna27141enn.pdf. Accessed 14 Mar 2016.

Convention on Biological Diversity. (2010). Decision X/20, 2010 on Cooperation with other conventions and international organizations and initiatives; Decision X/32, 2010 on Sustainable use of biodiversity.

Convention on Biological Diversity. (2012). Decision XI/6, 2012. Cooperation with other conventions, international organizations, and initiatives.

Costanza, R. (2008). *Natural capital. The Encyclopaedia of the Earth*. Cleveland: Cutler.

Costanza, R., de Groot, R., Sutton, P., van der Ploeg, S., Anderson, S., Kubiszewski, I., Farber, S., & Turner, R. K. (2014). Changes in the global value of ecosystem services. *Global Environmental Change, 26*, 152–158.

Council of the European Union. (2014). Charter of Rome on Natural and cultural capital. Brussels 16540.14.

Daniel, T. C., Muhar, A., Arnberger, A., Aznar, O., Boyd, J. W., Kai, M. A., Chane, K. M. A., Costanza, R., Elmqvist, T., Flint, C. G., Gobster, P. H., Grêt-Regamey, A., Lave, R., Muhar, S., Penker, M., Ribe, R. G., Schauppenlehner, T., Sikoro, T., Soloviy, I., Spierenburg, M., Taczanowska, K., Tam, J., & von der Dunk, A. (2012). Contributions of cultural services to the ecosystem services agenda. *Proceedings of the National Academy of Sciences of the U. S. A, 109*(23), 8812–8819.

Emerson, K., Nabatchi, T., & Balogh, S. (2012). An integrative framework for collaborative governance. *Journal of Public Administration Research and Theory, 22*(1), 1–29. Oxford University Press.

ESMERALDA. (2016). *European Research Project on Enhancing ecosystem services mapping for policy and decision-making*. www.esmeralda-project.eu. Accessed 14 Mar 2016.

European Environmental Agency. (2016a). *EIONET Pan-European ecosystem map for terrestrial, freshwater and marine ecosystems*. Copenhagen: European Environmental Agency. http://projects.eionet.europa.eu/eea-ecosystem-assessments/library/draft-ecosystem-map-europe. Accessed 14 Mar 2016.

European Environmental Agency. (2016b). *Conservation status at the national biogeographical level and the EU biogeographical level*. Copenhagen: European Environmental Agency. http://www.eea.europa.eu/publications/state-of-nature-in-the-eu. Accessed 14 Mar 2016).

European Environmental Agency. (2016c). *Facts and figures on the state and trends of the different biodiversity and ecosystem components*. Copenhagen: European Environmental Agency. http://www.eea.europa.eu//publications/eu-2010-biodiversity-baseline-revision. Accessed 14 Mar 2016.

European Environmental Agency. (2016d). *EIONET Methodological framework for natural capital accounting*. Copenhagen: European Environmental Agency. http://projects.eionet.europa.eu/ecosystem-capital-accounting/library/reference-document-natural-capital-accounting. Accessed 14 Mar 2016.

FAO. (2013). *Indigenous peoples' food systems and well-being*. Rome: FAO.

Fernández, H. R. (2015). From an informed public to social learning for water management: Is Argentina cast adrift? *International Journal of Social Science and Humanities Research, 3*(2), 66–70.

Folke, C., & Berkes, F. (1992). *Cultural capital and natural capital interrelations*. Beijer discussion paper series no.8. Beijer International Institute of Ecological Economics. The Royal Swedish Academy of Sciences.

Hill, B. H., Kolka, R. K., McCormick, F. H., & Starry, M. A. (2014). A synoptic survey of ecosystem services from headwater catchments in the United States. *Ecosystem Services, 7*, 106–115.

Holling, C. S. (1973). Resilience and stability of ecological systems. *Annual Review of Ecology and Systematics, 4*, 1–23.
Kiker, B. F. (1966). Historical roots of the concept of human capital. *Journal of Political Economy, 29*, 481–499.
MAES. (2013). *Mapping and assessment of ecosystems and their services. An analytical framework for ecosystem assessments under action 5 of the EU biodiversity strategy to 2020.* Brussels: European Union.
MAES. (2016). *Mapping and Assessment of Ecosystems and their services website*. Brussels: DG Environment. http://biodiversity.europa.eu/maes. Accessed 14 Mar 2016.
Maes, J., et al. (2013). *Mapping and assessment of ecosystems and their services. An analytical framework for ecosystem assessments under action 5 of the EU biodiversity strategy to 2020.* Luxembourg: Publications office of the European Union.
Martin-Ortega, J., Ferrier, R. C., Gordon, I. J., & Khan, S. (2015). *Water ecosystem services. A global perspective*. Cambridge: Cambridge University Press.
Mauerhofer, V., Hubacek, K., & Coleby, A. (2013). From polluter pays to provider gets: Distribution of rights and costs under payments for ecosystem services. *Ecology and Society, 18*(4), 41.
Millennium Ecosystem Assessment. (2005). *Ecosystems and human well-being: Synthesis*. Washington, DC: Island Press.
Nelson, J. G. (2009). *Places: Linking nature, culture and planning*. Calgary: University of Calgary Press.
OpenNESS. (2016a). European research project on ecosystem services concepts used in the River Basin management plans of the water framework Directive in five case studies in Europe: Italy, Scotland, Romania, Belgium and Portugal. http://www.openness-project.eu/. Accessed 14 Mar 2016.
OpenNESS. (2016b). *European Research Project on Operationalisation of natural capital and ecosystem services*. www.openness-project.eu. Accessed 14 Mar 2016.
Operas. (2016). *European Research Project on Ecosystem Science for Policy & Practice*. www.operas-project.eu. Accessed 14 Mar 2016.
Paavola, J., & Hubacek, K. (2013). Ecosystem services, governance, and stakeholder participation: An introduction. *Ecology and Society, 18*(4), 42.
Pahl-Wostl, C., Craps, M., Dewulf, A., Mostert, E., Tabara, D., & Taillieu, T. (2007). Social learning and water resources management. *Ecology and Society, 12*(2), 5.
Pearce, D., Markandya, A., & Barbier, E. B. (1989). *Blueprint for a green economy*. London: Earthscan.
Porras, I., Barton, D. N., Miranda, M., & Chacón-Cascante, A. (2013). *Learning from 20 years of payments for ecosystem services in Costa Rica*. London: International Institute for Environment and Development.
Posey, D. (1999). Culture and nature: The inextricable link. In D. Posey (Ed.), *Cultural and spiritual values of biodiversity*. London: UNEP/ITDG Publishing.
Spray, C., & Blackstock, K. (2013). *Optimising water framework directive river basin management planning using an ecosystem services approach*. CREW, The James Hutton Institute, The Scottish Government, Scotland.
Takashi, G., Sidle, R. C., & Richardson, J. S. (2002). Understanding processes and downstream linkages of headwater systems. *Bioscience, 52*(10), 905–916.
Throsby, D. (1999). Cultural capital. *Journal of Cultural Economics, 23*, 3–12.
Turnhout, E., Bloomfield, B., Hulme, M., Vogel, J., & Wynn, B. (2012). Listen to the voices of experience. *Nature, 488*, 455.
United Nations. (2003). *UN decade of education for sustainable development*. Resolution 57/254. New York. http://www.un-documents.net/a57r254.htm
Vlachopoulou, M., Coughlin, D., Forrow, D., Kirk, S., Logan, P., & Voulvoulis, N. (2014). The potential of using the ecosystem approach in the implementation of the EU Water Framework Directive. *Science of the Total Environment, 470–471*, 684–694.

Chapter 21
Water Conservation Business Arising from Company's Environmental Responsibility

Mia Suominen

1 Introduction

Vapo Oy is a Finnish, non-listed, joint-stock company owned by the Finnish State (50.1%) and Suomen Energiavarat Oy (49.9%). It was established in 1940. Vapo has 1000 employees and the annual group turnover is about 500 M€. It produces energy and environmental peat, forest fuel and pellets for heat, steam and electricity customers in Finland, Sweden and the Baltic countries. Peat energy fulfills 4–5% of the total energy demand in Finland, although only 0.7% of the total area is used for energy purposes. Finland has a remarkable area of peatland: 1/3 of the total area (9.3 million hectares) of which 13% is under conservation, 52% is in forest land and 32% is still untouched (Anon 2011). Peatland area is a valuable energy storage with energy content of 13,000 TWh (Virtanen et al. 2000).

The ecological sustainability of utilisation of peat reservoirs is guaranteed by legislation. Peat production can be started only in such areas that are no longer in a natural state. Beginning peat production requires natural value assessment of the swamp, environmental impact assessment (EIA) in large projects, and always environmental permitting. Peat production is regulated by an environmental protection act and decree, a law and decree of the EIA process, nature preservation act and decree, water act, waste act and decree, land use and building act, act on water and sea resources management, law of environmental impairment liability insurance, law of handling of dangerous chemicals and explosives and safety, law of handling flood risks and reindeer management act (Anon 2011).

M. Suominen (✉)
Vapo Ventures, Vapo Oy, Helsinki, Finland
e-mail: Mia.suominen@vapo.fi

© Capital Publishing Company, New Delhi, India 2017

J. Křeček et al. (eds.), *Ecosystem Services of Headwater Catchments*,
DOI 10.1007/978-3-319-57946-7_21

Table 21.1 Vapo's additional commitments to responsible peat production 2011–2016

Commitment	Realised	In progress
Commitments for the period 2011–2014		
Vapo is committed to building water treatment systems using the best available water treatment technologies (BAT) at all of its peat production areas by the end of 2014.	✓	
Vapo continues the enhanced monitoring of the effectiveness of water treatment structures at peat production areas.	✓	
On its website, Vapo publishes information on the contact persons for each production area, as well as reports on emissions and the monitoring of waterways produced for the authorities by independent consulting agencies.	✓	
Vapo increases the use of continuously operating measurement devices in monitoring the load on watercourses and publishes the results online.	✓	
Vapo is committed to producing peat only in production areas where the natural state has been altered, generally meaning ditched peatlands.	✓	
Vapo continues to actively exchange mires with significant nature value owned by the company to ditched peatlands, or to sell such mires for use for conservation purposes.	✓	
Commitments 2016–		
From 2016, the solid matter and humus load of new peat production areas will be lower than that of the same bog before peat production.		✓
All of Vapo's peat production areas will be subject to emissions monitoring during the peat production season, and half of the production areas will be subject to year-round emissions monitoring.		✓
Areas owned by Vapo and released from peat production will be in active after-use within 2 years of the end of production operations.		✓

Source*:* Vapo annual report 1.5.2014–30.4.2015

2 Environmental Turnaround Programme

Vapo started to produce energy peat in the 1970s. Very soon it was noticed that there is a water protection need related to peat production and the need was tackled. Vapo has built thousands of water treatment solutions for its own needs. In recent years, demand arose to increase the level of environmental responsibility far beyond the legislation requirements. That is why Vapo started the environmental turnaround programme (responsibility programme 2011–2016) in order to achieve better social acceptance of peat production. It has realised nine additional actions (Table 21.1).

In its new environmental programme, Vapo committed to responsible peat production. The company built water treatment systems using the best available water treatment technologies (BAT) at all of its peat production areas by the end of year 2014. That meant 9000 hectares with new water systems with a total cost of 30 M€. It enhanced monitoring of the effectiveness of water treatment structures at peat production areas. About 14,000 water samples are taken each year and all the production areas (over 400 areas) are inspected at 2-week intervals during the production season. On its website, Vapo published information on the contact persons for each production area as well as reports on emissions and the monitoring of waterways produced for the authorities by independent consulting agencies. The company increased the use of continuously operating measurement devices in monitoring the load of watercourses. Thirty measurement stations automatically measure the outgoing water every hour, 365 days per year.

Vapo also committed to producing peat only in production areas where the natural state has been altered, generally meaning ditched peatlands. It also actively exchanged mires with significant nature value owned by the company to ditched peatlands, or sold such mires for use for conservation purposes. It even promised that from the year 2016 the solid matter and humus load of new peat production areas will be lower than that of the same bog before peat production. All of Vapo's peat production areas will be subjected to emissions monitoring during the peat production season, and half of the production areas will be subject to year-round emissions monitoring. Areas owned by Vapo and released from peat production will next be in land use (forest, field, wetland) within 2 years of the end of production operations. With the concrete results of the environmental turnaround programme, Vapo has now achieved better general acceptance and citizens' trust for responsible peat production.

3 Rise of a Corporate Start-Up *Clean Waters*

Unique know-how of water treatment and conservation has accumulated to the organisation throughout the years. In autumn 2012 the idea arose of offering the know-how more widely outside the own organisation. A corporate start-up called "*Clean Waters*" was created to productise the know-how. Today, Clean Waters is a business line at Vapo offering natural water treatment solutions to different land users, mainly for forest water conservation, but also for the agriculture and to the urban environments for stormwater control. In forest water conservation projects, Clean Waters has implemented comprehensive water protection design, restoration of streams and construction of forest wetlands where special know-how and tailoring to local conditions is required. One example is the Lake Pyhäjärvi catchment area, where planning included negotiations with landowners and engineering of suitable water treatment structures in 22 outlets of the project area (Suominen et al. 2016). A total of 33 water protection structures were proposed to be built and in that way marked load reduction could be achieved in both suspended solids and P load.

Clean Waters uses natural treatment methods that are based on nature's own biological, chemical and physical processes. Such methods are e.g., sedimentation ponds, flow control dams, overland flow fields, vegetation fields, biofiltration and wetlands. In special cases, chemical treatment can be used. Natural treatment methods are based on retention, sedimentation, filtration, absorption and evaporation. Investment in water treatment is an investment in sustainability and productivity. Water treatment methods can result in economic advantages like increased yields or better flood control, improved biodiversity and recreational values and, of course, better water quality.

4 Water Emissions in Finland

Pollution and eutrophication are threatening the quality of water systems worldwide. Degradation of natural waters is caused by the loading of different land use on waters. According to the Finnish Environmental Institute (Tattari et al. 2015), there are many sources for water emissions in Finland, including also natural emission (Table 21.2). With natural water treatment actions it is possible to improve and maintain good quality of natural waters.

The increase of total organic carbon (TOC) emission is a challenge in Scandinavia, the UK and North-America. Climate change and decrease of acid fallout are thought to be the main causes of this change (Löfgren et al. 2003). There is a growing need to use the existing methods in larger areas as well as to develop totally new water protection methods.

Forests cover more than 70% of the land area of Finland. Measured by the proportional share of forest land, Finland is the most forested country in Europe. It is a significant source of welfare for Finnish people. Forest management operations like clear-cuttings, fertilisation and ditching also cause water load. The water load is greatest in the first years after operation and continues for 2–10 years (Anon 2011). The amount of water load is affected by the operation type, date and area size, topography, soil type, closeness of the groundwater and water basin. Water protection measures can reduce such loads considerably like protective buffer areas, ditch breaks, weirs and sedimentation pool.

The forest water conservation is even more important in the future because of climate change and increased forest biomass harvest in order to maintain lakes and rivers in a good/excellent ecological state (European Water Framework Directive). In Finland, recommendations for water protection measures in forestry were published already in the 1980s, but realisation in the field has been rather slow. The water structures (sedimentation pots and basins) have often been undersized and therefore they have filled rapidly with solid matter and lost their effectiveness. The water flood may also be too high and solid matter continues to the downstream water basin. In the future, it is important to develop and use water conservation management practices with more detailed and tailored water conservation. The target should be the use of the best available water treatment methods with good effectiveness in forest management operations.

Table 21.2 Water emissions in Finland. Survey carried out by Finnish Environment Institute

	Area	Solid matter		Total phosphorus		Total nitrogen		TOC	
	Million ha	kg $ha^{-1}a^{-1}$	1000 kg/a	kg $ha^{-1}a^{-1}$	1000 kg/a	kg $ha^{-1}a^{-1}$	1000 kg/a	kg $ha^{-1}a^{-1}$	1000 kg/a
Diffuse load									
Natural wash-off	30.4	5.1	155,000	0.05	1.500	1.3	39,500	57	1730.000
Forestry	0.28	255	71,000	0.47	130	5.8	1600	77	21,600
Agriculture	2.2	610	1,340,000	1.10	2400	13.0	33,000	220	484,000
Deposits in water	3.5			0.10	350	3.0	10,500		
Scattered settlement	1.07	5.2	5600	0.37	400	2.5	2700		
Drainage waters	0.69	297	205,000	0.39	270	4.7	3200		
Point load									
Municipal wastewater					160		10,900		9800
Industry			15,400		170		3700		26,000
Fish farms					90		780		
Fur farms					45		430		
Peat production	0.065	53	3400	0.30	20	8.1	524	146	9500
Total					**5535**		**106,834**		

Survey carried out by: Finnish Environment Institute Client and source: Finnish Energy industries, environmental pool http://energia.fi/sites/default/files/syke_maankayton_vesistovaikutukset_raportti_3132015.pdf
TOC total organic carbon

5 New Sustainable Businesses in the Pipeline

Business development has been an integral part of operations at Vapo, and the development of existing businesses will continue in much the same vein as before. A new approach for the company is starting up new businesses based on existing strengths and competencies. In its new strategy, Vapo has highlighted renewal as its key objective. The Vapo Ventures business has been harnessed as the engine of this renewal. The dream behind this renewal is the sustainable use of natural resources by increasing their refining rate to create energy products and new consumption goods. This constitutes a great opportunity for Vapo.

Developments and new innovation rarely happen on their own and at random, as an offshoot of basic operations. Because of this, renewal must be a managed and controlled process of assessing and analysing new opportunities that the organisation's assets, competencies or other competitive advantages make it well-equipped to take advantage of. An entrepreneurial approach helps keep the goal clear in the mind, whether that goal is identifying new business opportunities or improving the core business. Creating something new is not easy, and you may have to wait for results for a long time. Processes, technologies and product features have to be tested. Also hitting the right window of time is a major success. Stumbles and failures should not be a cause for concern, as they are part of the nature of the venture business. To create future growth, risks have to be taken (Heikkinen and Kerr 2014).

Vapo Ventures is an ongoing process of assessing business ideas based on the developed operating model. The operating model proceeds from an idea database, using the gate model and evaluation criteria, to preliminary assessments and R&D, and from there, to business plans and ultimately to actual business operations. The inputs for the Venture idea database include not only the company's current research activities, but also past R&D activities from throughout Vapo's history, all Vapo personnel, as well as the network of customers and partners. Continuous new business development guarantees the company's survival in the changing world. It may also result in possible future success stories like, for example, the case of Clean Waters.

6 Conclusion

Vapo has implemented a large environmental turnaround project since 2011. In addition to enhanced environmental responsibility, Vapo has succeeded in generating new ecologically sustainable business from its corporate responsibility. The first Vapo startup is Clean Waters, which provides solutions for the treatment of natural waters. This example shows that it is possible to create new, ecologically sustainable business opportunities from traditional business areas when know-how, network and assets are evaluated from a different point of view. Today, Clean Waters

Ltd. serves customers broadly in different segments from agriculture and forestry to city stormwater purification, and is starting to look to opportunities also abroad.

References

Anon. (2011). Ehdotus soiden ja turvemaiden kestävän ja vastuullisen käytön ja suojelun kansalliseksi strategiaksi. Työryhmämuistio MMM 1, pp. 10, 21–22, 88 (in Finnish).

Heikkinen, K., & Kerr, W. (2014). *What Finnish companies can do to become better in innovation*. NAG Research note 11.

Löfgren, S., Forsius, M., & Andersen, T. (2003). *The colour of water: Climate induced water colour increase in Nordic lakes and streams due to humus*. SLU, Suomen Ympäristökeskus, NIVA.

Suominen, M., Varis, L., & Pulkkanen, M. (2016). *Control of suspended solids and nutrient load in Lake Pyhäjärvi catchment area in Finland*. In European Forest Institute EFI's series "What Science can tell us" (manuscript submitted).

Tattari, S., Puustinen, M., Koskiaho, J., Röman, E., & Riihimäki, J. (2015). Valuma-alueen eri lähteistä tulevan vesistökuormituksen arviointi ja vähentämismahdollisuudet. Finnish Environmental Institute (in Finnish).

Vapo. (2015). *Vapo*. Annual report, 1.5.2014–30.4.2015 (in Finnish).

Virtanen, K., Hänninen, P., Kallinen, R.-L., Vartiainen, S., Herranen, T., & Jokisaari, R. (2000). *The peat reserves in Finland. Geological survey of Finland*. Report of Investigation 156 (in Finnish).

Chapter 22
Natural Protective Services in Mountain Catchments: Provision, Transaction and Consumption

Florian Rudolf-Miklau

1 Introduction

The term '*natural protective effect*' is commonly associated with the capacity of ecosystems to balance or buffer the action of natural hazard processes, such as floods, droughts, snow avalanches, mass-movements or soil erosion (Gray and Leiser 1982; Stöcklin et al. 2007; Wehrli et al. 2007; Salzman et al. 2001; Nairz et al. 2015). They are generated by certain elements in mountain catchments: the topography, soils, vegetation cover and forests, and waterbodies. Protective effects in ecosystems are inseparably linked with the emergence and impact of the hazard processes themselves and often originate from the same ecosystem elements. While in the normal case, natural forces driving or buffering hazard processes are well balanced in mountain catchments, excessive natural events (e.g. extreme precipitation, thick snow cover, strong earthquakes) may severely unbalance the ecosystems and trigger natural catastrophic events. However, antagonistic natural effects in ecosystems exist, that counterbalance the emergence, displacement and action (impact) of hazard processes (Rudolf-Miklau et al. 2012). Natural protective effects emerge and subsist by nature, respectively without human creative act. They are bounded in their capacity of protection, either seasonally or when reaching a critical threshold. As a rule, the sustainability of natural protective effects requires a certain treatment or maintenance of the ecosystem.

The most prominent natural protective effects are related to flooding. Ecosystems may buffer the formation of floods in a manifold way, in particular by soil water storage, by retention effects of depressions, lakes, wetlands or inundation areas (flood plains) or by the evapotranspiration of the forest canopy (Kundzewicz 1999; Salzman et al. 2001; Rudolf-Miklau et al. 2012). Especially mountain forests retain

F. Rudolf-Miklau (✉)
Federal Ministry for Agriculture, Forestry Environment and Water Management, Department III 5, Torrent and Avalanche Control, Vienna, Austria
e-mail: florian.rudolf-miklau@bmlfuw.gv.at

© Capital Publishing Company, New Delhi, India 2017
J. Křeček et al. (eds.), *Ecosystem Services of Headwater Catchments*,
DOI 10.1007/978-3-319-57946-7_22

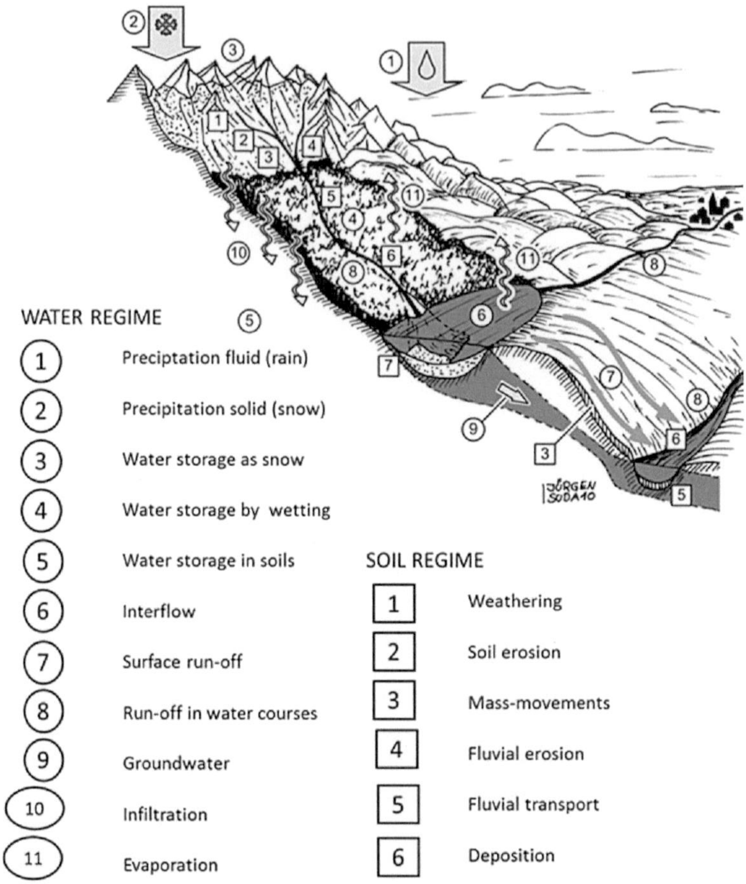

Fig. 22.1 Natural protective effects in mountain catchments: flow regime and displacement of solids on slopes

large quantities of precipitation and reduce (decelerate) surface run-off (Pearce 2001; Hegg 2006; Markart and Kohl 2009). Figure 22.1 depicts the variety of effects in mountain catchments, balancing the flow regime or displacement of solids. On steep slopes, mountain forests fulfil multiple protective functions related to gravitational hazards, such as the stabilization of slopes by fixing and dewatering them with the root system (Gray and Leiser 1982; Markart and Kohl 2009), the support of instable snow cover to prevent avalanche release (Meyer-Grass and Schneebeli 1992; Margreth 2004; Nairz et al. 2015) or the dissipation of kinetic energy by rock fall (Berger et al. 2002; Stoffel et al. 2006; Wasser and Perren 2014) landslides or debris flows (Binder and Endres 2012). Furthermore, the vegetation cover has important shielding effects to prevent soil erosion by surface run-off (Boardman and Poesen 2006; Morgan 2009), wind drift (Chepil 1945; Fryrear 1985; Morgan 2009) or shearing force of snow creep (Nairz et al. 2015).

In general, ecosystems contribute in multiple ways to the sustainable mitigation of risks for human life and the basic functions of existence (Rudolf-Miklau 2009), hence, *natural protective services* can be characterized as typical *ecosystem services*.[1] Daily (1997) and Salzman et al. (2001), for instance, identify the 'mitigation of floods' among the most essential services of ecosystems. Depending on the taxonomy applied, protective services may be classified as 'ecosystem disturbance regulation' (Costanza et al. 1997), ecosystem services, that 'are essential for human survival' (Holmlund and Hammer 1999) or just 'stabilizing processes' in the environment (Daily and Dasgupta 2001). By analogy with the principle, that ecosystems don't yield services until they are used by people (Ruhl et al. 2007), natural protection services only materialize if human beings (and their living space) are at risk by natural hazards or frequent endangered areas.

1.1 Objective of Investigation

The article deals with the *provision*, *transaction* and *consumption* of natural protective services by people and demonstrates the relationship between the owners of the land, from where the effects originate (*providers of protective service*), and the beneficiaries in areas, where effects evolve (*consumers of protective service*). Natural protective effects in mountain catchments are systematically traced in order to clarify the framework conditions and limits for provision of these public services. In this article a critical discussion is conducted, whether or not natural, protective service may be subject of a private property system and take on 'market values'. However, regarding the complexity of origin and effect of protective functions in nature, it seems questionable if an economic transaction may occur at all and if a monetary compensation can be gained by the land owner. Though, a sustainable preservation of protective effects in nature—in competition with other land use interests—presupposes the perception and appreciation of these ecosystem services by land owners, beneficiaries and the public. A hypothesis is argued, that the legal nature of natural protective services as public goods or even common pool resources justifies the establishment of a legal framework and state management system that limit the unrestricted use and exploitation of private (land) property.

[1] **Definitions:** "The benefits people obtain from ecosystems." (Millennium Ecosystem Assessment 2005); "The conditions and processes through which natural ecosystems, and the species that make them up, sustain and fulfil human life" (Daily 1997); "Flows of materials, energy, and information from natural capital stocks which combine with manufactured and human capital services to produce human welfare" (Costanza et al. 1997).

2 Tracing Natural Protective Effects for Human Population: Theoretical Approach and Practical Assessment

2.1 *Basic Model of Natural Protective Services*

For the human perception and valuation of the natural protective effects (as ecosystem services), a distinction between *direct* and *indirect* use—in specific (according to Costanza et al. 1997) between service benefits,

(a) used indirectly through 'structural' components derived from ecosystems (e.g. provision of flood protection for downstream areas by wetlands or inundation areas), or
(b) used directly through the dynamic action of ecosystem functions (e.g. erosion control by forests on steep slopes)

—is of utmost importance. For the perception of consumers and beneficiaries, natural protection effects can appear either *clear* and *traceable* or *oblique* and *complex*. In practice, the complex chains of effects are barely comprehensible and difficult to assess.

Hence, the assessment and evaluation of protective services for legal, political and economic decisions presuppose a *fundamental concept*, that visualizes the relation between source and consumer, the path of delivery (transaction) as well as the functional chain of effects. For practical purposes, it is useful to trace the protective service in three phases, from provision via transaction to consumption, by structuring it into a simple model. This *model* (Fig. 22.2)—adapted according to Ruhl et al. (2007)—contains seven components: In between the '*provision source*' (comp. A) and the '*service consumer(s)*' (beneficiaries) (comp. F) we can assemble the '*timing*

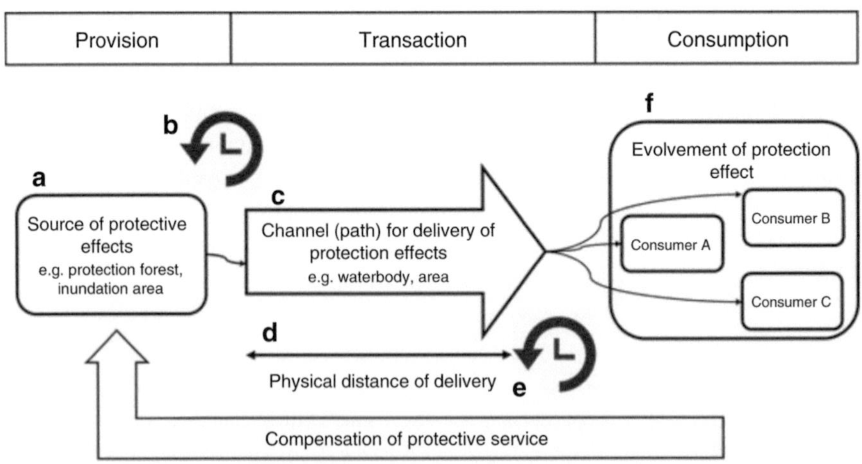

Fig. 22.2 Basic model to visualize the provision, delivery (transaction) and consumption of natural protective effects (Adapted according to Ruhl et al. 2007)

of the provision' (comp. B), the '*channel of delivery*', closely related to the '*physical distance over which delivery occurs*' (comp. D) and the '*timing of the delivery of the service ready to be used*' (comp. E). Whereas the latter—depending on delivery distance, latency of the protective service benefits, and user decisions—may not correspond with the timing of the service provision. A *compensation gain* (comp. G) for the beneficial use of the protective services by beneficiaries shall flow back to the provider ('owner').

2.2 Occurrence and Alteration of Protective Effects in Ecosystems

The *patterns of occurrence* of natural protective services in ecosystems range from being relatively *predictable* and *easy to identify* (map), to highly *variable in location and time*. The natural flood retention effects in a river basin, for example, provide a clearly allocated protective function for a small town situated downstream on the river bank, while for an adjacent large town outside the river basin the benefits from natural flood protection—protective effects, such as operability of transport and supply lines, public health or sense of people's security, water supply—might be vaguer and less perceptible to the public, but eventually even more effectual (Ruhl et al. 2007).

The *scale* of ecological properties providing protective effects is defined by the *spatial* and *temporal* dimensions at which the property has the most coherence (Limburg et al. 2002). The spatial scale associated with a particular protective ecosystem property is a function of numerous factors, including the area of influence of disturbances (e.g. the forest area destroyed by a storm or wildfire), the distance of displacement of material (e.g. of avalanche snow, mudflows or bedload in a river) or the efficiency range (e.g. of the retention effect of a flood plain downriver) (Millennium Ecosystem Assessment 2003). Likewise, the temporal dimension of natural protection functions plays a major role when tracing the effects from source to consumer, ranging from *quick* and *immediate* (e.g. trees trapping falling rocks) to *slow* and *retarded* (e.g. forest stands slowing the slope creep) effects.

Moreover, natural protective effects in ecosystems are subject to a constant *process of change*. Slower processes, though generally more stable than fast processes over short time frames, do change, and over time, may become more difficult to steer back to a previous course (Ruhl et al. 2007). *Resistance* and *resilience* of natural protective effects are by no means infinite. At some point, the change may exceed a threshold of irreversibility setting the entire ecosystem and its protective effects onto a course towards a completely different set of dynamic equilibria (Holling and Gunderson 2002).

Natural or *man-made alterations* of ecosystems inevitably may also enhance or decrease, even eliminate the natural protective functions. Furthermore, inevitable trade-off impacts and synergies for other ecosystem services have to be considered while managing natural protective functions (cf. Daily and Dasgupta 2001). Indeed, the Millennium Ecosystem Assessment (2004a, b) suggests that in many cases substantial changes (usually degrading) in the hazard regulating services (e.g. soil erosion, increase of surface runoff) are a direct result of management activities in these ecosystems, primarily for the benefit of the yield functions (e.g. forestry, agriculture, site development: usually to enhance them). When managing and consequently altering any ecosystem in order to sustain the protection of one property, these activities may cause effects of change for property of third party, usually reducing the protective effect or increasing the risk. These negative trade-off impacts and synergies have to be considered, as well for the relation of providers and consumers of protective effects, as for the property rights and interests of other parties involved.

This theoretical approach to tracing natural protective services shows, that they cannot simply be understood as a singular mitigating effect acting on hazard processes—compared to technical protection measures (e.g. flood retention dam, supporting structures for instable slopes)—but have to be treated as risk-reducing consequence of a complex interaction between the natural hazard process and the buffering effects of the ecosystem.

2.3 Assessment of Protective Effects in Mountain Ecosystems

If specific natural protective services in mountain ecosystems are investigated with respect to an economic transaction, the major challenges are associated with the assignability of the function to certain (groups of) providers or consumers (beneficiaries), the quantification of the protective effect—including the extent, number of providers and consumers and distribution of the share of effects among them—and the delineation of the area, where the protective effect enfolds (mostly equal to a hazard zone) (Salzman et al. 2001; Borsky and Weck-Hannemann 2006; BAFU 2014). In order to assess and describe the properties of protective effects, specific criteria—such as *effect relation*, *perceptibility* and *traceability*, *distance of effect enfoldment* (related to source area), *predictability*, *time of occurrence* and *possibility of delineation* (mapping)—are needed. For the purpose of an economic assessment of protective effect transaction, it is essential to express their benefits in monetary terms and as economic risk. However, only parts of the benefits are tangible (e.g. protection for real estates, houses, infrastructure) and assignable to a certain proprietor, while the majority of natural protective effects generate only societal (social) benefits or, hence, are rather intangible (e.g. protection of human life and health, environment and cultural heritage) (Maniak 2001; Romang 2004; Rudolf-Miklau 2009).

Usually, protective effects are presented in *hazard (indication) maps* or *hazard zone plans*, showing, inter alia, the area where protective effects enfold their benefits (Rudolf-Miklau 2011; Rudolf-Miklau et al. 2014; ÖROK 2015). The delineation of hazard zones presupposes that the area, where protective services take effect, can be contoured relatively exact and with respect to the boundaries of land property (land register or cadastre). Furthermore, the risk-reducing effects of ecologic protective measures ought to be assessed in a quantitative manner and integrated in the risk assessment methods applied (Borsky and Weck-Hannemann 2006; BAFU 2014). In concrete terms, it is necessary to know the measure of risk reduction (related to the reduced intensity of the hazard process) as well as the reduced frequency of damaging (catastrophic) events. Usually, this information is only available for directly, closely and immediately effective protective services.

In general, protective effects related to gravitational (on slopes: rock fall, landslides, erosion and snow avalanches) and torrential hazards (debris flow, fluvial bedload transport, drift wood, bank erosion) usually are traceable and sufficiently precise assignable to certain beneficiaries or providers. In practice, these hazard processes can be assessed applying numerical models and outlined in hazard maps (Jakob 2005; Rudolf-Miklau 2011; Rudolf-Miklau et al. 2014; ÖROK 2015). The related protective effects usually originate from forests or vegetation cover, infrequent also from soil and topography. Similarly, for flood related hazards the assessment applying numerical models and the outline in hazard maps is usually well-developed (Schanze et al. 2007; de Moel et al. 2009; Hagemeier-Klose and Wagner 2009; Rudolf-Miklau 2011), while the assignment of natural protective effects to certain property parcels (beneficiaries) is much more complex, especially if traced in larger river basins and in remote distance to the area of origin. Table 22.1 contains an overview of the most important protective effects in Alpine catchments and a rough assessment according to the criteria stated above.

In general, the opportunities for transaction of natural protective effects and the assignment of benefits to certain consumers is a function of the *threshold of detectability* in causal, spatial and temporal context. It seems obvious that only forest related protective effects on slopes are appropriate, to some extent, for an economic valuation and potentially are subject to economic transactions between definite providers and beneficiaries. In contrast, protective effects by topography, soils or waterbodies rather unfold conceptional benefits for large groups of beneficiaries or society as a whole and tend to be intangible in valuation. Altogether, it seems to be a principle, whichever natural protective service is assessed, that the perception and appreciation of the benefits is strongly divergent between providers and beneficiaries.

3 Legal Nature of Natural Protective Services

3.1 Common Pool Resource or Property Ownership Right?

From the legal perspective, protective effects in headwater catchments, that emerge from natural conditions or processes in ecosystems, can be characterized as positive economic effects, that are spreading from its source area, either directly to adjacent properties (parcels) or even to remote spaces further downstream. While the source

Table 22.1 Natural protective effects in mountain catchments

Natural hazard	Effect originating from				Effect relationship		Perceptibility		Traceable		Distance of effect		Predictability		Time of occurrence		Delimitable	
Natural protective effect	Forest, vegetation	Soil	Topography	Water body	Direct	Indirect	Clear	Oblique	Traceable	Complex	Close	Remote	Yes	No	Immediate	Retarded	Exact	Diffuse
Floods																		
Evaporation, Transpiration	x					→		→		→		→	←			→		→
Storage capacity of soils		x				→		→		→		→	←			→		→
Retentions by depressions			x			→		→		→		→	←			→		→
Retention by interflow		x				→		→		→		→	←			→		→
Retention by lakes and wetlands				x		→		→		→		→	←			→		→
Retention by flood plains			x			→		→	←			→	←			→		→
Sediment retention			x			→	←		←		←		←		←		←	
Debris flow																		
Stabilization of release area	x				←			→	←		←		←		←		←	
Drainage effect	x					→		→		→	←		←			→		→
Filtering of sediments	x				←		←		←		←			→	←		←	

Breaking effect	x		x		←		←		←		←			→	←		←	
Shielding and deflecting effect					←		←		←		←			→	←		←	
Erosion																		
Protective cover of soil	x				←		←		←		←		←		←			→
Armoring by roots	x				←		←			→	←		←		←			→
Wind shield	x		x		←		←			→	←			→	←			→
Shield against tractive force		x			←			→	←		←			→	←			→
Rock fall																		
Stabilisation of rock formation	x				←			→	←		←		←			→	←	
Broking effect (energy dissipation)	x				←		←		←		←			→	←		←	
Shield and deflecting effect			x		←		←		←		←			→	←		←	
Shallow Landslide																		
Protective cover of solids	x					→		→			←		←			→	←	

(continued)

Table 22.1 (continued)

Natural hazard	Effect originating from				Effect relationship		Perceptibility		Traceable		Distance of effect		Predictability		Time of occurrence		Delimitable	
Natural protective effect	Forest, vegetation	Soil	Topography	Water body	Direct	Indirect	Clear	Oblique	Traceable	Complex	Close	Remote	Yes	No	Immediate	Retarded	Exact	Diffuse
Armoring by roots	x					→		→		→	←		←		←		←	
Drainage effects	x					→		→		→	←		←			→	←	
Shielding or deflecting effects			x		←		←		←		←			→	←		←	
Deep-seated Landslide																		
Drainage effects	x					→		→		→	←		←			→		→
Shielding or deflecting effects			x		←		←		←		←			→	←		←	
Snow avalanche																		
Supporting in the release zone	x					→		→	←		←		←			→	←	
Wind shield	x					→		→		→	←		←			→		→
Braking effect (Energy dissipation)	x		x		←		←		←		←			→	←		←	
Shielding of deflecting effects			x		←		←		←		←			→	←		←	

area of these effects, as a rule, is clearly assignable to landed property, the area where the benefits take effect is usually spacious and difficult to demarcate. The delivery (transaction) of protective effects from provider to consumer is, with little exception, a transboundary resource flow generating multiparty legal relationships.

According to the prevailing opinion in European states (de Moel et al. 2009), protective services related to natural hazards are considered as *public goods*[2] or even *common pool resources*[3] (Ostrom 1990; Ostrom et al. 1999; Schanze et al. 2007; Ruhl et al. 2007). Public services start at a stage where the sphere and the capacity of an individual party is exceeded. In contrast, private goods (services) are property of single persons or certain groups of individuals if others can be (partly) excluded from consumption (of the effects) (Rudolf-Miklau 2009). In general, the natural protective services addressed in this article are related to catastrophes and serve the safety of a multitude of persons or even of a not delimitable part of society. Hence, it seems conclusive to treat them as *public protective services* that include the opportunity for everybody to consume the effects 'free of charge' (Weck-Hannemann and Thöni 2006).

As ecosystem services obviously provide economic values to humans, the gain from the land, where the natural capital that produces them is located, and of the land, where the services are delivered and effects are consumed, ought to reflect those values (Ruhl et al. 2007). According to the natural origin of protective services in ecosystems, private rights of use and gains are related to a specific treatment or maintenance (in order to preserve the protective effect) that restricts the free use of land and causes substantial costs. Hence, theoretically the consumer of protection owes the provider a monetary compensation. The concern is not who owns the land, where the natural protective service is produced, or the area where the protective service benefits are consumed—that is in general clarified—but rather, what the respective owners can do with their property and can reasonably expect from the consumers (respectively the public) in return. In turn, consumers (beneficiaries) that are charged a monetary compensation for the provision of protective effects must be able to expect a reliable and enforceable delivery of service, which must not be altered by the owner's utilization of the source area.

An established system of remuneration of protective services presupposes cultural norms, a public consensus esteeming the value of the service and rules of consumption enforced by law. Concerning ecosystem services '*in the public interest*', in most European countries (Schröter et al. 2005; Ruhl et al. 2007; Maes et al. 2012) the state has set up a *rule- and compensation-based system* that regulates the transaction (supply and demand) of ecosystem protection services and enables both, providers and consumers, to enforce their interests and rights through injunction or

[2]**Definition:** A *public good* is a good that is both, non-excludable and non-rivalrous in that individuals cannot be effectively excluded from use and where use by one individual does not reduce availability to others (Demsetz 1970).

[3]**Definition:** *Common pool resource* (CPR) is a type of good consisting of a natural or human-made resource system, whose size or characteristics makes it costly, but not impossible, to exclude potential beneficiaries from obtaining benefits from its use (Ostrom 1990).

compensation remedies. One mean of resolving the matter is to devise a rule that allows one property owner to force the other to change behaviour in land use; an alternative mean of resolving the competing interests is to establish liability principles that define a compensation, if the owner of the source area uses the land incompatibly to the protective function or if the consumer of protection alters the claims for safety by increasing the risk (Ruhl et al. 2007). The establishment of such a rule- and compensation-based regulatory system comprises also limitations to the open access to consumption of protective effect, that particularly prevent 'free-rider' use without any compensation. This requires—according to the definition of 'common pool resources'—that the state sets up rules that regulate the public access to natural protective services and takes into account the explicit agreement of the owner to the use of the service. Furthermore, the system must also provide incentives to the owner of the protective resource to invest in, use and maintain his property (the ecosystem) in favour of the provision of the protection function (Ostrom et al. 1999).

3.2 Hypotheses: Advantages of Group-Owned Property (Cooperative) or State Governed Regimes

The transaction of protective effects may be qualified as *transboundary resource flow* in the delivery path. In order to fully account for all the relevant property interests, rights must be assigned not only in the resource stock (natural protective function in our case) but also in the resource flow (protective effect as ecosystem services in our case). And as the flow of ecosystem services often, if not usually, transcends the boundaries of defined private property parcels, this requires a mechanism for distributing the rights among a broad spectrum of property owners, all of whom can reasonably make some claim to them (Bouckaert 2000; Ostrom et al. 1999). As a rule, owners of parcels from where natural protection functions originate, are inseparably bound to owners of parcels where protective effects are delivered and consumed, and the latter are all inseparably bound to each other. These complex legal provider-consumer relations require a regulatory system that is appropriate to balance private (individual) interests and public welfare. Hence, alternatively *group-owned property*[4] regimes for natural protective services may overcome the problems of multilateral legal relations (Ruhl et al. 2007). According to Cole (2002) in common property arrangements the ownership group holds the right to exclude others from the use, whereas within the group each individual has both rights and duties with respect to use of and investment in the resource. When the members of the group do not all agree on how to deal with the common interests, however, the law should provide means of deciding whose—as a rule the majority in the

[4]Property law as well as public law provide many arrangements for group ownership, known in law as common property or concurrent property, such as co-tenancy, partnerships and corporations.

group—view prevails. A voluntary formed corporation or legally formed partnership is 'owned' by its shareholders, and thus protective services to which the corporation holds title is a form of '*common property*'. When disagreements arise among the group members about how to manage and maintain the protective service, a legally enforced decision-making apparatus can be engaged to decide the dispute.

Another alternative is a *public property system* for protective services, where the cooperative effort merely replaces the individual owners with the state, and the state defines the rights of each citizen with respect to the protection service and the rules and liabilities enforced against them (Ostrom et al. 1999). Such a system can take various forms, either simply by a state ownership of source areas of protective functions (e.g. protection forest estates, retention areas) or by a system regulated by law, that delegates the task of management and control of areas with protective function to a government institution. If the state manages and regulates the provision and consumption of protective services on private land, usually a *lawful regulatory system* exists enforcing the owner's obligations to tolerate, the owner's entitlement for financial compensation and the official instrument of land expropriation for pushing through public interests of protection (Rudolf-Miklau 2009). The major advantage of state governed systems for protective services is based on the fact that governments can make policy decisions, such as preserving natural protection functions of ecosystems or protecting natural inundation areas, that produce economic values not appreciated by the public and recognized in the market. In particular, the government needs not to compare these ecosystem service values using the same private discount rate that market participants use in assessing alternative uses of a resource. In fact, a *'social' discount rate* should be adopted that reflects a longer view toward sustainable provision of protection, greater aversion to risk, and a greater public appreciation of the complex nature of ecosystems (Ruhl et al. 2007).

4 Economic Evaluation and Compensation of Natural Protective Services: Possibilities and Limits

4.1 Gainful Transaction or Positive Externality?

Ecosystem services, in general, are for the most part, *free* for the taking; however, this does not mean they are without value. As already stated above, these services are close in economic behaviour to *public goods* (Ruhl et al. 2007; Blankart 2008).

In general, natural protective effects of ecosystems are hardly captured in markets or adequately quantified in economic terms; furthermore, they are often given too little weight in policy decisions. Owners of natural capital resources—natural protection function in our case—perceive them as *positive* externalities[5] for which

[5] **Definition:** According to Simpson (1998), an *externality* is defined as any kind of cost or benefit from the production or consumption of a good or service that is not borne or enjoyed by the producer or consumer but is borne or enjoyed by a third party.

the owner derives no gain. Even when the owner of the source area of natural protective functions can 'regulate' the volume and timing of this services flowing to the beneficiaries (consumers), often the following problems will impede the ability of the owner to engage in transactions with users from which some economic gain could be derived:

- information costs,[6]
- transaction costs, and
- open access: consumption by 'free riders'.

In a very revealing example, dealing with the downstream benefits from natural flood control effects of wetlands, Ruhl et al. (2007) demonstrate the complexity and limitation of economic evaluation of protective services from ecosystems: Given the assumption that the owner of the wetland is able to control the provision of this natural protection effect, this raises the question "*whom would the person charge for the service, and how? The challenge for an owner of a wetland area, for example, is that he cannot practicably charge a price to dictate which downstream landowners derive the flood control benefits of the wetland area, and the downstream owners cannot crowd each other out for access to the benefits. Consequently, while the wetland is privately owned and as such is a private good, the ecosystem service of flood control it provides is a public good and one from which the wetland owner can derive little direct benefit. There is no way for the demand for flood mitigation services from wetland parcels to become an effective demand in the market to which wetland owners can respond.*"

Generally, the owners of resource areas of protective service—in default of a possibility of economic exploitation—has no interest to take into account the positive externalities associated with ownership of natural capital resource of protection and rather seeks for alternative and gainful options in land use, such as the development of building land or profitable agricultural/silvicultural cultivation.

4.2 *Obstacles for Establishment of Economic Transaction of Natural Protective Services*

One obstacle for the establishment of compensation-based, gainful transactions between providers and consumers seems to be the *problem of information (costs)* and joint comprehension/appreciation of protective services. Due to their complexity this presupposes intensive research on a deeper understanding of the protection function for the specific case of application, which would cause—except for the simple case of direct protective effects (for neighbouring parcels)—tremendous costs and be disproportionate time consuming.[7] As a rule, the evaluation of protective

[6] Research costs for the assessment of protection effects and valuation of the benefits for the consumer.

[7] Costs for studies and expert opinions on the transaction value of protective effects could exceed the compensation costs and are hardly to afford by a single party.

effects can be done best on the basis of a hazard or risk map provided by state services. Sufficient information on the economic value of the protective service should promote negotiations between providers and beneficiaries on a financial compensation for a sustainable supply of protection.

Another obstacle seems to be the *transaction costs*[8] for delivery of protective services. Especially with a multitude of providers and beneficiaries it seems inefficient to monitor the extent of protective effect over time and the actual distribution among parties involved; as well challenging would be the recurrent enforcement of the claim. For the example of the flood protection effects of wetlands, Ruhl et al. (2007) once again 'get to the heart of the dilemma': "*For example, flood control benefits of wetlands may often be experienced for long stretches of river systems. There may be numerous owners of different wetland parcels responsible for the supply of the benefits, as well as numerous owners of different benefited parcels. Even to initiate the kind of transaction in mind, a sufficient number of wetland parcel owners would need to identify each other, agree to threaten to eliminate the wetlands (assuming no legal restriction), evaluate the flood control benefit values, identify the flood control beneficiaries, develop a strategy for negotiating with them, and devise a method for allocating any payments received among the group.*"

Not least the fact, that protective services behave like an '*open-access resource*', it is not possible to exclude parts of potential beneficiaries from the consumption and, de facto, the benefit is available to anyone. For this so-called '*free-rider problem*' Ostrom et al. (1999) pronounce the problematic issues of overuse and lack of contributions by beneficiaries to maintenance and improvement of the service, if they are able to consume the benefits without an effective regulations limiting access and defining rights and duties. Generally, in the absence of the ability to control the distribution of protective services, these free rider problems can be overcome only through collective acts of compensation by the consumers of the resource, whether voluntary or coerced.

4.3 Assessment and Realization of Financial Compensation for Public Protective Services

The obstacles stated above for a fair compensation-based transaction system are mainly caused by market failure, thus it should be intrinsically a prerogative of the state, to regulate and manage the provision and transfer of protective services. For this purpose, it is necessary to assess the appreciation of the public for protective effects of ecosystems (e.g. the protection function of forests).

[8] **Definition:** Transaction cost is a cost incurred in making an economic exchange (restated: the cost of participating in a market). These include the costs of acquiring sufficient information for contracting, bargaining, and consummating the transactions as well as attendant costs of monitoring, policing, and enforcing transactions and agreements after the fact.

In practice, this is commonly done by assessing the *willingness to pay* (WTP) (Weck-Hannemann and Thöni 2006) of the public (beneficiaries) for several options of protective services (choice-based conjoint, CBC) including a "none" choice option (Olschewki 2013). However, one critical issue is, how people's willingness to pay (WTP) for various types of protective effects is related to their geographic and cognitive distance from them (Ruhl et al. 2007). In general, the public appreciation for protective services—except for the instance of a catastrophic situation—will be expectedly low. Hence, the state has to establish, a regulatory system in order to realize financial settlement and compensatory payment by the beneficiaries. De facto, the fund-raising for compensation of benefits from protective services is frequently realized by a fixed distribution key (determined by an official decree), especially in group-owned property systems (e.g. water cooperatives). These public compensation systems are mainly based on 'social' discount rates and suppress also a potential price formation mechanism that might take into account a special preference of the beneficiaries for the protective service.

5 State-Governed Management of Natural Protective Services: The Austrian Example

In the political perception natural protective services are part of a comprehensive risk management system. On European level, there is neither a specific policy existent dealing with protective functions in ecosystems, nor are these environmental benefits explicitly part of a general ecosystem services strategy, even though they are addressed in several legal norms: e.g. the preservation of natural retention areas in the European Flood Directive (EFD)[9] or the maintenance of the protection function of forests in the "Mountain Forest Protocol"[10] to the Alpine Convention. Nevertheless, a political awareness of these ecosystem services can be clearly detected on closer inspection in most national legal policy systems. The established Austrian system shall serve as a demonstrative example for a rule- and compensation-based governmental risk management system, comprising also natural protective services. (Rudolf-Miklau 2009).

According to the prevailing opinion, the provision of protective services is a governmental responsibility and, consequently, understood as part of public safety policies and basic public services (Weck-Hannemann and Thöni 2006; Kerschner 2008; Rudolf-Miklau 2009). However, neither the Austrian constitution nor substantive law grant a subjective right of protection. It is exclusively in the political decision of the state, to which extent public services for risk prevention and catastrophe management are provided and which precautionary actions are left to the

[9] Directive 2007/60/EC of the European Parliament and of the Council of 23 October 2007 on the assessment and management of flood risks.

[10] Protocol on the Implementation of the Alpine Convention relating to the Mountain Forests, enforced on the 27th of February 1996

initiative of society or the responsibility of the individual (*own precaution*) (Weber 2004). In particular, the 'neighbourhood law' in the Austrian General Civil Code defines the principle of land owner's *self-responsibility* to protect his own property. More concretely, this means that he has no entitlement against his neighbour to claim defensive measures to inhibit impacts from natural hazards (that originate from the neighbouring estate). Damages by purely natural events are classified as '*force majeure*' (vis major) and are exclusively included in the sphere of risk of the aggrieved party. Though the Civil Code also determines an obligation of the land owner to use his estate in a risk preventive manner and to take 'acceptable' measures against the risk originating from the use of his property (Karner 2011). This means specifically that the owner of a neighbouring estate only has a 'no-fault entitlement' for injunction and elimination, if the owner of the estate, from which the hazard originates, has actively increased the risk (Wagner and Jandl 2013). Hence there is no legal base in Austrian civil law for an enforceable claim against a land owner to provide a certain protective service for other neighbouring or parties situated downstream. Conversely this means that owners of real properties that provide natural protective effects can invoke a right of consent and an entitlement of compensation, if natural protective services are demanded in the public benefit or the security requirements of other parties.

These principles stated above are in line with the regulations in the Austrian Forest Act concerning the treatment and maintenance of *protective forests*. The owner is obliged to take forest management measures only to the extent which he can cover by the income from yield of wood; in the case of *object-protecting forest*, only to the extent which he can cover from public subsidies and compensation by beneficiaries of the protective service. In the particular case of *protective forests stipulated by decree*,[11] the Forest Act determines an entitlement to compensation for the forest owner, if he has to accept pecuniary prejudice by the governmental restrictions. The compensation has to be paid by the beneficiary—in case of a multitude of beneficiaries shared proportionally to the individual benefits or averted losses. These regulations in the Austrian Forest Act may be clearly interpreted as compensation rule for a natural protective service provided by forest ecosystems. In practice, these regulations are well enforceable as, in the case of protective forests, there are clear relations between providers and consumers (beneficiary) of the protective service, both owners (user) of neighbouring or adjacent real estates.

An explicit duty requiring preventive actions against natural hazards is determined in the Austrian laws concerning traffic infrastructure. In the interest of transport user's (passenger's) safety, the owner of a real estate neighbouring roads, railway lines or cableways, are obliged to use their property in a risk-reducing manner or take 'acceptable' precautions against natural hazards, that originate from their real estate. This obligation also comprises the treatment of the area, forest stands and soils in favour of the natural protective effects (Hattenberger 2004, 2005; Wagner and Jandl 2013). A comparable duty requiring protective measures by the land owner is regulated in the 'Ordinance for Slope Protection' of the city of

[11] German term in Austrian Forest Act: 'Bannwald'.

Salzburg for the benefit of house owners at the bottom of the rocky slopes (Rudolf-Miklau and Kanonier 2016). In all these cases the Austrian state has defined a strong public interest and regulates the entitlement and consumption of natural protective services, based on 'social' discount rates (resp. officially regulated reference rates), by law.

Concerning other natural protective effects with more remote areas of benefit and more complex channels of delivery, especially effects related to flood retention and soil erosion, the Austrian law (especially the Water Act or Soil Protection Acts) does not contain comparable regulations for provision and compensation. However, flood retention or sedimentation areas have to be outlined in the hazard maps according to Forest or Water Act. Furthermore, these zones have to be taken into account, according to applicable provincial law, in the land use and development planning on the municipal level. The absence of a legally based entitlement for protective flood retention services opens a market for financial compensation according to supply and demand of available land. In Austria, in the competition of land use interests, the value (price) of (for) natural flood retention areas has been significantly increasing, especially in regions with limited availability of usable land (particularly, in Alpine regions). When fostering the natural retention function in flood control policy the Austrian state will have to adapt the management for provision of inundation areas, either by establishment of a stricter rule- and compensation-based regulatory system or a public property system (by acquisition/purchasing and public management of appropriate real estates) for this important public interest. At the moment, the disproportionately high cost for information (verification management) and transfer of flood protective effects on rivers as well as 'free-rider' use of the benefits completely suppress a fair compensation system based on contractual legal relationships. Hence, the Austrian state has to close this gap by public compensation payment and lawful restrictions of use.

Furthermore, the Austrian state also provides extensive, publiclysubsidized services and incentive systems for the management and maintenance of forests and catchment areas with protective function. Financial support from the national disaster relief fund and provincial budgets is available for projects in protection forest rehabilitation and watershed management. The consumption of these subsidies by the beneficiaries requires, again, own contributions equivalent to the extent of benefits or averted losses (Rudolf-Miklau 2009). The Austrian Water Act determines the constitution and regulations for *Water Cooperatives*,[12] that comprise all beneficiaries benefiting from services provided by protection forests, inundation areas or mountain catchments. The law regulates in detail the constitution of these cooperatives, the internal fund raising and burden sharing, the mediation of conflicts and achievement of decisions as well as the shared liability of the cooperative members (Rudolf-Miklau et al. 2014). This institution—actually only realized in the province of Salzburg—is a perfect example for a group-owned property regime for protective services. In other parts of Austria, this cooperative system is suppressed by the

[12] Legal nature: Statutory cooperative body under public law.

safety investments of municipalities in place of their citizens or only existent on the level of *Regional Water Associations.*[13]

Although the Austrian Assessment Report on Climate Change (2014) clearly states the urgent need for a stronger public awareness and appreciation for ecosystem services—including natural protective effects—in view of a rapidly changing environment, no Austrian national strategy was implemented so far in order to foster this important target. Nevertheless, it has been demonstrated that some efficient and well-established instruments already exist related to the protective function of forests that fulfil to a large extent the requirements of ecosystem service management and fairly balance the benefits and burdens related to natural protective effects.

6 Conclusions

Protective effects of ecosystems represent essential public services for the mitigation of natural risk, that are hardly perceived and appreciated by society and political decision-makers, in particular by the beneficiaries and consumers. Furthermore, it was shown in this article, that the 'market' fails in regulating the economic transaction from providers (land owner of estates, where the protective effect originates) to consumers (land owners and beneficial owner, that consume the protective service) and that guarantees a gainful compensation for protective services according to the principles 'supply and demand'. Facing the risk of loss of essential protective effects in ecosystems, when land owners seek for more gainful exploitation of their real estates (e.g. building land development, intensive agricultural use, profit-oriented forestry), it seems inevitable that the state has to close this gap by establishing a lawful rule- and compensation-based system. According to the principles of protective services treated as 'common pool resources' the state should set up rules that regulate the public access to protection and take into account the explicit agreement of the owner to the use of the service.

From the investigations in this article, a clear preference for a 'group-owned' (institutional cooperation by the beneficiaries) or a 'public-property' (state owned protective ecosystems or governmental risk management services) system, concerning the provision and transaction of protective services, may be derived. These schemes seem to be most appropriate to overcome the problems of information costs (verification management), transaction costs, and free riders related to the public use of protective services. A lawful compensation system should mainly be based on 'social' discount rates (regulated rates) and takes into account the special preference of the beneficiaries ('willingness to pay'). For the scenario of an impending loss of protective functions in ecosystems, a combination of incentive instruments and proscriptive regulation or legal constraints on the use of property should be available in order to steer land use towards a risk-adapted and sustainable manner.

[13] Members of regional water associations are mainly municipalities or water management enterprises.

The Austrian example shows clearly that policies for the comprehensive management of natural protective services are still uncommon, although partially realized in the legal system for the case of protective forests. Even if natural flood retention services are explicitly addressed in the European Flood Directive, an integrated compensation system for the provision and transaction of these services is still missing and seems to be persistently inhibited by deficits in traceability, assignability of beneficial use to certain individuals and general public appreciation (willingness to pay). Regarding the complexity of protective services in ecosystems only far sighted strategies on the European and science-based management concepts will solve this policy issue, that soon might attain paramount importance in a rapidly changing environment.

References

Austrian Panel on Climate Change (APCC). (2014). *Austrian assessment report on Climate Change (AAR14)*, Vienna.

Berger, F., Quetel, C., & Dorren, L. K. (2002). Forest: A natural protection mean against rock falls, but with which efficiency. *International Congress Interpraevent, 2*, 815–826.

Binder, F., & Endres, U. (2012). Bergwaldbewirtschaftung und Schutzfunktionen. *LWF aktuell, 91*: 49 (in German).

Blankart, C. B. (2008). *Öffentliche Finanzen in der Demokratie* (7th ed.). München (in German): Verlag Vahlen.

Boardman, J., & Poesen, J. (2006). *Soil erosion in Europe*. Chichester: Wiley.

Borsky, S., & Weck-Hannemann, H. (2006). *Sozio-ökonomische Bewertung der Schutzleistung des Waldes vor Lawinen*. alpS Projekt C.2.5, Innsbruck (in German).

Bouckaert, B. (2000). Original assignment of private property. In B. Bouckaert & G. DeGeest (Eds.), *Encyclopedia of law and economics. Vol. II, Civil law and economics*. Cheltenham: Edward Elgar.

Bundesamt für Umwelt (BAFU). (2014). PROTECT Bio – Methodik zur Beurteilung der Wirkung biologischer Schutzmaßnahmen gegen Naturgefahren als Grundlage für ihre Berücksichtigung bei Risikoanalysen. Synthesebericht Phase III, Bern (in German).

Chepil, W. S. (1945). Dynamics of wind erosion: I. Nature of movement of soil by wind. *Soil Science, 60*(4), 305–320.

Cole, D. H. (2002). The meaning of property rights: Law versus economics. *Land Economics, 78*(3), 317–330.

Costanza, R., d'Arge, R., de Groot, R., Farber, S., Grasso, M., Hannon, B., Limburg, K., Naeem, S., O'Neill, R. V., Paruelo, J., Raskin, R. G., Sutton, P., & van den Belt, M. (1997). The value of the world's ecosystem services and natural capital. *Nature, 387*, 253–260.

Daily, G. C. (1997). *Nature's services: Societal dependence on natural ecosystems*. Washington, DC: Island Press.

Daily, G. C., & Dasgupta, S. (2001). Ecosystem services, concept of. In S. Levin (Ed.), *Encyclopedia of biodiversity* (Vol. 2). San Diego: Academic Press.

de Moel, H., van Alphen, J., & Aerts, J. C. J. H. (2009). Flood maps in Europe—Methods, availability and use. *Natural Hazards and Earth System Science, 9*(2), 289–301.

Demsetz, H. (1970). The private production of public goods. *The Journal of Law and Economics, 13*(2), 293–306.

Fryrear, D. W. (1985). Soil cover and wind erosion. *Transactions of the ASAE, 28*(3), 781–784.

Gray, D. H., & Leiser, A. T. (1982). *Biotechnical slope protection and erosion control*. New York: Van Nostrand Reinhold Company Inc..

Hagemeier-Klose, M., & Wagner, K. (2009). Evaluation of flood hazard maps in print and web mapping services as information tools in flood risk communication. *Natural Hazards and Earth System Sciences, 9*(**2**), 563–574.

Hattenberger, D. (2004). Rechtliche Aspekte betreffend Lawinenschutzbauten (Teil 1). *Baurechtliche Blätter* 7. Jg. Vienna: Springer (in German).

Hattenberger, D. (2005). Rechtliche Aspekte betreffend Lawinenschutzbauten (Teil 2). *Baurechtliche Blätter* 8. Jg. Vienna: Springer (in German).

Hegg, C. (2006). *Waldwirkung auf Hochwasser*. Berichte der Bayerischen Landesanstalt für Wald- und Forstwirtschaften, *55*: 29–33 (in German).

Holling, C. S., & Gunderson, L. H. (2002). Resilience and adaptive cycles. In L. H. Gunderson & C. S. Holling (Eds.), *Panarchy: Understanding transformations in human and natural systems*. Washington, DC: Island Press.

Holmlund, C. M., & Hammer, M. (1999). Ecosystem services generated by fish populations. *Ecological Economics, 29*(2), 253–268.

Jakob, M. (2005). Debris-flow hazard analysis. In M. Jakob & O. Hungr (Eds.), *Debris-flow hazards and related phenomena*. Berlin/Heidelberg: Springer.

Karner, E. (2011). Schutz vor Naturgefahren und Haftung. *Zeitschrift für Verkehrsrecht (ZVR)* 2011/60 (in German).

Kerschner, F. (2008). Handbuch Naturkatastrophenrecht: Vorsorge, Abwehr, Haftung und Versicherung bei Naturkatastrophen. *Schriftenreihe Recht der Umwelt*. Manz: Vienna.

Kundzewicz, Z. W. (1999). Flood protection—Sustainability issues. *Hydrological Sciences Journal, 44*(4), 559–571.

Limburg, K. E., O'Neil, R. V., Costanza, R., & Farber, S. (2002). Complex systems and valuation. *Ecological Economics, 41*(3), 409–440.

Maes, J., Egoh, B., Willemen, L., Liquete, C., Vihervaara, P., Schägner, J. P., & Bouraoui, F. (2012). Mapping ecosystem services for policy support and decision making in the European Union. *Ecosystem Services, 1*(1), 31–39.

Maniak, I. U. (2001). *Ökonomische Grundlagen und Bewertungsmaßstäbe von wasserwirtschaftlichen Systemen*. Berlin/Heidelberg: Springer (in German).

Margreth, S. (2004). Die Wirkung des Waldes bei Lawinen. Forum für Wissen, 21–26 (in German).

Markart, G., & Kohl, B. (2009). Wie viel Wasser speichert der Waldboden? Abflussverhalten und Erosion. *BFW-Praxisinformation, 19*: 25–26 (in German).

Meyer-Grass, M., & Schneebeli, M. (1992). Die Abhängigkeit der Waldlawinen von Standorts-, Bestandes-und Schneeverhältnissen. *International Congress Interpraevent, 92*: 443–455 (in German).

Millennium Ecosystem Assessment. (2003). *Ecosystems and human well-being: A framework for assessment*. Washington, DC: Island Press.

Millennium Ecosystem Assessment. (2004a). *Conditions and trends assessment*. Available at http://www.millenniumassessment.org/en/products.chapters.aspx

Millennium Ecosystem Assessment. (2004b). *Scenarios assessment*. Available at http://www.millenniumassessment.org/en/products.chapters.aspx

Millennium Ecosystem Assessment. (2005). *Ecosystems and human well-being: Synthesis [1]*. Washington, DC: Island Press.

Morgan, R. P. C. (2009). *Soil erosion and conservation*. London: Wiley.

Nairz, P., Mears, A., Sauermoser, S., Gabl, K., Stoffel, M., & Margreth, S. (2015). Avalanches: Evolution and impact. In F. Rudolf-Miklau, S. Sauermoser, & A. Mears (Eds.), *The technical avalanche protection handbook*. Wiley.

Olschewki, R. (2013). How to value protection from natural hazards – A step-by-step discrete choice approach. *Natural Hazards and Earth System Science, 13*, 913–922.
ÖROK (Österreichische Raumordnungskonferenz) (ed.) (2015). Risikomanagement für gravitative Naturgefahren in der Raumplanung. – Wien. (= ÖROK-Schriftenreihe 193) (in German).
Ostrom, E. (1990). *Governing the commons. The evolution of institutions for collective action*. Cambridge: Cambridge University Press.
Ostrom, E., Burger, J., Field, C. F., Norgaard, R. B., & Policansky, D. (1999). Revisiting the commons: Local lessons, global challenges. *Science, 284*(5412), 278–282.
Pearce, D. W. (2001). The economic value of forest ecosystems. *Ecosystem Health, 7*(4), 284–296.
Romang. (2004). Wirksamkeit und Kosten von Wildbach-Schutzmaßnahmen, Geografica Bernensia Reihe G 73, Universität Bern (in German).
Rudolf-Miklau, F., Skolaut, C., & Sauermoser, S. (2014). Avalanche hazard assessment and planning of protection measures. In F. Rudolf-Miklau, S. Sauermoser, & A. Mears (Eds.), *The technical avalanche protection handbook*. Berlin: Wiley.
Rudolf-Miklau, F. (2009). *Naturgefahren-Management in Österreich*. Vienna: LexisNexis Orac (in German).
Rudolf-Miklau, F. (2011). Naturgefahrenkarten und –pläne. In J. Suda & F. Rudolf-Miklau (Eds.), *Bauen und Naturgefahren: Handbuch für konstruktiven Gebäudeschutz*. Vienna: Springer/Ambra. (in German).
Rudolf-Miklau, F., & Kanonier, A. (2016). Gravitative Naturgefahren: Risiken managen. *Recht der Umwelt*. Manz: Vienna (in print).
Rudolf-Miklau, F., Hübl, J., Fiebig, M., Formayer, H., Jaritz, W., Mölk, M. Kleemayr, K. (2012). Naturgefahrenprozesse und –szenarien. Suda, J. und Rudolf-Miklau, F. Bauen und Naturgefahren: Handbuch für konstruktiven Gebäudeschutz. Springer/Ambra Vienna. (in German).
Ruhl, J. B., Kraft, S. E., & Lant, C. L. (2007). *The law and policy of ecosystem services*. Washington, DC: Island Press.
Salzman, J., Thompson Jr., B. H., & Daily, G. C. (2001). Protecting ecosystem services: Science, economics, and law. *Stanford Environmental Law Journal (SELJ), 20*, 309.
Schanze, J., Zeman, E., & Marsalek, J. (2007). *Flood risk management: Hazards, vulnerability and mitigation measures*. New York: Springer.
Schröter, D., Cramer, W., Leemans, R., Prentice, I. C., Araújo, M. B., Arnell, N. W., & Anne, C. (2005). Ecosystem service supply and vulnerability to global change in Europe. *Science, 310*(5752), 1333–1337.
Simpson, R. D. (1998). Economic analysis and ecosystems: Some concepts and issues. *Ecological Applications, 8*(2), 342–349.
Stöcklin, J., Bosshard, A., Klaus, G., Rudmann-Maurer, K., & Fischeret, M. (2007). Landnutzung und biologische Vielfalt in den Alpen–Fakten. Perspektiven, Empfehlungen. vdf-Verlag Zürich (in German).
Stoffel, M., Wehrli, A., Kühne, R., Dorren, L. K., Perret, S., & Kienholz, H. (2006). Assessing the protective effect of mountain forests against rockfall using a 3D simulation model. *Forest Ecology and Management, 225*(1), 113–122.
Wagner, E., & Jandl, C. (2013). *Steinschlagschutz: Rechtliche Aspekte, Schriftenreihe Umweltrecht und Umwelttechnikrecht 4*. Trauner Verlag: Linz. (in German).
Wasser, B., & Perren, B. (2014). Wirkung von Schutzwald gegen gravitative Naturgefahren-Protect-Bio. *Schweizerische Zeitschrift für Forstwesen, 165*(9), 275–283. (in German).
Weber, K. (2004). Grenzen des Rechts: Erwartungshaltung versus Einlösbarkeit im Recht des Naturgefahrenmanagements. In Fuchs, S., Khakzadeh, M.L., & Weber, K. (eds) Recht im Naturgefahrenmanagement, Studienverlag Innsbruck (in German).

Weck-Hannemann, H., & Thöni, M. (2006). Rechtliche Instrumente des Naturgefahrenmanagements aus ökonomischer Sicht. In: Fuchs, S., Khakzadeh, M. L., & Weber, K. (eds), Recht im Naturgefahrenmanagement, Studienverlag Innsbruck (in German).

Wehrli, A., Brang, P., Maier, B., Duc, P., Binder, F., Lingua, E., & Dorren, L. (2007). Schutzwaldmanagement in den Alpen: eine Übersichtl. *Schweizerische Zeitschrift für Forstwesen, 158*(6), 142–156. (in German).

Chapter 23
Citizens Participation in Ecosystem Services

Claude Poudrier

1 Introduction

Human activities and climate change are increasing the vulnerability of headwater catchments and water resources (Křeček et al. 2012). Community participation is an important factor in effective management of mountain watersheds. Often, this fact is still neglected by decision makers and practitioners whose focus is exclusively upon the physical or economic environments. Ecosystem services can capture a wider set of costs and benefits, not traditionally valued in economic analysis (TEEB 2010). The challenge for technical, economic and development aspects of mountain watersheds is how to engage community partners and how to spread understanding of the technical problems of applied watershed management into the community sufficiently that local communities are able to develop as informed, empowered and active catchment citizens, capable of acting as critical friends in watershed management decision making (Ewing et al. 2000). Ferrier and Jenkins (2009) emphasized the responsibility of the public in catchment management. But, achieving this goal involves community-based environmental education and the mobilization of community stakeholder groups according to Blackstock and Richards (2007).

This chapter concerns an environmental and citizenship education training programme, appropriate to ecosystem services (their formulation and implementation). In Québec, the Program in Environmental Education and Citizenship (PEEC) was developed to train students, teachers or decision makers to become actors in community problem solving. The idea of catchment citizenship has been developed particularly in western regions of US. As an example, the Los Angeles & San Gabriel Rivers Watershed Council was founded, in 1996, because of the five different water agencies managing water in this region were failing to adequately exchange information even among themselves, let alone with the public (Council for Watershed

C. Poudrier (✉)
Program Environmental Education and Citizenship (PEEC), Trois-Rivières, QC, Canada
e-mail: claude.poudrier@p2ec.ca

© Capital Publishing Company, New Delhi, India 2017

J. Křeček et al. (eds.), *Ecosystem Services of Headwater Catchments*,
DOI 10.1007/978-3-319-57946-7_23

Health 2011). Now rebranded as the Council for Watershed Health, this NGO has become the centre for practical watershed education, research and analysis in Southern California, focusing on the watersheds of the Los Angeles Basin. The Watershed Council is uniquely situated at the intersection of research and policy to drive applied research to improve policy and practice. Elsewhere, in Baltimore, American citizens use urban waterways like the Patapsco River as sources of drinking water and for a variety of activities including boating, fishing and swimming. Of course, cleaning up, restoring and sustaining the qualities of this water resource is essential to protecting the local community's health and to improving their overall quality of life. Revitalizing such urban waterways also help reconnect citizens to open spaces and their natural habitat, as well as having positive economic impacts on local businesses, tourism and property values. It could also spur private investment and job creation in these communities.

2 Environmental Education and Citizenship

Since 1993, PEEC, the Québécois Program in Environmental Education and Citizenship has been building an effective strategy for active environmental citizenship. The PEEC citizenship concept is that defined by Jane Jenson, Director of the Canadian Policy Research Network's Family Network, as both a status and a relationship (between state and citizens. Her citizenship concept has three dimensions: (1) rights and responsibilities, (2) access and (3) belonging. She also explores how citizens' formal equality can be turned into real participation.

The PEEC programme is based on an educational model known as *"Action Research for Community Problem Solving"* (AR:CPS), which encourages learners to become actors in, for and with their communities (Poudrier 2011). The approach was developed in the United States by William B. Stapp and his students (Wals et al. 1990), where model was often used to change the conventional school environment and to improve teaching and learning conditions in schools.

The AR:CPS approach advocates a citizenship education learning strategy that leads actors of all ages to:

- Identify a problem of concern to them in their community.
- Analyze the problem by considering its different issues.
- Identify potential solutions.
- Select the one that best fits the situation.
- Develop and implement an action plan.
- Evaluate the process and its outcome.

The AR:CPS model use thirteen steps (Table 23.1) for developing community projects that integrate or linked to the social, physical or biophysical environment. The teacher leads the first two steps. The students then become partners in the decision-making processes and gradually take on all or part of the project

Table 23.1 AR:CPS thirteen steps

1.	Process planning by the teacher, in collaboration with all concerned partners: principal, colleagues, parents, community members, etc.
2.	Diagnosis evaluation of students' abilities in solving problems, in order to spot their "deficiencies". As the project goes along, the teacher will have to try to develop required abilities through appropriate teaching and exercises.
3.	Students' awareness of community problems. An exploratory visit of immediate environment is an excellent way to discover these problems and become sensitive to them.
4.	Listing of noticed problems. The brainstorming technique is then useful. Problems may be classified and their interrelations underscored.
5.	Identification of criteria for the selection of a problematic situation to be solved.
6.	Selection of problematic situation.
7.	Search for information on this problematic situation. It may be documentary researches, interviews, observations, etc. A lot of information comes from community environment itself.
8.	Clear definition of problem inherent to the analyzed situation.
9.	Search continuation on the problem now clearly defined.
10.	Examination of possible solutions (again, brainstorming session).
11.	Development of criteria for the selection of preferred solution.
12.	Development and implementation of action plan.
13.	Evaluation of action considering expected effects and follow-up.

management. The teacher continues to work in the role of guide, facilitator and a resource person.

Throughout, the AR:CPS process leads students to take meaningful action in their communities. Its approach is geared toward solving real problems by developing and then implementing action plans. The students lead the decision making, step by step, making their way through a democratic process. Throughout, they log everything they experience in an experiential learning journal. They record their findings, thoughts and even feelings. Their journals describe what they have learnt, their new insights on their problem, changes in their awareness and understanding, their experimentation with participation strategies, and their perseverance and struggle in the face of set-backs and adversities. Through reflection on these accounts, the approach not only leads young people to solve community problems but also highlights new opportunities for the development of learning and even solutions for school-related teaching/learning problems. As Wals (1994) discovered in Detroit, while these activities may target just one environmental problem, the method actively engages both students and teachers in an interdisciplinary learning process that seeks improvement in their local biophysical and/or social environments.

Programme evaluations conducted thus far suggest that the approach is popular with students, teachers, parents and school administrators and that all are very satisfied with its application and its outcomes. Especially, it seems, learners prefer this approach to traditional teaching models. The reason is that the model opens up the community to the school. The immediate environment, as seen by the actors, becomes a source and tool for learning. The students select and develop projects based on a problem-solving approach. The students identify various problems in their school or neighbourhood, such as vandalism, vacant lots, rundown buildings, bullying, waste of energy or drinking water, etc., and take steps towards their resolution. It offers real benefits to all involved. It helps educators discover their students' potential, renews their skills and generates community goodwill. It helps learners develop self-esteem, teamwork skills and self-awareness of their individual power as agents within, for and with their community. It promotes reflective practice alongside the development of critical thinking and communication skills. Of course, for the educational institutions as a whole, the project is valuable because it helps enhance the school-community relationship and encourages school staff to be more deeply involved with their surrounding community.

3 Discussion

The PEEC citizenship programme is rooted in the Action Research process of participants solving real life problems in, with, and for their community. Currently, PEEC engages 15 or so school boards, two private schools, colleges, as well as some Day Care and Seniors Centres. In 2008, there were 70 schools, some pre schools and universities, involved in the programme. PEEC involves Elementary and Secondary School teachers, Education consultants and also Adult Education and Pre-school Educators.

PEEC is the only one recognized by the Ministry of Education in Quebec (MEQ), so the MEQ is engaging with researchers providing teacher training, while the PEEC project itself offers training for learners of all age groups from 4 years old upwards. Moreover, a Canadian premiere in Environmental Education was achieved when the National Film Board of Canada has produced the documentary Hope Builders (2010). This film shows students of 11–12 years old applying the AR:CPS Thirteen Steps for solving the problem of vandalism in, for and with for their community. From Hope Builders, we can see the students who need to consider different aspects related to the problem they want to solve as biophysical, social, economic, politics, social, historic or cultural. The evaluations of the skills developed by the students has shown a high sense of critical thinking, an efficient capacity in communication and qualities related to entrepreneurship profile as perseverance and strategy.

This investment is thought worthwhile because this AR:CPS model proves inspirational; its success, as observed by teachers, is due to the great motivation and perseverance that its student-citizen actors show in their efforts to improve a situation or solving a real problem that affects their families and community. In March 2005, the Canadian Policy Research Network (CPRN) published the results of an online survey of 200 or so young Canadian adults aged from 18 to 25 years.

The survey was designed to identify young adults' priorities in relation to six themes predetermined by an advisory committee made up of people of the same age. However, the key survey finding was that young people truly want to be active citizens and to be involved in community decision making. This echoed the results from an earlier CPRN revealed the same hunger for participation and confirmed Canadian citizens' desire to participate in bringing about changes in their communities.

Thus far the evaluations of AR:CPS projects have demonstrated that student participation in community problem solving strongly promotes the development of their sense of accountability and community belonging because these young people are the leaders of the projects they create, albeit in collaboration with a guide (teacher or other person) (Poudrier 2005). This methodology contributes to developing students' sense of empowered personal responsibility toward and engagement within their communities (cf Haigh 2010). The AR:CPS approach engages students in developing partnerships with members of their communities. They learn that collaboration, team work and coordination are critical and that, today, no actor can succeed alone.

In 2010–2012, the PEEC project included collaboration with the Earthwatch Institute on the active citizenship role in revitalisation of headwater catchments affected by the acid atmospheric deposition (Křeček 2014). As the CPRN's Pat MacKinnon also reports, Canada's citizens are keen to play a larger role in the democratic process and welcome greater opportunities to engage with decision makers on issues that affect their collective quality of life. The PEEC Project strongly supports this goal as does that of the later CPRN 'Focus on Youth' programme (2004), whose goals include the identification of effective approaches for engaging young people in their community's democratic process. One of the effective methods, this suggests is the AR:CPS model.

Equally, the Conseil supérieur de l'éducation du Québec (1998) states that no citizenship education programme can yield tangible results if the students do not have real opportunities to apply the values, rights, responsibilities and civic involvement they are taught in class. This link between teaching and practical engagement is crucial for successful citizenship education and the advantages of the PEEC programme include students' gaining frequent opportunities to exercise their citizenship learning. The AR:CPS model creates multiple opportunities for students to apply their citizenship education skills because it spans an entire school year and builds skills that are key to active participation in democratic society, such as critical thinking, communication, persistence and collaboration (Poudrier 2005). The Conseil supérieur de l'éducation du Québec (1998) also argues that citizenship goes hand in hand with the development of community roots. The PEEC programme's AR:CPS approach engages students in community service and encourages them to develop their sense of social responsibility.

4 Conclusions

The PEEC citizenship concept seems to be a critical toolkit supporting the self-sustainable ecosystem service design in headwater environments. This approach is something that gives meaning to school learning, and the development of critical thinking skills and creativity as well as for citizens who apply the model. It gives learners the chance to acquire experience outside the conventional framework and in the real world contexts of their community and encourages their emergence as active democratic citizens. The key problem in the protection of mountain watersheds is finding ways of integrating the needs and perspective of the technical, scientific and economic agents in these regions with those of the community of inhabitants (Haigh 2010). To achieve this goal, it is essential that environmental and citizenship education in these regions should work together to produce the kind of informed, engaged, effective and reflective change agents that are capable of making informed decisions for their communities. This PEEC programme and the AR:CPS model that underpins it has wider significance than just the needs of Canadian society. Its uniqueness lies in the fact that the PEEC can be applied to all age groups, from early child 4 years old until the adult sector of education.

It is a model that should be widely applicable to other regions of the world and anywhere that seeks to involve citizens, regardless of age, in the decisions that concern them. In so doing, this empowering style of environmental and citizenship education will be able to contribute to improving the quality of life and environmental management by building, not only the requisite foundations of knowledge but also, and above all, by developing the behaviours and attitudes that are key to educated citizenship.

References

Blackstock, K. L., & Richards, C. (2007). Evaluating stakeholder involvement in river basin planning: A Scottish case study. *Water Policy, 9*(5), 493–512.

Council for Watershed Health. (2011). *Announcing the Council for Watershed Health! 8/1/2011*. Available from http://watershedhealth.org/Default.aspx. Accessed Jan 2016.

Ewing, S. A., Grayson, R. B., & Argent, R. M. (2000). Science, citizens, and catchments: Decision support for catchment planning in Australia. *Society and Natural Resources, 13*, 443–459.

Ferrier, R. C., & Jenkins, A. (2009). *Handbook of catchment management*. Chichester: Wiley-Blackwell.

Haigh, M. (2010). Headwater control: An agenda for the future. Chapter 1. Landcon keynote. In M. Zlatic (Ed.), *Global change: Challenges for soil management, Advances in GeoEcology 41*. Reiskirchen: Catena Verlag.

Křeček, J. (2014). *International field project on management of mountain watersheds: A challenge for field science and multi-disciplinary education*. In International strategies in higher education, Proceedings of Esterházy Károly College, Eger (Hungary).

Křeček, J., Haigh, M., Hofer, T., & Kubin, E. (2012). *Management of mountain watersheds*. Dordrecht: Springer.

Poudrier, C. (2005). Environmental education and active citizenship. *Education Canada, 5*, 51–53.

Poudrier, C. (2011). *AR:CPS the pedagogical model of action research for community problem solving*. Available from http://www.P2EC.ca. Accessed Jan 2016.

TEEB. (2010). *The economics of ecosystems and biodiversity. Ecological and economic foundations*. London: United Nations Environment Programme, Earthscan Ltd.

Wals, A. (1994). Action research and community problem-solving: Environmental education in an inner-city. *Educational Action Research, 2*, 163–182.

Wals, A. E., Beringer, A., & Stapp, W. B. (1990). Education in action: A community problem-solving program for schools. *Journal of Environmental Education, 21*, 13–19.

Chapter 24
Watershed Ecosystem Services and Academic Programmes on Environmental Education

Erika Pénzesné Kónya

1 Introduction

Watersheds are exceedingly intricate and inter-dependent complexes of ecosystems. The watershed has emerged as the logical unit for managing surface water and to support the management of groundwater. Watersheds are ecosystems composed of a mosaic of habitat patches that are connected by a line network of streams and patch network of lakes (Millennium Ecosystem Assessment 2005). There are some fundamental features which characterize the environmental education tasks in connection with watershed ecosystems:

- Aquatic systems are linked with terrestrial environment; riparian area along rivers and streams represents this connection.
- Riparian areas are vital to ecological and biological health of watershed ecosystems—provide more than 90% of instream nutrients in the aquatic food cycle
- Aquatic ecosystems are dynamic: slow water flow in the flood, store organic matter, give opportunity for growing biodiversity. In ecologically healthy watershed ecosystems the most important component is the degree to which native riparian biodiversity survives spatially and temporally (Stapp 2000).

Not all freshwater ecosystems will be affected in the same way by climate change. The type of climate change will vary by region, and even across segments of a single basin (Pajtók-Tari et al. 2011). The uneven nature of climate change impacts means that we must understand the different climate vulnerability, sensitivity, and hydrological importance of different aspects of a basin in order to prioritize management responses. In effect, climate change will lead to a tapestry of different risks across freshwater systems (Křeček et al. 1996). Particular elements of the ecological system will be at risk at certain local points on the Earth.

E. Pénzesné Kónya (✉)
Department of Botany and Ecology, Károly Eszterházy University, Eger, Hungary
e-mail: konya@ektf.hu

© Capital Publishing Company, New Delhi, India 2017

J. Křeček et al. (eds.), *Ecosystem Services of Headwater Catchments*,
DOI 10.1007/978-3-319-57946-7_24

Scientists traditionally play an important role in Hungary in facilitating decision making related to the conservation and environmentally sound utilization of freshwater resources. Freshwater resources are also of special concern to various environmental NGOs. More than 90% of the total drinking water demand in Hungary is satisfied from subsurface resources because of the big rivers. Signs of excessive use of these resources have appeared in some regions. Approximately 92% of the dwellings are connected to public drinking water network. The quality of water supplied by public utilities fulfils general national health requirements. However, meeting the relevant standards is a serious problem, especially, in light of the EU-accession related requirements. Freshwater resources have been of special concern for state authorities, local governments and citizens for reasons other than for domestic consumption.

2 Impacts of Climate Change on Water Resources

Analysis of long-term observations shows a decreasing tendency in precipitation amounts and average soil moisture content. Winter and spring precipitation amounts show a significantly decreasing trend. Drought frequency has increased, primarily in the last two decades. It is expected that one of the possible consequences of anticipated global climate change will be an average further decrease of precipitation levels for the next decades in the Hungarian region by approximately 50–100 mm/°C annually. This could cause severe water supply problems in semiarid areas and dry lands. Another consequence of the climate change might be the increasing occurrence of extreme weather conditions causing also floods due to extraordinary heavy rains and fast thawing of the snow in the catchment areas including in mountains of the neighbouring countries. A Special Scientific Committee was established to deal with possible consequences of climate change. The Committee is an important advisory body elaborating long-term strategies and response measures to mitigate the adverse impacts.

Changes in the wet and dry periods show a higher risk of flooding and dry periods in Europe, mainly in the Carpathian Basin. Bartholy and her colleagues (2013) showed the estimated seasonal change in MDS (Mean Dry Spell) for the end of the twenty-first century. Considering the average longevity of the dry periods there will be considerable growth in summer: the value of the index can grow from 6 to 8 days. What about the wet periods and precipitation? On the basis of complex, corrected regional climatic simulations it can be stated that summers become drier and in autumn and also in winter the number of the days with great amount of rain or other form of precipitation will grow. In watershed regions the risk of flooding will grow in winter and in autumn and the amount of water falling out of the atmosphere within a short period will grow.

3 Higher Education in Hungary: A Detailed Examination of Educational Programmes

Hungarian higher education has a long history, dating back to the fourteenth century. The country's first university was founded in 1367 in Pécs, around the same time as other central European universities, such as those in Prague, Krakow and Vienna. Due to the expansion of higher education in the last two decades, the number of enrolled students and the capacity of the institutions have increased considerably. From 1990 to 2010 the student population in higher education quadrupled, from 90,000 to around 400,000. Hungary has been taking part in the Bologna Process since 1999, whose most important goal is the creation of the European Higher Education Area. In force from September 1, 2006 the new Bologna regime three-cycle degree system has been introduced. The first degree programmes (3–4 years) lead to Baccalaureus/Bachelor's degrees, while second degree programmes (1–2 years) lead to Magister/Master's degrees. Unified, undivided, long-term Master's degree programmes (5–6 years) are offered in some fields of study, e.g. in human medicine, dentistry, pharmacy, veterinary medicine, architecture, law and in programmes of art and education. All Bachelor's and Master's degrees grant access to the market of human resources. Based on a Magister/Master's degree or its foreign equivalent, PhD or Doctor of Liberal Arts (DLA) degrees as third cycle degrees can be awarded after 3 years of study.

In addition to the degree programmes described above, higher education institutions offer non-degree programmes: short cycle programmes, higher-level vocational trainings, postgraduate specialist training courses based on a first or second degree as well as summer universities and partial trainings in the framework of lifelong learning. Higher education programmes may be offered in full-time study, part-time study or distance learning courses. The Higher Education Act currently in force is in accordance with the principles of the Bologna process—three cycle degree structure (BA/BSc, MA/MSc, PhD/DLA).

In Hungary environmental education and within that studying watershed ecosystem ecology is not a complex programme but it can be found at several levels of academic programmes (Fig. 24.1). What is the task in connection with watershed ecosystem services? After completing a certain academic programme, the student, applicant should know the basic biotic and abiotic components of watersheds, the basic natural processes and interrelationships occurring in watersheds, and how watershed structure and functions may vary in time and space (Naiman 1992). Some background in the life sciences is helpful for comprehending this material, but not required. Some geographical, physical and chemical knowledge can be also recommended for these programmes. Let's see BSc and MSc programmes within the area of Natural Sciences. Geography BSc is the first educational programme complex which can support sustainability, watershed description and structure studies for Hungarian areas, for Europe and for the Earth (Haigh 2008). There is a special Geography MSc programme with specialization of Environmental Risk Management at the Faculty of Natural Sciences of Eszterházy Károly University in Hungary.

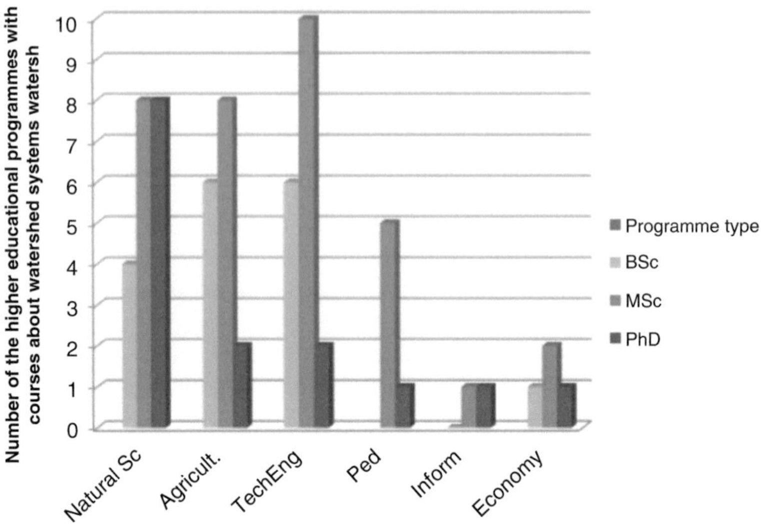

Fig. 24.1 The Hungarian higher educational programmes (According to the National Law of Higher Education, 2015)

Within the area of agricultural studies MSc in Environmental Management is designed to develop your undergraduate knowledge and improve it through application and research. The field of Environmental Management is broad and the programme reflects this diversity, with emphasis on Natural Resource Management, Environmental Impact Assessment, Environmental Technologies, Environmental Informatics, which are the key research areas of the Water and Environmental Management.

4 Biodiversity in Education Programmes

Hungary also declared that nature protection is an important aim and has to be regulated. The National Nature Conservation Master Plan is not only a part of the National Environmental Programme, but is also organically connected to other national plans and strategies, and determines the related sectoral policies that are completed in this planning period, such as the National Agri-Environmental Program, the National Forestry Strategy, the National Regional Planning and the National Development Program. The implementation of these plans greatly influences the status of nature, so they must represent the interests of nature conservation in a complex way, constituting an organic system. Hungary's flora and fauna are very rich and have significant biodiversity. This uniqueness and nature resource value resulted in the formation of a special Hungarian biogeographic region, the Pannonicum, with various types of rocks, formed in different geological eras under

Fig. 24.2 Palingenia longicauda, an ephemer indicator invertebrate species of big rivers in Hungary, Duna and Tisza (Photo: Pénzesné)

different formation conditions; karst areas are the most valuable among the geological and geomorphological resources. There is wide variety of habitats and wildlife with natural forests, lowland and hilly grasslands, nature-like floodplains along our large rivers, traditionally used meadows, vineyards and orchards in hilly and mountainous regions, traditionally cultivated farming areas, and peculiar natural landscapes with unique beauty. One of the symbolic animals of European wet areas and watershed areas is mayfly, *Palingenia longicauda* (Fig. 24.2) which is protected by the Convention on the Conservation of European Wildlife and Natural Habitats—Bern Convention. It is a good indicator species and the biggest habitat of the species within Europe can be found in Hungary.

Unfortunately, and as a familiar global trend, agricultural practice destroyed several wetland habitats like moorlands, fern woods, gallery forests, etc. Terrestrial and aquatic ecosystems have characteristic trophic (feeding) patterns that organize the flow of energy into, through, and out of the watershed ecosystem and support the growth of organisms within the system. Food "chains" are rarely linear, hence the term food web, often used to describe the trophic interactions of organisms in an ecosystem. Within ecology as a discipline, watershed ecology and the role of biodiversity is a special topic (Fig. 24.3) which appears in also numerous programmes in higher education, not only in connection with hydrogeological systems, but as a part of biology (Noss 1990). The ecology or biodiversity studies is also a special topic in Hungary, as Hungary is situated in the Carpathian basin having a unique biogeographical region, the Pannonian region. Environmental education of NGO's with Primary and Secondary school education programmes deal with mainly biodiversity conservation, selective waste collection, waste recycling and energy use. Green Infrastructure is a holistic tool to tackle biodiversity loss while contributing to smart and sustainable socio-economic growth. It provides us with solutions for a better quality environment, a healthier society and a more prosperous economy. Green

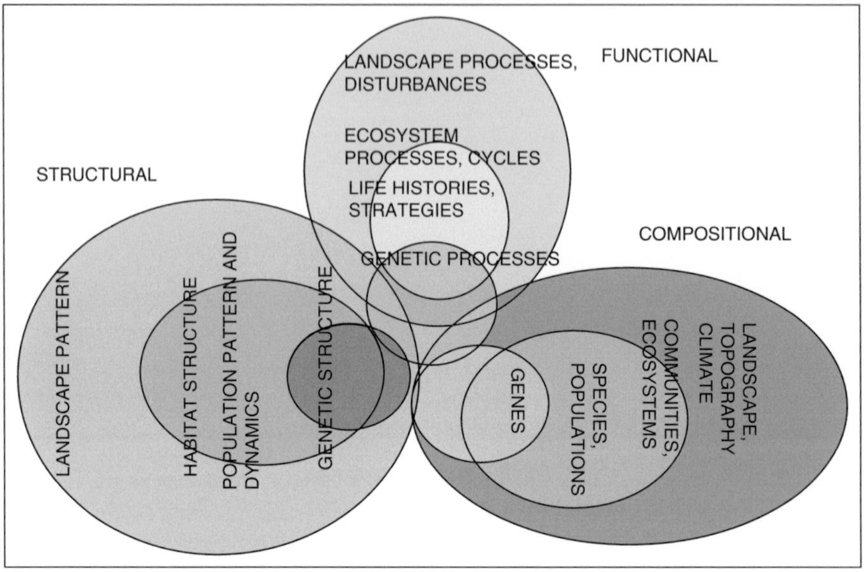

Fig. 24.3 Watershed ecosystems and processes in biodiversity

Infrastructure is especially important due to its potential to help us preserve and enhance ecosystem services and adapt to climate change in a multi-stakeholder setting. Under Target 2 of the 2020 Biodiversity Strategy, the EU committed itself to deploy green infrastructure and restore a minimum of 15% of Europe's degraded ecosystems.

4.1 Focusing the Tasks: Integrating New Data and Research Result in Academic Programmes

In addition to organised educational programmes there are other ways of life-long learning forms: summer universities and National Excellence programme of students in Hungary. One of the Program of Excellence in Higher Education is connected to the Duna Strategy of Europe started in June 2011. Within the programme, university conferences and research applications have been started to involve students into actual research projects on environmental risk management and sustainable energy use in Duna region.

4.2 *In the Teachers' Training and Postgraduate Programmes on Environmental Education of Hungarian Universities*

Students focus on an environmental question, problem, or issue requiring background research and investigation. They learn more about the issue through classroom instruction, the collection of data, conducting experiments and by talking to experts and reviewing important publications. They also reflect on their personal experiences and values related to the issue. This process should be appropriate with practices growing in complexity and sophistication across the grades, starting with educator guided investigation and progressing to student-led inquiry. As students mature, the level and complexity of inquiry will likewise progress.

4.2.1 Outdoor Field Experiences

Students participate in one or more outdoor field experience sufficient to collect the data required for answering the research questions and informing student actions. The outdoor field experiences should be made mainly by students actively involved in planning the investigation, taking measurements, or constructing the project within appropriate safety guidelines. These experiences can take place off-site and on the school grounds.

4.2.2 Action Projects

Students participate in a project during which they take action to address environmental issues at personal or societal level. These projects provide students a better understanding of the actions that they can take to protect and conserve natural resources, and allow them to have a sense of control over the outcome of environmental issues. Action projects should be student directed and can take the form of on-the-ground restoration projects on school grounds or in their community, or can be focused on increasing engagement. Students make conclusions based on research, experiences and data analysis, and consider alternate hypotheses. Students should synthesize and communicate results and conclusions to an external audience such as other classrooms, schools, parents, or the community.

There is an Eco-school system in Hungary for the primary schools which plan additional environmental education programmes for children and also a Forest School system was introduced in primary education in 2011; 543 schools own the Eco-school (ES) title in Hungary. These programmes need prepared teachers to use the environmental education systems in an appropriate way which is the task of the academic level. The PhD programme "Educational aspects of cultural and natural values" of Eszterházy Károly University and other educational aspects of ecosystem research can be the way to introduce good educational methodology and get data for using them in environmental education, also on watershed ecosystems.

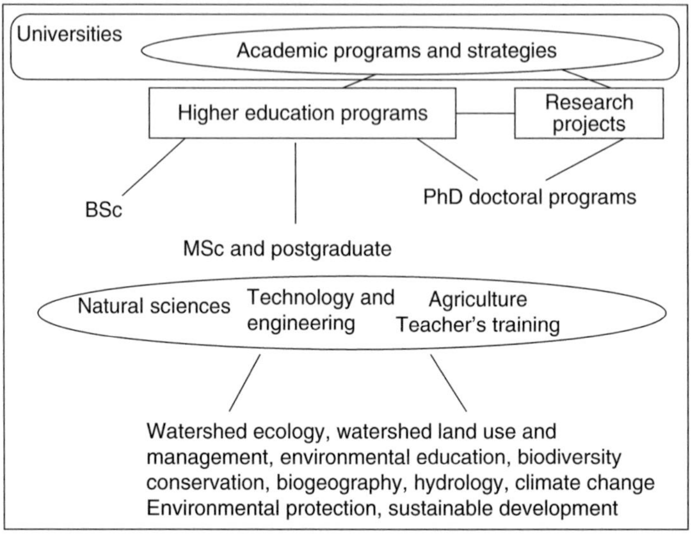

Fig. 24.4 The summary of processes in higher education when research and the results of watershed ecosystem services get into higher education and environmental education programmes

5 Conclusions and Discussion

Watershed ecosystems are essential for human being and to maintain the natural resources on Earth. Some groups of people, particularly those living near wetlands, are highly dependent on these services and are directly harmed by their degradation. Climate change can damage watershed ecosystems very rapidly so the application of new research results in higher education is necessary. On the basis of the analysis of Hungarian educational programmes from bachelor to PhD level the results show that Natural sciences, Engineering and Technology and Agriculture areas are the most relevant areas to study watershed ecosystems. Teacher's training programmes are also important because there are a lot of eco-schools and other ecological programmes for biodiversity conservation and environmental protection (Fig. 24.4).

References

Bartholy, J., Pongrácz, R., & Hollósi, B. (2013). Analysis of projected drought hazards for Hungary. *Advances in Geosciences, 35*, 61–66. doi:10.5194/adgeo-35-61-2013.

Haigh, M. (2008). Internationalisation, planetary citizenship and higher education Inc. compare. *A Journal of Comparative Education*, *38*(4): 427–449 [ISSN 0305–7925, Online ISSN: 1469–3623]. doi:10.1080/03057920701582731

Křeček, J., Rajwar, G., & Haigh, M. (Eds.). (1996). *Hydrological problems and environmental management in highlands and headwaters*. Rotterdam: CRC Press.

Millennium Ecosystem Assessment. (2005). *Ecosystems and human well-being: Biodiversity synthesis*. Washington, DC: World Resources Institute.

Naiman, R. J. (Ed.). (1992). *Watershed management: Balancing sustainability and environmental change*. Paris: Springer.

Noss, R. (1990). Indicators for monitoring biodiversity: A hierarchical approach. *Conservation Biology, 4*(4): 355–364. http://www.un.org/esa/agenda21/natlinfo/countr/hungary/Hungarywater04f.pdf

Pajtók-Tari, I., et al. (2011). Moments of school subjects promoted by climate change. In *Responsibility, challenge and support in teachers' life-long professional development* (pp. 49–66). Brussels: ATEE.

Stapp, W. B. (2000). Watershed education for sustainable development. *Journal of Science Education and Technology, 9*, 183–197.

The World Water Development Report Series. *World Water Assessment Programme*. UNESCO. http://www.unesco.org/new/en/ natural-sciences/environment/water/wwap/wwdr/

Conclusion: Headwater Control

Introduction

Headwater areas are those lands that shed water and receive water only from precipitation. These zero to first order watersheds form the uplands of every catchment. Within Europe (countries of EU15 plus Poland, Estonia, Latvia, Lithuania, Czech Republic, Slovakia, Slovenia, Hungary, Bulgaria, Romania, Bosnia and Herzegovina, Albania and Macedonia), Paracchini et al. (2000) identified the area of 1,000,000 ha of headwater catchments (i.e. 27% of the total land surface). These upland watersheds present an important practical problem for environmental managers. They provide the recharge zones for both surface and ground waters. They also form the margins of hydrological (and often socio-economic) systems. For many years, they have been granted little value, recognised mainly as areas with a potential for land use conversion but, frequently, they are also frontiers of development. At almost every scale, headwaters can become wracked by conflict concerning the exploitation of natural resources, tourism, nature protection and political control. This is why adverse conditions in headwater regions can have dramatic effects on both the environmental and political stability of areas downstream.

More recently, it is being recognised that headwaters provide many valuable environmental services. Headwaters provide society with water, peat, timber, grazing land, and locally aesthetic, cultural, recreational and educational benefits to local communities; and they can help mitigate hazards (flood flows, sediment loads, landslides, and/or avalanches). They may also have a role in buffering climate change, through carbon sequestration (Cannell 1999). The physical costs and benefits of environmental management and different land use practices have been discussed by many previous publications on headwaters (e.g., Taniguchi et al. 2009; Beheim et al. 2010; Křeček et al. 2012).

© Capital Publishing Company, New Delhi, India 2017

J. Křeček et al. (eds.), *Ecosystem Services of Headwater Catchments*,
DOI 10.1007/978-3-319-57946-7

Nairobi Declaration for the International Year of Freshwaters 2003

The problem of undervaluing of distant headwater catchments was identified by delegates at the *International Conference on Sustainable Management of Headwater Resources* (The Fifth International Conference on Headwater Control, Nairobi, 2002). The major output of this conference was the *Nairobi Declaration for the International Year of Freshwaters 2003,* which was ratified by five United Nations agencies (Jánský et al. 2003). This provided the mandate for the creation of new management structures at all levels, which should be designed to improve the coordination, cooperation and empowerment of all stakeholders in headwater regions (Box 1, item 16).

Box 1: Nairobi "Headwater" Declaration for the International Year of Freshwater 2003

We, the participants in the International Conference on Sustainable Management of Headwater Resources, held in Nairobi (Kenya) on 5–8 September 2002,*

1. Acknowledging with gratitude the United Nations General Assembly Resolutiton No. 55/196 to declare the year 2003 as the International Year of Freshwater, thus drawing the world's attention to the need to foster sustainable development and management of freshwaters;
2. Noting the outcome of the recent World Summit on Sustainable Development in Johannesburg 2002, where commitments were made "to increase access to clean water and proper sanitation, to increase access to energy services, to improve health conditions and agriculture, particularly in drylands, and to better protect the world's biodiversity and ecosystems";
3. Recognizing that headwater regions are sensitive environments, source areas for both surface and groundwater resources, and lands that affect the quality of freshwater supplies;
4. Keeping in mind that headwater regions lie at the margins of both watersheds and, often, social and economic systems;
5. Recognizing also the critical environmental functions of headwater regions and their importance for the livelihoods of both their inhabitants and for those who inhabit lands downstream as evoked in Chapter 18 *"Protection of the quality and supply of freshwater resources: application of integrated approaches to the development, management, and use of water resources"* of Agenda 21 adapted at the United Nations Conference on Environment and Development (1992), and also as

(continued)

Box 1 (continued)

stipulated in its Chapter 13 entitled *"Managing fragile ecosystems: sustainable mountain development"*;

6. Affirming our concern to mitigate the consequences of the increasing human impact in headwater regions caused by competing demands for water, forestry, agriculture, energy production, tourism, transport and urban development, which continue to affect the environment adversely, not least with respect to the provision of clean water supplies and the maintenance of other hydrological functions;
7. Noting with further concern that policies can impair, seriously and inadvertently, the course of headwater resources management, and that this can create problems downstream for the quality, quantity and distribution of available freshwater resources;
8. Recognizing that sustainable management of headwater regions needs a holistic, integrated approach which respects the needs of all stakeholders in the regions, values and empowers the headwater inhabitants, and which recognizes their central role in the stewardship of headwater systems;
9. Affirming that the sensitive and scientific management of natural resources, supported by improved access to the high quality data required is essential for fostering development that is not only sustainable, but ideally self sustaining;
10. Conscious that unsustainable management has negative impacts on the health, productivity, social and economic welfare and ecosystems of headwater regions;
11. Aware of the increasing demand for potable waters that will be required for human health, welfare and well-being, and of the crucial role that headwater regions will play in meeting this demand;
12. Conscious also of the potential negative interactions between the inhabitants of headwater regions and those downstream, caused by competition for the limited resources available in the regions, and aware also that headwater areas accommodate and provide for livelihood of a large number of populations, whose activities and resources consumption may have significant effects on the well-being of those who live downstream;

Declare that:

13. Sustainable development should be the baseline for all environmental policy, planning, management practice, education and law in headwater regions;
14. UN agencies should continue their work with all stakeholders to appraise their situations, to identify gaps in knowledge, needs and constraints, and to support them in their efforts to resolve their problems and undertake

(continued)

Box 1 (continued)

practical action towards more self-sustaining and environmentally sensitive development;

15. An 'international commission' for headwater management should be established in order to provide direction and continuity for headwater issues and to create an awareness of headwater concerns at governmental level;
16. Priority should be given to the creation of new management structures at all levels, which should be designed to improve the coordination, cooperation and empowerment of all stakeholders of headwater regions, not least to enhance the participation of women, disadvantaged social groups and minority communities, and to tap and develop the full spectrum of local indigenous knowledge relating to watershed planning and management;
17. Greater effort should be devoted to the refinement of methods for generating and sharing the appropriate and reliable information needed for environmental research, planning and management and also for the transfer of appropriate low cost technologies, especially with respect to 'cushioning' the impacts of environmental hazards for human populations;
18. Greater attention needs to be paid to the special roles and hydrological functions of headwater wetlands and peat lands, which should be a special focus for future headwater workshops, and also to the impacts of anthropogenic processes on watershed functions in headwater regions;
19. The quality of life for the inhabitants of headwater regions should become a primary concern, including the basic needs for a healthy environment and the regeneration of degraded headwater habitats where required;
20. Greater attention should be paid to applied environmental education aimed at building capacity for headwater management and changing social attitudes against wasteful and polluting uses of headwater resources;
21. NGOs (community-based non-governmental organizations devoted to environmental and/or social uplift) should be empowered to play a greater role in the planning, regeneration and management of headwater habitats, by promoting more efficient mechanisms for financial support for effective NGOs;
22. Greater attention should be given to management of headwaters in arid and semi-arid lands, especially with respect to groundwater management and improved accessibility of potable waters to headwater inhabitants, while one of the main focuses should be to reduce the time wasted in carrying water to households from distant water sources;
23. Attention should also be paid to alternative measures that would reduce the dependence of downstream areas on the resources of headwater areas,

(continued)

Box 1 (continued)

including reducing wastage and increasing the efficiency of resource utilization, not least of water;

24. The equitable distribution and use of headwater resources remain a major concern, and planning and management of headwater regions needs to be integrated within the broader framework of watershed management that addresses the concerns of both headwater inhabitants and those downstream.

We therefore call upon UNU, UNESCO, UN-HABITAT, FAO, UNEP, UNDP and other concerned international and national organizations, governments of both developed and developing countries, corporations and NGOs, to facilitate headwater research, monitoring, capacity-building, self-sustaining sustainable development, and better management of the headwater environments, and to help create linkages and synergies in this regard among environmental managers, scientists, communities, policy/decision-makers, practitioners and the general public.

*This conference, which took place from 5 to 8 September 2002 at the United States International University – Africa in Nairobi, Kenya, was jointly organized by the United Nations University, Tokyo, Japan, UNESCO-Nairobi, the United Nations Centre for Human Settlements (UN-HABITAT), the United Nations Environment Programme (UNEP), USIU and Kenyatta University, Nairobi, Kenya, in collaboration with International Association for Headwater Control (IAHC); International Association of Hydrological Sciences (IAHS); and the World Association for Soil and Water Conservation (WASWC).

Environmental Benefits of Headwaters

Among the environmental services of mountain catchments, Körner and Ohsawa (2005) include water resources recharge (quantity and quality), recreation, timber, habitat, and beautiful scenery. According to Reed (2008), the formulation and solution of environmental services requires flexible and transparent decision making processes that embrace a diversity of knowledge and values. For example, instead of promoting 'sustainable timber production', there is the concept of multi resource forestry, which seeks the simultaneous production of several interdependent substances and services and that concentrates on maintaining forests as self-sustaining biophysical systems (Dudley and Stolton 2003). The FAO (2006) has reported several positive examples of collaborative watershed management based on action research, which have employed diagnostic studies, field testing and the validation of improved technologies, as well as social and environmental impact assessment. Recently, the symbolic term 'water towers' has been widely adopted to express the impact of mountains on freshwater supplies in downstream areas (FAO 2006). On the global scale, Viviroli et al. (2007) analysed the potential significance of mountain water resources by including the actual population in the adjacent lowlands and its water needs: 7% of global mountain area provides essential water resources, while another 37% delivers relatively important supportive supply. Significance of mountain water resources is still increasing with expected changes of the global climate (IPCC 2007). Moreover, to identify the above benefits, unfortunately, the estimates of hydrological characteristics (water yield, runoff timing, probability of extreme events, water quality, ecological discharge) are still affected by serious uncertainties, namely in mountain catchments.

The aim of this book is to discuss impacts of human activities on the environmental services provided by headwater catchments with a special emphasis on multi-criteria assessment tools. Stern and Dietz recognise three environmental value systems. 'Egoistic' values predispose people to protect environmental attributes that affect them personally. 'Altruistic' values subsume concern for environment within the welfare of human society. 'Biospheric' grant primacy to all life, including that part which is human. The headwater literature tends to conceive human communities as integral parts of the watershed system and often argues that human welfare is best served by serving the needs of Nature (Berry 1999). Changes in headwater and mountain watersheds have major impacts on both the sustainable development of these headwaters regions and on those who live downstream (Miklau 2017; Palán and Punčochář 2017; Sato 2017).

Several authors emphasize the need to communicate the environmental message about headwaters and highlands to communities that, at present, merely consume its services and protest about the consequences of poor headwater management during times of crisis following extreme events such as torrential floods or landslides (Marui 2017; Kostadinov 2017). Other discussion concerns the creation of new legislative and planning frameworks and the spread of watershed consciousness through education (Poudrier 2017).

Headwater Ecosystem Services

In 2010, Haigh (2010) proposed that *"Tasks for the future include: more effective integration of environmental science with the socio-economics of livelihoods and ways of life in headwater societies; greater exploration of ethics both professional and as a force in environmental decision making; and doing more to get the environmental management message through to the public through education, local media, and participatory/community engagement"*.

This new book takes this agenda forward by focussing on the problems of paying for the environmental security and services that headwater and highland regions provide (Miklau 2017; Palán and Punčochář 2017). Generally, the concept of environmental services is based on the link between flows of value to human societies as a result of the state and quantity of natural capital (MEA 2005). Assessment of ecosystem changes in relation to human well-being concentrated on the enhancement of adequate conservation and sustainable management practices.

The Millennium Ecosystem Assessment (MEA 2005) grouped ecosystem services into four categories: provisioning, regulating, supporting and cultural services. Concerning the impacts of development in headwater catchments (Haigh and Křeček 2000), the provisioning services reflect namely drinking water supply, timber and fisheries; regulating services refer to carbon and bio-geo-chemical cycles, water quality genesis, runoff timing and flood occurrence; supporting services contribute particularly to biodiversity; and cultural services address recreation, tourism and aesthetic experiences. The concept of ecosystem services is based on means by which all different types of environmental effects and their societal impacts can be evaluated together in a systematic manner. Ecosystem services can capture a wider set of costs and benefits including many that are not traditionally valued by economic analysis (TEEB 2010). Since the 1970s, there have been attempts to evaluate the economic impacts and cost-benefits of controlling degradation of headwater regions. However, these aspirations have been limited by the rather broad spectrum of uncertainties in relations between the processes (degradation, restoration) and effects. There are still large gaps in the understanding of the nature and value of degradation and restoration impacts. There is also, still, a conceptual mismatch between the relatively short-term time-frames used for economic analysis and the longer perspectives needed for effective environmental management. Therefore, current economic-evaluation models do not adequately account environmental benefits.

This volume explores the notion of ecosystem services as a means for creating a holistic overview that can be explained in economic, social and political, as well as scientific, terms (cf. Van Haveren 2000). The fact remains that many downstream problems in catchments have their origins in, sometimes distant, headwaters, that few of these problems respect the artificial boundaries set by academic and administrative structures, but that many tend to be out of the public eye ahead of times of extreme event or resources shortages (Miklau 2017; Schreier 2017; Haida et al. 2017).

An abiding concern is the increasingly popular concept of Ecosystem Services and its associated methodologies for environmental valuation (del Campo et al. 2017; Haida et al. 2017; Haigh 2017; Lievois et al. 2017), especially with reference to meeting the costs of adaptation to climate change in different parts of the world (Canada – Schreier 2017; Japan – Marui 2017; Southern Spain – del Campo et al. 2017; Hungary – Pénzesné Kónya 2017; Eastern Alps – Haida et al. 2017; North Bohemia – Křeček and Palán 2017). In several cases, public participation and environmental education are seen as key to the future (Haida et al. 2017: Pénzesné Kónya 2017; Poudrier 2017; Zingari 2017). In some chapters, authors tackle more traditional Headwater Control issues, sometimes from the Ecosystem Services perspective, but more usually focussing on the land itself and the management of land degradation processes such as forest degradation that affect water supply (Priezgalski 2017). Specific issues addressed include the legacies of 'acid rain' (Křeček and Palán 2017), shifting cultivation (Rawat 2017), forest clear-cutting (Kubin and

Křeček 2017; Görücü 2017); the management of agricultural terrace systems (Fullen et al. 2017), soil erosion sensitivity (Kertesz 2017); coal-mining (Haigh 2017); peat extraction and water conservation (Suominen 2017); water supply (Sato 2017; Gophen 2017; Balek 2017); deforestation and flood hazard (Palán and Punčochář 2017); and torrent and landslide control (Kostadinov et al. 2017; Marui 2017).

Some Key Findings and Recommendations

The role of headwaters as 'water towers' (Viviroli et al. 2007) receives much attention in this volume. Sato (2017) aims to manage the impacts of climate changes that have raised the average annual air temperature in Japan by 1.2 °C since 1910 and may add a further 3 °C by the close of the century. The problem is that this will reduce snowfall and hence the snow-melt runoff in rivers, which is critical for the water supply during the crop growing season in affected regions. Planning for deforestation is the concern of Palán and Punčochář (2017) in Brazil, where deforestation of 30% in the most forested headwater areas is expected to lead to a 98% increase in 2-year peak flows and 117% increase in 20-year peak flows, with specific discharge increases of 380% (2-year) and 400% (20-year). The authors go on to cost the increased insurance risk for affected municipalities. Elsewhere, in the gently sloping peat-lands of Finland, clear cutting decreased evapotranspiration by 22% and so water yield increased by 35%, mainly in connection with flood peak flows, which increased by 62–103%. The removal of the spruce stands also resulted in nitrogen concentrations increasing by up to 100%, effects which persisted in ground waters 25 years after the clear cut and 10 years after natural regeneration had replaced the forest (Kubin and Křeček 2017). In southern Spain, Del Campo et al. (2017) suggest that, with climate change and advancing aridity, the hydrological aspects of forests should be considered direct and tangible benefits rather than indirect and forest managed accordingly.

The holistic integration of ecology, economy and society within programmes and policies is also the recommendation of Rawat (2017), faced with the task of sustainable environmental management in the Kiliki basin of Nagaland, where almost two thirds of the land is managed by an increasingly stressed system of shifting cultivation (*jhum*). In a different context, Marui (2017) also argues that comprehensive watershed management, including the formulation of a disaster emergency management plan, is the only possible way of coping with the unpredictable nature of large-scale landsliding in the mountains of Japan. To this end, in Poland, Priezgalski (2017) recommends ecological planning to mitigate environmental damage and proposes several simple 'house-keeping' measures for headwater management. These include the prevention of clear-cutting on slopes >40%, reforestation of slopes >15%, rehabilitation of skidding roads after logging, and multi-species ecological silviculture to reduce the damages caused by disease and wind-throw. Observing the recovery of headwater catchments in North Bohemia from the devastation caused by acid rain in the 1980s, Křeček and Palán (2017) commend five main priorities for mountain foresters addressing issues of water resources control and drinking water supply. These five are the protection of surface waters by buffer strips, the conservation of biodiversity, especially with respect to wetlands, soil protection, especially for slopes >30%, fog drip maintenance at higher altitudes (>900 m), and evapotranspiration management across the whole catchment.

Several further chapters in this volume deal with the role of lessons from the past management of mountain for planning the future management of environmental services in headwaters. Among these, Görücü (2017) emphasises the need for engagement with the political process to ensure that measures necessary to support ecosystem services are put in place and sustained. Similarly, Lievois et al. (2017) note that the protective role of forests has only recently entered territorial planning documents and that this has happened because of decades of research that have supplied policy makers with the tools that they need to frame laws and disburse funding. In Israel, Gophen argues the need to find 'the golden path' between the competing demands on Lake Kinneret, which include water supply, agriculture, sustainable lake ecology, the problems of pollution and water level fluctuation, and impacts on local residents and an emergent ecotourism business. In China, Fullen et al. (2017) consider the role ecotourism may play in the conservation of a World Heritage historical artefact, the Yuanyang Terraces of Yunnan, PR China, where the successful harmonization of high-value organic agriculture and eco-tourism may provide the key to sustainable development.

Concerning the ecosystem service approach, Zingari (2017) argues that the lesson to be learnt is the need to reconcile the demands of conserving both natural and cultural capital by bringing together local experience and scientific knowledge which, in policy terms, means that planning must include constructive dialogue between local stakeholders and other interested parties. This poses new challenges for environmental education and the promotion of active citizenship. Here, Poudrier (2017) with Pénzesné Kónya (2017) highlight the role of environmental education in ensuring that local stakeholders have both the knowledge base and commitment necessary to contribute to such dialogue. Of course, the economic valuation aspect of the Ecosystem Service concept evokes squeamishness in some contributors. For example, exploring the impacts of reclaimed opencast coal-mines in the headwaters of South Wales, Haigh (2017) criticises the general notion of Ecosystem Services as encouraging the commodification of Nature, but finally agrees, that where finance is required for the restoration of degraded lands, the economic approach may have real benefits. Miklau (2017) goes further by discussing the qualities of the 'compensation gain,' that should go back to the providers of headwater ecosystem services. However, Schreier (2017) points out that neither Market Value methods (replacement or restorative value, preventative expenditures), nor Shared Preference methods (non-market valuation, contingent valuation-willingness to pay, conjunct choice comparison) nor Revealed Preference methods (hedonic pricing, benefit transfer) provide a wholly satisfactory basis for decision making. He agrees that determining restorative cost methods is a realistic approach but worries that many of the ecosystems services of undisturbed environments will not, perhaps cannot, be restored and that problems, such as resilience to emerging climatic changes, will not be addressed.

Management of Headwater Catchments

In recent years, EU initiated several studies addressing the problems of definition, theory, typology and a measuring approach of ecosystem services in Europe. Unfortunately, these do not recognise the watershed concept. Watersheds are key natural functional land systems and central to the assessment of the environmental roles of headwater ecosystems. Watershed outputs (including water yield, runoff timing, water chemistry and biota, flood regime, sediment yield) serve as key indicators of the health of headwater ecosystems and of their land use sustainability.

The modern concept of 'Headwater Control' emerged in the late 1980s, in parallel with the transformation of the former Socialist countries. It arose from the shared concerns of applied environmental scientists, especially field operatives in forestry, soil and water conservation and water resource management, who work in uplands and on the margins of surface water systems. It contained the search for a more holistic, hands-on and sustainable approach to watershed management.

The 'Headwater Control' initiators, however, had aspirations that extended beyond the academy and a wish to engage with the real world of environmental problem solving on the ground. In particular, they were concerned about the way that headwater regions are managed. They were concerned that, often, the management of headwater regions was effected by outside interests and often by agencies whose interest was also restricted to the exploitation of a single resource. This resulted in blinkered, partial management, often to the disadvantage of the local environment, economy and population. The solution was to promote a more holistic approach based on the integrated locally-empowered, synthesis of lands and livelihoods (Bunch 1982).

For the future, it is hoped that more may be done to develop the role that could and should be played by environmental education for sustainable development because, over recent meetings, it has become obvious that improved watershed management, like all environmental management, demands a change in social attitudes. This includes a shift in emphasis from granting primacy to short-term economic gains and away from belief in the, still current, myth that it is desirable for technology, routinely, to replace the functions of nature (Berry 1999). It is hoped that such headwater thinking may contribute to the agenda of the United Nations 2030 Agenda for Sustainable Development, which was adopted on 25 September 2015 at the United Nations Sustainable Development Summit (United Nations 2015a). Among its 17 Sustainable Development Goals are: 6: Ensuring the availability and sustainable management of water and sanitation for all, whose targets include to "implement integrated water resources management at all levels, including through transboundary cooperation as appropriate; protect and restore water-related ecosystems, including mountains, forests, wetlands, rivers, aquifers and lakes, to expand international cooperation and capacity-building support; and to support and strengthen the participation of local communities in improving water and sanitation management". In addition, Goal 7 concerns clean energy, 8 concerns inclusive economic development, 13 concerns combatting climate change, 16 concerns promoting peaceful and inclusive societies for sustainable development and, of course, 15 concerns the protection and restoration of terrestrial ecosystems, sustainable forest management, reversing land degradation and the conservation of biodiversity (United Nations 2015b). All of these concerns are central to the mission of Headwater Control and to the chapters in this volume.

The aim of effective headwater management should be to maximise the benefits of headwater catchments to their stakeholders. Effective headwater management continues to require better technologies, better policy frameworks, changes in land husbandry patterns and direct engineering intervention. Research needs to be reoriented to focus on the tolerances, exchanges, checks and balances within headwater landscapes and the impact of changes downstream. There is a need for new data, especially from longer-term environmental monitoring and also for new models of headwater management and there is a need for greater sharing and utilisation of existing information. There is a need to focus on the creation of truly self-sustainable structures, both socioeconomic and environmental. Much watershed management aims to solve problems either by constraining Nature or by taking the functions of Nature into human control. Sometimes this is called 'sustainable development'. However, in practice, 'sustainable' often means no more than capable of being kept going through repair, maintenance and management—a perpetual concern, cost and responsibility for any host community. Headwater control strives for self-sustainability and to create systems that can look after themselves—either because their support is inherent in normal pattern of land-use or because environmental management is returned to the self-sustaining hand of Nature. Effective management of headwater wetlands in the frame of integrated watershed planning also demands some assessment of the role of key components such as riparian buffer zones, runoff genesis, and water supply as well as a better system for community education and more effective participatory management processes.

This new collection of research and discussion papers includes many valuable applied scientific contributions that have practical value to the better management of headwater regions. In addition, the hope is that this new collection will contribute to the future development of a more holistic headwater control that is both better valued by and better integrated with stakeholders of all kinds.

References

Beheim, E., Rajwar, G. S., Haigh, M., & Křeček, J. (Eds.). (2010). *Integrated watershed management: Perspectives and problems*. Dordrecht: Springer.

Berry, T. M. (1999). *The great work*. New York: Bell Tower.

Bunch, R. (1982). *Two ears of corn: A people-centered approach to agricultural development*. Oklahoma City: World Neighbors.

Cannell, M. G. R. (1999). Growing trees to sequester carbon in the UK: Answers to some common questions. *Forestry, 72*, 237–247.

del Campo, A. D., González-Sanchis, M., Lidón, A., García-Prats, A., Lull, C., Bautista, I., Ruiz, G., & Francés, F. (2017). Ecohydrological-based forest management in semi-arid climate. In J. Křeček, M. Haigh, T. Hofer, E. Kubin, & C. Promper (Eds.), *Ecosystem services of headwater catchments* (pp. 45–58). Cham: Springer.

Dudley, N., & Stolton, S. (2003). *Running pure: The importance of forest protected areas to drinking water*. Gland: Research Report of the World Bank and WWF Alliance for Forest Conservation and Sustainable Use.

FAO. (2006). *The new generation of watershed management programmes and projects, FAO forestry paper 150*. Rome: Food and Agriculture Organization of the United Nations.

Fullen, M., Zhu, Y., Wu, B., Chengyun, L., Mei, L. Y., Xin, A. T., & Colinet, G. (2017). Agro-environmental sustainability of the Yuanyang rice terraces, in Yunnan Province, China. In J. Křeček, M. Haigh, T. Hofer, E. Kubin, & C. Promper (Eds.), *Ecosystem services of headwater catchments* (pp. 117–126). Cham: Springer.

Gophen, M. (2017). Headwaters management constraints within the Kinneret ecosystem services. In J. Křeček, M. Haigh, T. Hofer, E. Kubin, & C. Promper (Eds.), *Ecosystem services of headwater catchments* (pp. 79–88). Cham: Springer.

Görücü, Ö. (2017). Ecosystem services supporting water supply systems. In J. Křeček, M. Haigh, T. Hofer, E. Kubin, & C. Promper (Eds.), *Ecosystem services of headwater catchments* (pp. 15–22). Cham: Springer.

Haida, C., Geitner, C., Hama, M., Hastik, R., Höferl, K. M., & Schneider, K. (2017). Prioritizing adaptation needs for ecosystem services: A regional case study in the Eastern Alps. In J. Křeček, M. Haigh, T. Hofer, E. Kubin, & C. Promper (Eds.), *Ecosystem services of headwater catchments* (pp. 151–168). Cham: Springer.

Haigh, M. (2010). Headwater control: An agenda for the future. In Zlatič, M. (ed.), Global change: Challenges for soil management. *Advances in GeoEcology*, *41*: 1–12.

Haigh, M. (2017). Impacts of reclaimed opencast coal-land on headwater ecosystem services. In J. Křeček, M. Haigh, T. Hofer, E. Kubin, & C. Promper (Eds.), *Ecosystem services of headwater catchments* (pp. 183–206). Cham: Springer.

Haigh, M., & Křeček, J. (Eds.). (2000). *Environmental regeneration in headwater areas*, *NATO science series 2: Environmental security 68*. Dordrecht: Kluwer.

IPCC. (2007). *Climate change: Synthesis report*. Genéva: World Meteorological Organization.

Jánský, L., Křeček, J., & Haigh, M. J. (2003). *Sustainable management of headwater resources (Nairobi "Headwater" declaration for the International Year of Freshwaters 2003). Environment and sustainable development Programme*. Tokyo: United Nations University.

Körner, C., & Ohsawa, M. (2005). Mountain systems. In R. Hassan, R. Scholes, & N. Ash (Eds.), *Ecosystems and human well-being: Current state and trends*. London: Island Press.

Kostadinov, S., Olivera Košanin, O., Petrović, A., & Dragićević, S. (2017). Extreme climate events and erosion control in headwater catchments of Serbia. In J. Křeček, M. Haigh, T. Hofer, E. Kubin, & C. Promper (Eds.), *Ecosystem services of headwater catchments* (pp. 215–224). Cham: Springer.

Křeček, J., & Palán, L. (2017). Environmental services in mountain catchments affected by acid atmospheric deposition. In J. Křeček, M. Haigh, T. Hofer, E. Kubin, & C. Promper (Eds.), *Ecosystem services of headwater catchments* (pp. 169–182). Cham: Springer.

Křeček, J., Haigh, M., Hofer, T., & Kubin, E. (Eds.). (2012). *Management of mountain watersheds*. Dordrecht: Springer.

Kubin, E., & Křeček, J. (2017). Ecosystem services in headwaters of the boreal environment. In J. Křeček, M. Haigh, T. Hofer, E. Kubin, & C. Promper (Eds.), *Ecosystem services of headwater catchments* (pp. 23–32). Cham: Springer.

Lievois, J., Gauquelin, X., Dubois, A., & Evans, A. (2017). Account of the French progression for a better integration of the protective function of mountain forests. In J. Křeček, M. Haigh, T. Hofer, E. Kubin, & C. Promper (Eds.), *Ecosystem services of headwater catchments* (pp. 69–78). Cham: Springer.

Marui, H. (2017). Mitigation of natural hazards in mountain watersheds of Japan. In J. Křeček, M. Haigh, T. Hofer, E. Kubin, & C. Promper (Eds.), *Ecosystem services of headwater catchments* (pp. 89–104). Cham: Springer.

MEA. (2005). *Ecosystems and human well-being: Synthesis*. Washington, DC: Millenium Ecosystem Assessment, Island Press.

Miklau, F. R. (2017). Natural protective services in mountain catchments: Provision, transaction and consumption. In J. Křeček, M. Haigh, T. Hofer, E. Kubin, & C. Promper (Eds.), *Ecosystem services of headwater catchments* (pp. 247–270). Cham: Springer.

Palán, L., & Punčochář, P. (2017). Environmental security and headwater control in Brazil. In J. Křeček, M. Haigh, T. Hofer, E. Kubin, & C. Promper (Eds.), *Ecosystem services of headwater catchments* (pp. 127–138). Cham: Springer.

Paracchini, M. L., Folving, S., & Bertolo, F. (2000). Identification and classification of European headwaters. In M. J. Haigh & J. Křeček (Eds.), *Environmental regeneration in headwater areas, NATO science series 2: Environmental security 68*. Dordrecht: Kluwer.

Pénzesné Kónya, E. (2017). Watershed ecosystem services and academic programmes on environmental education. In J. Křeček, M. Haigh, T. Hofer, E. Kubin, & C. Promper (Eds.), *Ecosystem services of headwater catchments* (pp. 279–288). Cham: Springer.

Poudrier, C. (2017). Citizens participation in ecosystem services. In J. Křeček, M. Haigh, T. Hofer, E. Kubin, & C. Promper (Eds.), *Ecosystem services of headwater catchments* (pp. 271–278). Cham: Springer.

Priezgalski, E. (2017). Technical and ecological methods to control the water cycle in mountain watersheds. In J. Křeček, M. Haigh, T. Hofer, E. Kubin, & C. Promper (Eds.), *Ecosystem services of headwater catchments* (pp. 7–14). Cham: Springer.

Rawat, M. S. (2017). Environmental management in the headwater catchments of Kiliki River, Nagaland, North East India. In J. Křeček, M. Haigh, T. Hofer, E. Kubin, & C. Promper (Eds.), *Ecosystem services of headwater catchments* (pp. 105–116). Cham: Springer.

Reed, M. S. (2008). Stakeholder participation for environmental management: A literature review. *Biological Conservation, 141*, 2417–2431.

Sato, Y. (2017). Climate trends and impacts on water balance in a snow-dominated watershed of northern Japan. In J. Křeček, M. Haigh, T. Hofer, E. Kubin, & C. Promper (Eds.), *Ecosystem services of headwater catchments* (pp. 33–44). Cham: Springer.

Schreier, H. (2017). Maintaining environmental services in mountain watersheds when climatic extremes, water demand and land use pressures are increasing. In J. Křeček, M. Haigh, T. Hofer, E. Kubin, & C. Promper (Eds.), *Ecosystem services of headwater catchments* (pp. 139–150). Cham: Springer.

Suominen, M. (2017). Water conservation business arising from company's environmental responsibility. In J. Křeček, M. Haigh, T. Hofer, E. Kubin, & C. Promper (Eds.), *Ecosystem services of headwater catchments* (pp. 239–246). Cham: Springer.

Taniguchi, M., Burnett, W. C., Fukushima, Y., Haigh, M., & Umezawa, Y. (Eds.). (2009). *From headwaters to the ocean: Hydrological change and watershed management*. Boca Raton: CRC Press, Taylor & Francis Group.

TEEB. (2010). *The economics of ecosystems and biodiversity. Ecological and economic foundations*. London: United Nations Environment Programme, Earthscan Ltd..

United Nations. (2015a). *Transforming our world: The 2030 Agenda for sustainable development.* Resolution adopted by the General Assembly on 25 September 2015, Seventieth Session, Agenda Items 15 and 116, A/Res/70/1. Retrieved July 15, 2016, https://sustainabledevelopment.un.org/?menu=1300

United Nations. (2015b). *Transforming our world: The 2030 agenda for sustainable development A/RES/70/1*. Retrieved July 15, 2016, https://sustainabledevelopment.un.org/?menu=1300. .

Van Haveren, B. P. (2000). Reconstructing headwater areas social and institutional considerations. In M. J. Haigh & J. Křeček (Eds.), *Environmental regeneration in headwater areas*, *NATO science series 2: Environmental security 68*. Dordrecht: Kluwer.

Viviroli, D., Dürr, H. H., Messerli, B., Meybeck, M., & Weingartner, R. (2007). Mountains of the world, water towers for humanity: Typology, mapping, and global significance. *Water Resources Research, 43*, W07447.

Zingari, P. C. (2017). Ecosystem services, European Union policies, and stakeholder's participation. In J. Křeček, M. Haigh, T. Hofer, E. Kubin, & C. Promper (Eds.), *Ecosystem services of headwater catchments* (pp. 225–238). Cham: Springer.

Index

© Capital Publishing Company, New Delhi, India 2017

J. Křeček et al. (eds.), *Ecosystem Services of Headwater Catchments*,
DOI 10.1007/978-3-319-57946-7

MIX
Papier aus verantwortungsvollen Quellen
Paper from responsible sources
FSC® C105338

If you have any concerns about our products,
you can contact us on
ProductSafety@springernature.com

In case Publisher is established outside the EU,
the EU authorized representative is:
Springer Nature Customer Service Center GmbH
Europaplatz 3, 69115 Heidelberg, Germany

Printed by Libri Plureos GmbH
in Hamburg, Germany